Marine Concrete

To Catriona and Colin
and to the memory of my parents, Ailie (Hill) and Alec Marshall

Marine Concrete

A L MARSHALL

Principal Lecturer
School of Environmental Technology
Sunderland Polytechnic

(Formerly Head, Offshore Engineering Unit)

Blackie
Glasgow and London

Van Nostrand Reinhold
New York

Blackie and Son Ltd
Bishopbriggs, Glasgow G64 2NZ
and
7 Leicester Place, London WC2H 7BP

Published in the United States of America by
Van Nostrand Reinhold
115 Fifth Avenue
New York, New York 10003

Distributed in Canada by
Nelson Canada
1120 Birchmount Road
Scarborough, Ontario M1K 5G4, Canada

16 15 14 13 12 11 10 9 8 7 6 5 4 3 2 1

First published 1990

British Library Cataloguing in Publication Data

Marshall, A.L.
Marine concrete.
1. Marine structures. Construction materials. Concrete
I. Title
620.4162

ISBN 0-216-92692-0

Library of Congress Cataloging-in-Publication Data

Marshall, A.L.
Marine concrete / A.L. Marshall.
p. cm.
Includes bibliographical references.
ISBN 0-442-30297-5
1. Concrete—Corrosion. 2. Sea-water corrosion. I. Title.
TA440.M29 1990
620.1′3623—dc20
89-70443
CIP

Phototypesetting by Thomson Press (India) Ltd, New Delhi
Printed in Great Britain by Thomson Litho Ltd, East Kilbride, Glasgow

Preface

Concrete is commonly regarded as a mundane, prosaic material whilst the sea is perceived as a fearsome environment, endowed with mystery. Mystery stems from lack of knowledge, and to that extent both concrete and sea have something in common—we fall a long way short of knowing enough about them. Fortunately we have learned enough from our investigations and experiences to be able to set the limits within which we should operate.

It is important for the engineer to seek to quantify the effects of the environment on materials and structures so that these can be made safe and adequately durable for their intended economic life. This is especially true for marine structures. Thus the primary purpose of this book is to provide a useful synthesis of the behaviour of concrete and concrete structures in the marine environment. An outline of the content of the book is provided in the latter part of the first chapter and so will not be anticipated here. The chief aim throughout, however, is to work as far as possible within a context of the appropriate governing physical phenomena, giving due consideration to the mathematical relationships between them. Moreover, without intending to be a design manual, an introduction is given to the sources of information which designers are likely to use, as well as to structural achievements. It is hoped that there should emerge an implicit integration between structure and constituent materials and the surrounding environment. Hopefully, too, the references cited provide useful entry points to the literature.

A body of literature is established by people, of course, and in referring to them an author is also expressing tacit gratitude. Warm thanks are expressed to the members of the community represented by their citations, although it is stressed that the presentation of data herein is the responsibility of the present author. Thanks are expressed to the publishers for their patience and understanding when, through the intervention of Nature and bodily infirmity, the enterprise was seriously delayed and could so easily have been lost. Finally, and by no means least, a manuscript must be made presentable for others to read and so particular thanks are due to Margaret Slaughter for her efforts in preparing the typescript.

ALM

Contents

4 Behaviour of plain concrete

1 Introduction

This book considers concrete in relation to the sea: how it can be affected as a material, the loads placed upon it, how it has been used, and how it might develop. Unravelling its behaviour necessitates following two main strands: the structural application and the nature of the environment in which it is placed.

Concrete of a sort was used in marine works by the Romans two thousand years ago and probably by the Greeks before that, so, in broad historical terms, we are not dealing with anything very new. On this time-scale it is really only quite recently that we have come close to a proper appreciation or understanding of the fundamental processes involved although, for all that, there remain still unanswered questions. Nevertheless, despite the 'newness' of this awareness, empiricism, fortunately, has enabled structures and vessels to be built over many years and adequate defences against the sea's ravages to be prepared. On the other hand, without identifying and quantifying the primary influences and principles, extrapolation on a basis of past experience alone is bound to be constrained and uncertain. Better understanding does not of itself bring greater freedom, of course, but it does imply that the bounds of possibility can at least be more clearly defined and hence how near they might be approached. At a minimum this should assist in more efficient and economic use of resources as well as in improving endurance. At the same time it can provide scope for stepping further in new directions by indicating where they might (or perhaps should not) lead.

1.1 Locating the sea

Retrospection always has been easier and more prevalent than prescience and this applies as much to perceptions of knowledge in general as to particular aspects of it. Erratically lurching from one state to another might be a more realistic assessment of its progress than the usual convenient and tidy simplicity of modulation. Culs-de-sac, obstacles and diversions occur frequently along the way and accident and serendipity have frequently played as much of a part as inspiration and dedication.

For man, the sea has always existed and is still capable of appearing as terrifying and mysterious as it did to our ancestors. Progress in our understanding of it has been no different from any other branch of knowledge, stuttering over the centuries from the first tentative journeys along the coast, through the great voyages of the Norsemen and the mariners of Western Europe, to the sophisticated studies of today with all the modern paraphernalia of scientific observation and measurement. Exploration and experience of landmarks and coastal features two and a half thousand years ago would be noted in an account known as a periplus but it was not until the beginning of the fourteenth century that records were preserved in charts (if there were earlier records, they have not survived). Based on confirmed observation, they are the first proper maps, for all their flaws. (Such a slow pace of improvement of record is not due solely to the rate of development of the specific technology, of course; it is a function of many complex factors including social change, the spread of literacy and better communications.)

1.2 The moving sea

As every British schoolchild used to know, the eleventh century (Danish) King Cnut of England demonstrated that even he could do nothing to thwart the inexorability of the rise and fall of the tide. The power of the sea has been well recognised since long before even that period but, while fear of the unknown might have deterred some, a combination of curiosity, boldness, and sometimes greed, has taken others where none dared go before. Not all returned but the gradual accumulation of knowledge and experience built up a more comprehensive picture of the oceans.

Most of the major surface current systems of the world had been discovered by fifteenth and sixteenth century navigators although, lacking some method of accurate location and measurement, quantification was necessarily very limited. Precise determination of longitude was the governing factor which, without reliable chronometers, was impossible. They did not come into general use until the nineteenth century: for instance, Cook's voyage to the Antarctic in the 1770s, a well supported expedition, had only one reliable instrument among the four with which they set out. The major exercises which were to establish, over a few decades, the circulation pattern of the oceans, did not begin until the latter part of the nineteenth century. Theory and observation were not married until the first quarter of the twentieth.

1.3 The marriage of observation and theory

Length of time or failure to reach accord might hinder but does not prevent progress. Indeed it is empirical success which has often led to the search for theory and principle notwithstanding its duration. For example, the ancient civilisations knew how to transfer water, and the Greeks, and particularly the Romans, mastered many of the techniques required without knowing anything about the theory of fluid flow as we understand it today. Archimedes discovered the basic principle of buoyancy in the first century but it was to be another fifteen hundred and more years before Newton and Leibnitz 'invented' calculus. The next one to two hundred years saw the foundations laid of modern fluid mechanics and hydrodynamics, the many celebrated contributors including now familiar names such as Lagrange, Euler, Laplace, Poiseuille and Reynolds. Nonetheless it was not until the Second World War that nineteenth century wave theories were successfully put together with observations, military needs spurring the drive towards successful wave forecasting. Subsequently oceanographic studies mushroomed, the same needs, to a large extent, dictating the pace of progress, but accelerated more recently by the development of offshore oil and gas resources. It is probably also fair to acknowledge the contribution made by the advent of the micro-chip and all the capabilities entrained with it.

The alliance of economic, commercial, and military pressures is a formidable one, but it has still needed the base formed by what is essentially scientific curiosity. Just as the combination of theorising and empiricism have helped to bring us to where we are, so has the coupling of intellectual enquiry and the cash nexus. Knowing the results without understanding why is no more satisfactory than postulation without verification. Seeking answers without the wherewithal to find them is equally frustrating.

1.4 Understanding the material

Even such a brief historical excursion as this should enable the work of recent years to be set in its proper context. It is the culmination—so far—of a very long chain of enquiry, and demonstrates how our most recent and burgeoning acquisitions of knowledge are dependent on the past. No less is this true of the behaviour of the material. While it can be conceptualised or idealised on a macro- or meso-scale, its behaviour ultimately is dependent on what takes place on the micro-scale. At that level, fundamental physical and chemical processes are involved and, unless we have due regard to them, it is unlikely that we can make the best use of what we have, let alone understand what is going on.

A historical roll call would once more produce familiar names marking a similarly erratic path of development, but in some respects it is less open to imaginative mis-interpretation; it is to a degree less vulnerable to, or conditioned by, everyday or cultural preconceptions. Preference in the discussion thus far has been given to how our knowledge of 'the sea' has arrived at its present status precisely for that reason. The experience of a fisherman, say, from an island in the Java Sea will be very different from that of the crew-member of a North Atlantic trawler, much as the bushman of the Kalahari will view 'hot' and 'cold' differently from the Eskimo. The units of temperature measurement may be the same but local conditions are decidedly not. Moreover, although local perceptions of climatic severity vary, the surface is still warmed by the same sources. On the other hand a lump of concrete is hard, wherever it is, and so is not so susceptible to varying perception: experience of it is less subjective.

1.5 Adopted approach

No text can be totally comprehensive so a certain amount of selection is inevitable. Apart from factors such as availability of information, relative significance and allowable space, it must be influenced to some extent by personal preference and interest. However, as far as possible, the same general approach has been aimed for throughout, where it seems appropriate. Any review of behaviour and the related evidence must be based on the underlying physical and chemical phenomena and processes. Assessment of the effects of varying some parameter or factor is dependent on the fundamental relationships which have been established, supplemented by the empirical evidence. For example, it is pointless to assume that a reaction will proceed twice as fast in Hong Kong as in Helsinki if physical chemistry suggests otherwise. Consequently, where relevant fundamental relationships can be identified, reference is made to them. This should have the benefit that, as new data becomes available, it may be more easily fitted alongside that which currently exists.

By the same token, it seems equally pointless to elaborate on what happens to materials and structures in the marine environment unless that itself is defined or set out: considering the likely effects of fatigue and impact loading from waves in a particular situation will be rather difficult unless the loads can be quantified. In other words, the context has to be established.

With this in mind two chapters are devoted specifically to the sea and the sort of physical and mechanical environment it provides. Unfortunately, some aspects are more difficult and less quantifiable than others with the result that parts of the text are not uniformly straightforward. Should this be exacerbated by idiosyncrasies of style and presentation, then apologies are due but some relief may be afforded by the fact that it is important not to consider any particular text in isolation: hopefully sufficient references are cited to guide the concerned reader to informed sources.

1.6 Bibliographical context

However hard one might try to avoid duplication and overlap, they are impossible to avoid entirely. Indeed a modicum of repetition is, perhaps, no bad thing since it might facilitate interlock with other texts. Although selection can be invidious, it might be helpful to the reader new to parts of the subject, to mention a few books (in English) which the author has found very useful. Most, if not all, are referred to at some stage in later chapters but a modest introductory bibliography makes them easier to pick out. Neither inclusion nor omission is intended to imply any sort of star or preferential rating, of course, the list simply means they are known to this author.

Less guidance is likely to be needed on concrete *per se* and there is a proliferation of sound references covering many aspects of the behaviour of the material, its structural design, and so on. Quoting three here may then suffice: Lea's 'Chemistry of cement and concrete'; Neville's 'Properties of concrete'; and the 'Handbook of structural concrete', edited by Kong et al (full bibliographic references are to be found in later chapters). In contrast, although there are many books on the sea, there are relatively few with comprehensive guidance on load calculation or the computation of wave behaviour. Wiegel's 'Oceanographical engineering' is a classic with, more recently, Sarpkaya and Isaacson's 'Mechanics of wave forces on offshore structures' and 'Water wave mechanics for engineers and scientists' by Dean and Dalrymple bringing matters in line with later developments as well as fundamental theory. On a more immediately practical note, the 'Shore protection manual' is well known while very recently Gerwick's 'Construction of offshore structures' addresses many of the problems of constructability, installation, and current practice. Although intended primarily for steel structures, 'Planning and design of fixed offshore platforms', edited by McClelland and Reifel, should not be ignored either. T.J.O. Sanderson's 'Ice mechanics, risks to offshore structures' is another important source, mentioned later.

Like all of the selected texts, with their variety of approach and scope, the main purpose of this present one can be summarised in a verse by Longfellow about the building of a ship:

> Build me straight, O worthy Master!
> Staunch and strong, a goodly vessel,
> That shall laugh at all disaster,
> And with wave and whirlwind wrestle!

If we can all succeed in helping to fulfill such a purpose we shall have made a useful contribution.

1.7 Content summarised

At the outset it was indicated that this book is concerned with concrete and the sea. Chapter 2 therefore deals with the marine environment, starting with the basic properties of temperature and salinity, before going on to aspects of behaviour such as tides, surges and currents. The remainder of the chapter is then devoted to waves since, as well as being important, they are more problematic, and an understanding of their behaviour is obviously essential to anyone involved with marine structures of any sort. Then, with Chapter 2 having set out part of the theoretical base line, Chapter 3 goes on to consider some of the sea's actions on structures. Beginning with dynamic pressure, forces due to drag and inertia follow, succeeded by a more extended discussion of the mathematical basis of diffraction theory which is rather less accessible than the others. Sources are given where the methods of attack on the (elaborate) numerical solution of problems can be followed up. In view of the sea's variability, a short discussion of stochastic approaches is provided, and wave slam and wave breaking are referred to since not all wave forces follow from such simplicities as linear theory. Some attention is paid to the inherent uncertainty of load calculation. Wind loading should not be ignored, nor should the modification or amplification of wave (and current) loading as a result of marine growth and fouling. The chapter concludes with a short discussion of ice loading—important in the light of developments in Arctic and sub-Arctic waters.

The primary objective of Chapters 2 and 3 is to provide guidance on the nature and selection of tools to quantify the sea's behaviour and its action on structures. The remaining chapters, by and large, then go on to examine the response of concrete to the sea and how structures might be designed to resist its action.

Plain concrete is the subject of Chapter 4 with durability its most important characteristic. In-service experience has been varied and is documented accordingly. Since permeability is regarded as probably *the* vital factor affecting durability, its underlying processes and relationships are outlined, and reference made to its measurement (potentially useful as a measure of durability, it is argued). Bleeding and pore pressures are then touched on before a more extended discussion of fatigue. Wave loading is clearly an ideal potential initiator of fatigue damage so it is essential that the response of concrete be properly understood, even if it is sometimes only necessary to the extent that durability is affected. Wave slam, collision, and accident also lead to the risk of impact damage. Finally, since the natural environment and storage of hot oil or cold liquids induce temperature effects, the chapter concludes with a consideration of thermal behaviour (climate can also be a major factor in this, especially above the water line).

In reviewing reinforced and prestressed concrete, Chapter 5 to some extent follows a pattern similar to the preceding one. With regard to durability, corrosion is a major contributor to structural deterioration, so the subject is discussed with some emphasis on its electro-chemical nature. This leads on to the subject of bond, prior to elaborating on fatigue. As in chapter 4, impact is incorporated before concluding with thermal behaviour.

Chapter 6 moves on to concrete marine structures, opening with comment on the design process and a note on the consequences of innovation.

Considerations of safety and risk lead naturally to the definition of limit states which feature prominently in the design of marine structures. Design guidance from various sources is then reviewed before the text discusses aspects of the design of oil and gas platforms and Arctic structures, concluding with 'miscellaneous' structures including sea defences.

Finally Chapter 7 deals with one or two items which did not quite fit with the earlier scheme of things. Floating concrete receives some attention although the book as a whole centres on fixed structures. A very important aspect of both types, however, is inspection, maintenance and repair, and no book on marine concrete should ignore it, particularly since it is considered to represent a growth market world-wide.

2 The marine environment

2.1 Introduction

Making successful use of concrete or, for that matter, any material, demands a proper appreciation and understanding of its environment. This is particularly true of the sea with which most of us can claim some familiarity, normally pleasurable. From an engineering point of view, however, we are likely to be concerned more with its extremes such as the highest level its surface will reach and the maximum load it will impose on a structure and the frequency of occurrence of such events. Additionally physical, chemical, and biological properties may be important: density, salinity, composition, potential marine growth, for example.

Since we are dealing with apparently random phenomena (waves generally are wind driven thereby depending largely on weather systems) and there may be a dearth of data for the immediate locality of a particular structure, computation and assessment of effect are made difficult and uncertain. A degree of idealisation is therefore required some of which is presented in this chapter. However, the initial approximations and assumptions can be all too easily obscured in the necessary attempt to apply mathematical, scientific and technological rigour. This can be exacerbated by the power and apparent precision of modern aids to calculation. It is necessary therefore to remember that judgement is more significant than superficial accuracy, particularly in the sea: the unexpected and disastrous does happen as we may be reminded by the loss of the Alexander Kjelland in the North Sea, the Ocean Ranger off Newfoundland, or the extent of subsidence of the Ekofisk field in the North Sea.

In the northern North Sea the design wave height is 30m. Assuming a sinusoidal surface (the normal assumption) and a wave length of 400m, this means that for every lm length of crest, the moving surface itself enfolds 6000 tons of water, the wave velocity being about 55 mph. One might therefore expect the dynamic forces to be appreciable. This has been shown to be the case for major concrete structures in the North Sea where load from a single, 25 m wave is about 50,000 tons for a Condeep structure (described in a later chapter) (Garrison, 1974). Clearly therefore, environmental loading can be

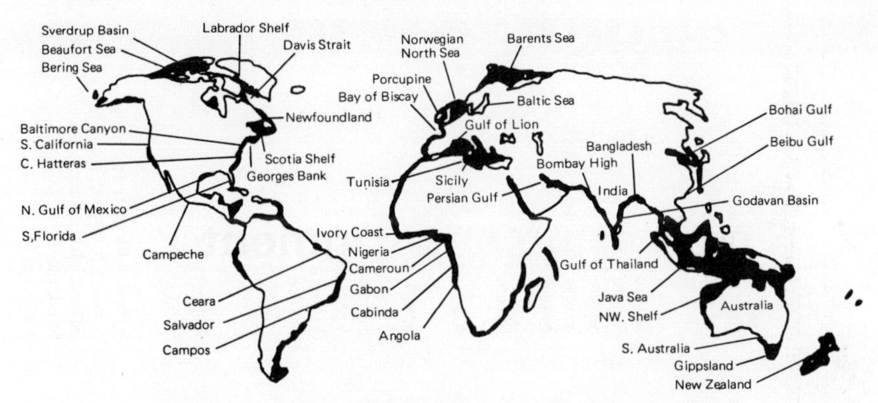

Figure 2.1 The World's principal offshore hydrocarbon basins

substantial, and it has been stated regarding steel platforms in the North Sea, that 'provision for environmental forces involved can account for nearly one third of the designed steel content of a large structure' (Dept. of Energy, 1980).

Load calculation is left to a subsequent chapter. First we present here an abbreviated account of some of the more typical aspects of the sea's properties and behaviour, at least as they are likely to relate to engineering structures, with particular attention being paid to waves.

While the world's seas and oceans take up about 71% of the surface area and the average depth is around 3800 m (the maximum exceeds 11000 m), for most engineering purposes we are concerned with the continental shelves. Conventionally they are defined by the 200 m depth contour (with a range, however, from 20 to over 500 m) and it is below the shelves and the continental slopes that most of the offshore hydrocarbon basins lie, as Figure 2.1 shows (King, 1983). While the small scale may be misleading it is clear that, in addition to the margins of the major continents, the shelves include a substantial part of the Arctic Ocean (the Barents Sea and the Beaufort Sea, for intance), the North Sea, The Baltic, the Mediterranean, the Java Sea, and a number of other important areas. The climatic range is thus complete, from equatorial to polar, although general environmental conditions for design purposes are not directly related, as Table 2.1 indicates (SDA, 1985).

2.2 Temperature

Ocean temperature variation is much less extreme than atmospheric. In moderate climates, as in the mid-latitudes, variations are not likely to be of major structural significance although account must be taken of them as in normal design for inland structures. However, in more extreme climates substantial vertical and lateral temperature gradients may exist. In the Arctic,

Table 2.1 Environmental conditions in major offshore provinces

Country	Offshore provinces—Total area to 200 m water depth (Mn. hectares) (Typical design wave height m)	Representative water depths (metres)		Water Depths		Weather conditions and Max distance to 200 m water depth or sector limit km	
				Current production (m)	Maximum exploration (m)		
Western Europe							
United Kingdom	47 (20–35)	North sea sector	30–200	<50–187	1370	Severe	300 km
		Celtic sea	50			Severe	70 km
Norway	30 (20–35)	North sea sector	30–200	<50–146	407	Severe	280
		Norwegian sea	100–500			Severe with ice hazards	20–50
Denmark	5 (20)	40–60		<50	<100	Severe	300
Ireland	4 (30)	80–500		90	495	Severe	
Netherlands	7 (20)	<70		<50	<100	Adverse	300
West Germany	4 (20)	<40		<30	<100	Severe	300
France	3 (30)	English channel	<200	—	<200	Severe	130
		Atlantic	100–200	—	221	Very severe	200
		Gulf de Lyon	>200	—	2012	Moderate	75
Spain	9 (30)	Atlantic deepwater		—		Severe	70
		Mediterranean prospects		95	1400	Favourable	80
Italy	12 (4–18)	<150m & deepwater prospects		<50–140	827	Favourable	10–20
Greece	4 (4–18)	<100 and deepwater		<50	200	Favourable	5–50
North America							
USA: Alaska	110 (20–30)	Beaufort sea	<100 m	—	<30	Severe with year round ice cover in areas,	65
		Bering Sea	⩽100 m	—	—	Severe with ice hazard	1200
		Gulf of Alaska	<200 m	—	<200	Moderate	270
Pacific	6 (20)	<200 m and deepwater prospects		<50–260	425	Favourable	50

Gulf of Mexico	28 (15–25)	Generally <100 m but also much deeper tracts	<10–312	655	Favourable but violent cyclonic storms	110
Atlantic	10 (30)	Most prospects deep and very deep water	—	1676	Severe-moderate	200
Canada						
Beaufort sea and Arctic Islands	65 (20)	Most prospects <100 m but some very deep water fields in Arctic Islands	—	—	Beaufort—severe, ice cover most of year Arctic Islands—very severe, multi year permanent ice cover	150 10–100
Hudson Bay	131	Shallow but no prospects identified	—	<200	Severe, permanent ice hazards in some areas	550
Eastern Seaboard	57 (35)	<200 & deepwater prospects	(90)	1486	Severe, iceberg hazard	375
West Coast	24 (20)	<200 m. no exploration to date	—	—	Moderate	30–100
Central America and Caribbean						
Mexico, Gulf	15 (20)	50–150 m	100	<200	Favourable but hurricanes	190
Pacific	11 (15)	Prospects <100	—	<100	Favourable	65
Trinidad	2 (20)	<200 m	<80	<200	Favourable	90
Other countries	32	Generally <200	—	—	Favourable	
South America						
Argentina	92 (15–25)	<100 with prospects over 100	<30	<200	Range from favourable to sub-arctic	75–850
Brazil	67 (15)	<50 and 100–200	<20–190	830	Generally favourable	
Chile	16 (15–25)	<100	<30	<200	Range from favourable to sub-arctic	20–30
Columbia	5 (15)	<100	—	201	Favourable	
Peru	11 (15)	<100 in general, potential to 200	50–130	<200	Favourable	30–140

Table 2.1 (Contd.)

Country	Offshore provinces—Total area to 200 m water depth (Mn. hectares) (Typical design wave height m)	Representative water depths (metres)	Water Depths		Weather conditions and Max distance to 200 m water depth or sector limit km	
			Current production (m)	Maximum exploration (m)		
Venezuela	9 (20)	< 150 m	—	< 200	Favourable but hurricanes	
Other countries	24	Generally < 100	—	—	Favourable	
Algeria	2 (15)	Mostly < 100 m	—	< 200	Temperate, favourable	20
Angola	5 (15)	< 100 m	20–70	< 200	Tropical, favourable	45
Cameroon	2 (15)	< 100 m	20–70	< 200	Tropical, favourable	50
Congo	0.3	100 and > 500	60–100	1340	Tropical, favourable	80
Egypt	7 (15)	Gulf of Suez < 100	15–90	< 200	Favourable	
		Red Sea-deepwater prospects	—	555	Favourable	25
		Mediteranean < 100	—	< 200	Favourable	
Gabon	2 (15)	< 100	< 40	< 200	Favourable	45
Ghana	1 (15)	< 100	—	892	Favourable	60
Ivory Coast	1 (15)	< 150 & deepwater	60	569	Tropical	12
Libya	9 (15)	< 200	148	300	Temperate, generally favourable but occasional storms	145
Morocco		50–200 & deepwater	—	242	Temperate, favourable	30–90
Nigeria	7 (10)	< 50	5–30	< 200	Tropical, favourable	60
South Africa	27 (20)	> 50	—	370	Generally favourable but subject to large storms	10–300
Tunisia	19 (15)	< 200	10–160	328	Favourable	330

Other countries	26	—	—	—	
Middle East					
Iran	11 (12)	<50	<50	<200	Very favourable
Kuwait and Neutral Zone	2 (12)	<100	<50	<100	Very favourable
Oman	4 (15)	<100 and possibly deepwater	—	<200	Generally favourable
Qatar	2 (12)	<100 m	10–50	<100	Very favourable
Saudi Arabia	6 (12)	<100	<60	<100	Very favourable
UAE	6 (12)	<100 m	10–60	<100	Very favourable
Far East & Australasia					
Australia	147 (15–25)	Generally less than 150	50–100	1373	Favourable to moderate
Brunei	1	<100	10–90	<200	Favourable
India	33 (25)	<100, wide shallow areas of continental shelf	60–90	<200	Generally favourable
Indonesia	133 (10–20)	Generally <100 but many deeper prospects	15–50	<200	Favourable but typhoons
Malaysia	24 (10–20)	Generally <100	30–80	<200	Favourable but typhoons
New Zealand	27 (25)	100–200	120	<200	Fair to moderate
Phillipines	19 (20)	50–150	90	350	Favourable but typhoons
Thailand	23 (15)	100 in general	70–90	<200	Favourable
Other countries	134	Various	—	—	
CPE's					
USSR: Arctic seas	222				Very severe
Black sea and Caspian	35				Moderate
Pacific areas	143				General severe
European areas	10				Moderate to severe
China	143 (10–20)	Anticipated <200	<30	<200	Moderate to favourable but with frequent typhoons

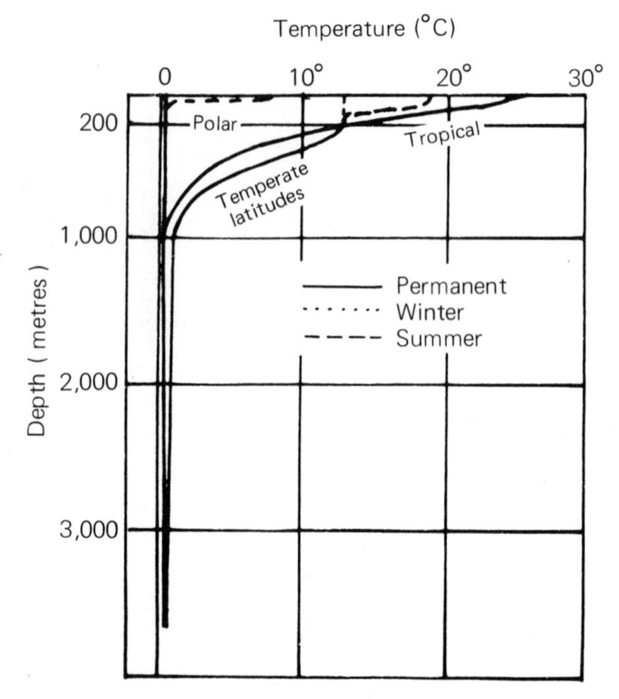

Figure 2.2 Typical oceanic temperature distribution

for instance, atmospheric temperatures below −50°C may cool the upper part of a structure while beneath the ice line the lower part is relatively warm at around 0°C. The converse effect occurs in tropical areas with an upper temperature of possibly 35°C (with higher surface temperatures due to radiant heating) and relative coolness beneath the sea. This is not the place to give world data, it is enough to mention the phenomenon and suggest the need for site-specific data.

Figure 2.2 (Myers, 1969) shows the typical vertical temperature distribution of the ocean. Some of the more interesting effects from an oceanographical point of view occur at too great a depth to concern us here, but the figure does indicate the potential vertical temperature gradient which may exist in deeper structures.

2.3 Salinity

Sea water is, of course, saline with a typical 'salt' content of 33–35‰ (parts per thousand). The predominant elements (in the form of sodium chloride) are chlorine, 19‰, and sodium, 10.5‰. The aggressive effects of this are well known and are discussed later but there is another, occasionally important aspect which should not be forgotten; namely, density.

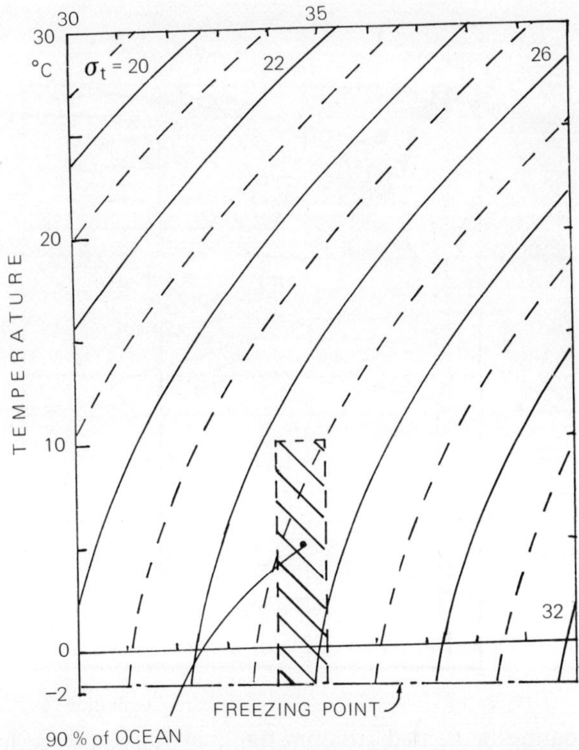

Figure 2.3 Relationship between salinity, density and temperature

2.3.1 *Salinity and density variations*

While affected also by temperature and pressure, the density of sea water is governed primarily by its salinity—Figure 2.3 shows the relationship between salinity, density and temperature (Pond and Pickard, 1978) from which it can be seen that, for example, 1‰ change in salinity gives about 0.7‰ change in density at 20°C. Variations in density mean, of course, that there will be variations in buoyancy, and this is well recognised in the seasonal and geographical differential requirements in the load lines of ships. In the open sea, surface water variations in any one location are normally insignificant in relation to structures (although they may be important oceanographically). However, in some coastal locations seasonal variations, due to snow and ice melt or substantial land precipitation and subsequent runoff, may be much greater. That is, there is a sizeable increase in fresh water supply. Additionally, global differences in salinity can be appreciable as Figure 2.4 indicates (Couper, 1983): the difference between the Mediterranean and the South China Sea is more than 4‰. Seasonal, and more transient, surface differences on the coast of SW Norway are of the order of 2‰ (Dir. Fish. Rsch., 1981). Off the West Coast of Africa values less than 20‰ have been recorded.

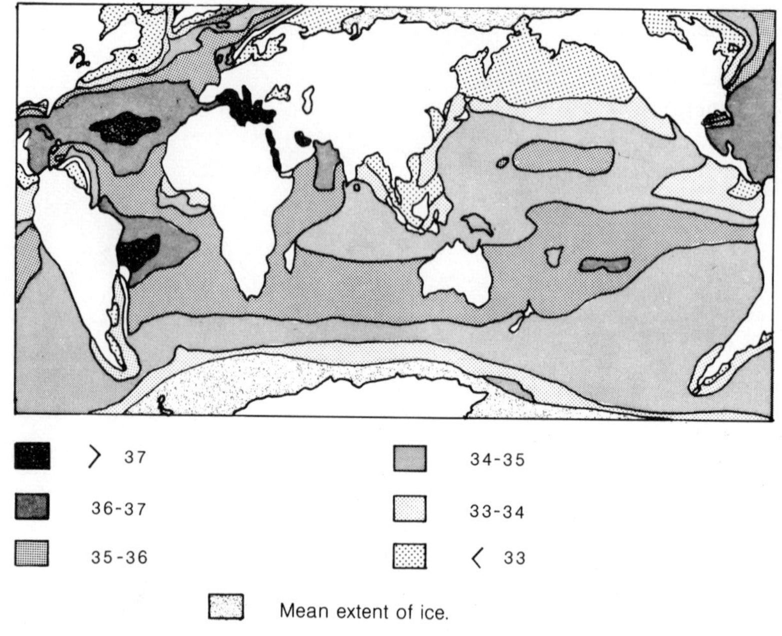

Figure 2.4 Global variations in salinity (in August)

Thus a floating or floated structure built in Italy may lie higher in the water when being installed offshore China, or a structure built in a Norwegian fjord may be lower in the water than anticipated when the deck is being mated in the spring or early summer. For ships and barge-mounted structures where there is a large water-line area the difference in level will be small, but for concrete structures of the Condeep and Sea-tank type, gravity or buoyant, the water-line area is small and appropriate allowance should be made (this is true also for other semi-submersible configurations).

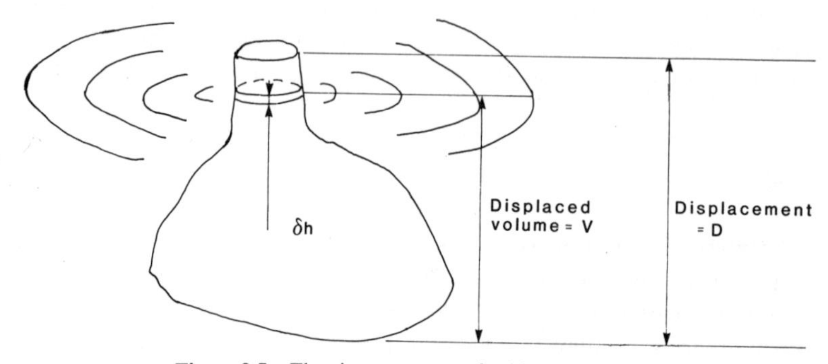

Figure 2.5 Floating structure of arbitrary shape

2.3.2 *Displacement*

Figure 2.5 shows a floating structure of arbitrary shape. Its weight (displacement) is D, V_1 is the volume of water displaced of salinity s_1 and density d_1, and V_2 is the displaced volume of salinity s_2 and density d_2. The change in depth of immersion is δh and A the water-line area.

Then

$$D = V_1 d_1 = V_2 d_2$$

and

$$\begin{aligned} V_1 - V_2 = A \cdot \delta h &= D \cdot \left(\frac{1}{d_1} - \frac{1}{d_2} \right) \\ &= D \cdot \frac{(d_2 - d_1)}{d_1 d_2} \\ &= D \cdot \frac{(s_1 - s_2)0.7}{(1.021)(1.022)} \text{ say} \\ &\approx \frac{2}{3} \cdot D \cdot \delta s \end{aligned}$$

Hence

$$\begin{aligned} \delta h &= \frac{2}{3} \cdot \frac{D}{A} \cdot \delta s \\ &= \frac{2}{3} \cdot \frac{1}{A} \text{ per 1000 tons displacement, for a 1‰ change in salinity.} \end{aligned}$$

If the structure has, say, three towers 12 m outside diameter at the water line then δh is 0.2 m per 100 000 tons displacement. Large displacements and greater variations in salinity can clearly be seen to result in significant changes in depth of immersion and freeboard. Obviously this is not always a problem but it must be checked for such configurations, which are not unusual in concrete offshore structures: a 250 000 ton structure subjected to a salinity change of 2‰ will have a change in draught (immersed depth) of 1 m.

2.4 Tides

Depth of immersion and freeboard are self-evidently important: over-topping under any other than extreme conditions is unlikely to endear the designer to the purchaser of a structure. Water level varies rapidly due to waves but the most obvious regular variation is that due to the tide. Fortunately, however, tidal motion is predictable well into the future and tidal records exist for coastal locations all over the world. Unfortunately similar records do not exist for the open oceans although an area such as the North Sea is well documented.

2.4.1 *Tide generation*

Tide generation is due to gravitational attraction of the moon and the sun, and the earth's rotation. Consequently when earth, sun and moon are in line (as at full or new moon) the principal tidal components due to the moon and sun are in phase and the highest (spring) tides occur. The components are out of phase at quarter and three quarter moon so the lowest (neap) tides occur then. There are several components of different periods but the two main ones are the so-called M_2 component due to the moon (the principal lunar semi-diurnal constituent, frequency 29 degrees per mean solar hour) and the S_2 component due to the sun (the principal solar semi-diurnal constituent, frequency 30 degrees per solar hour) (Hansen, 1962). The S_2 component is about 0.45 of the M_2. The semi-diurnal nature of these generally results in there being two high tides every 24 hours 50 minutes. Dependence on planetary motions means that spring and neap tides alternate at 14–15 day intervals. In the Atlantic the two high tides tend to be equal while in the Pacific they are not. There are also examples of purely diurnal tides.

2.4.2 *Tidal models*

Attempts have been made to model the tides for the world's oceans and Figure 2.6 (Cartwright, 1978) shows the M_2 tide. The co-tidal lines join points where high tide occurs at the same time and orthogonal co-range lines (omitted for clarity) join points where the tidal range is the same. The centres of the 'spider's webs' are amphidromic points (or amphidromes) or tidal nodes, which are points where the tidal range is zero. Moving radially out from such points along co-tidal lines it can be seen that the range increases or, to put it another

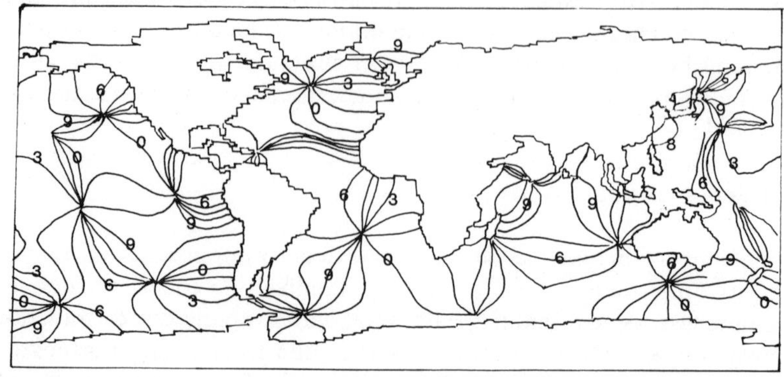

Figure 2.6 M_2 tide model showing co-tidal lines (—). Numbers refer to relating time (hrs) of occurrence of high tide

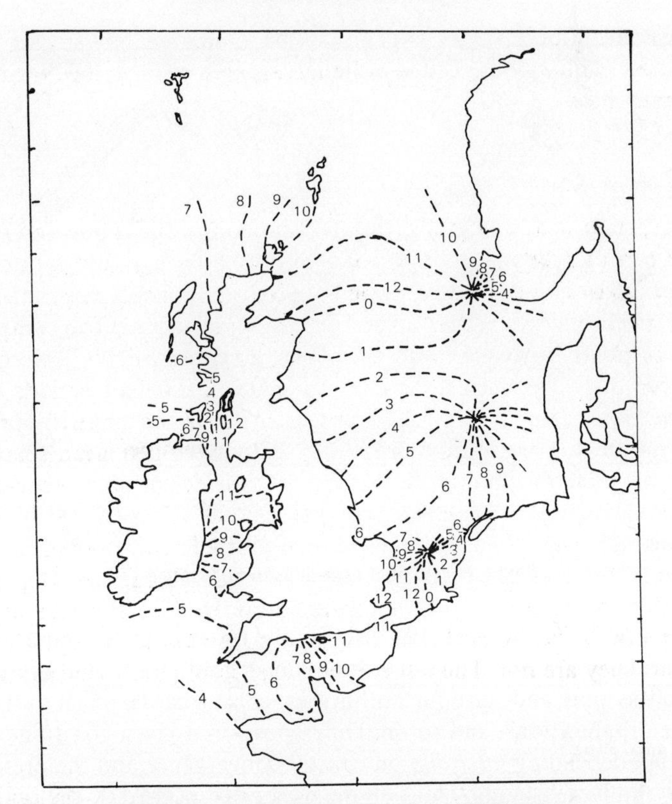

Figure 2.7 Co-tidal lines around the British Isles

way, there is a general tendency for tidal range to *decrease* as one moves out from the coast. Figure 2.7 shows the system around the British Isles (Dir. Fish. Rsch., 1981). From this it is apparent that the tide rotates anticlockwise around the amphidromic points or nodes and this is generally true in the northern hemisphere. In the southern hemisphere rotation is clockwise.

Figure 2.7 indicates that the flood tide (or tidal wave) progresses southwards down the east coast of the British Isles and northwards up the west coast of northern Europe. While generalisations can be made, care is required in applying them. In the South Atlantic for example the net effect is a tidal flow north westwards so that a point on the African west coast will have a high tide some hours ahead of a point on the same latitude on the east coast of South America.

2.4.3 *Variations in water level*

So far as structures are concerned the two primary effects of tidal variations are changes in water level and tidal currents. Changes in level—tidal range—can be considerable with a maximum of 16 m or so in the Bay of Fundy, Nova Scotia. In the Mediterranean, on the other hand, the change is only

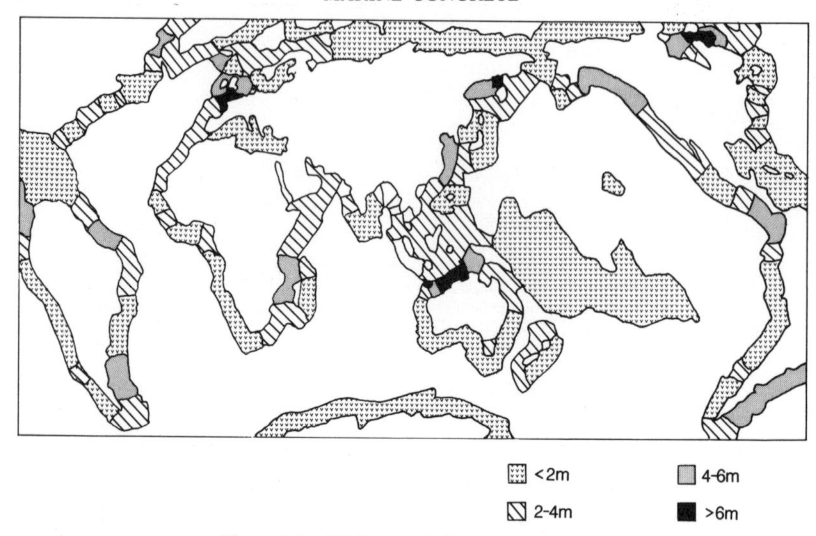

Figure 2.8 Global variations in tidal range

a few cm. Figure 2.8 (Couper, 1983) shows the order of global variations but clearly local, site-specific data must be obtained: guidance should be available from such as port and harbour authorities, coast guards, naval authorities, maritime organisations, and so on. Differences on a given coast line can be substantial, depending *inter alia* on coastal convergence and the presence of estuaries and large islands. Thus, on the west coast of the UK the range may be less than 1 m in the North Channel between Ireland and Scotland but over 11 m in the Severn estuary, only 200 miles or so distant.

2.4.4 *Tidal currents*

Tidal current velocities are generally fairly low (other than at particular locations, as in the Severn estuary), say 3 knots or less. In mid-ocean tidal streams are weak but they tend to increase as they approach the continental shelves and particularly as they close on coasts. Hence, over a large part of the North Sea the velocity is less than 1 knot but rises to over 3 knots off the East Anglian coast of the UK and the southern Dutch coast. Of course, while such velocities are low they, like other currents, will impose a load on the total structure and so their gross effect may be appreciable, particularly on more massive, large volume structures. Again, local guidance must be sought—in the UK for example, the Admiralty publishes appropriate charts. (In fact the Admiralty, via the Hydrographer of the Navy, produces the Admiralty Tide Tables for a substantial number of world ports. For the British Isles, predictions are also produced for around 200 'secondary ports', based on the predictions for 'standard ports'.)

2.5 Surges

Abnormal meteorological conditions can cause significant variations, known as surges, from the computed tidal level. Apart from their effect on deep-draught ships in shallow waters (negative surges) and their obvious influence on the immersed depth and freeboard of offshore structures, positive surges can result in serious coastal flooding. Areas particularly prone to such problems are the Netherlands, eastern England, the Atlantic seaboard of the US, the Gulf of Mexico, the Bay of Bengal, and the Pacific seaboard of Japan (Couper, 1983).

Changes in atmospheric pressure affect sea level resulting in the so-called 'inverted barometer' effect which produces about 0.01 m increase in tidal level for each millibar fall of pressure below the mean. In addition wind drag on the sea's surface produces a change (Pugh 1980). Computer models have been produced to predict these effects for the seas around the British Isles

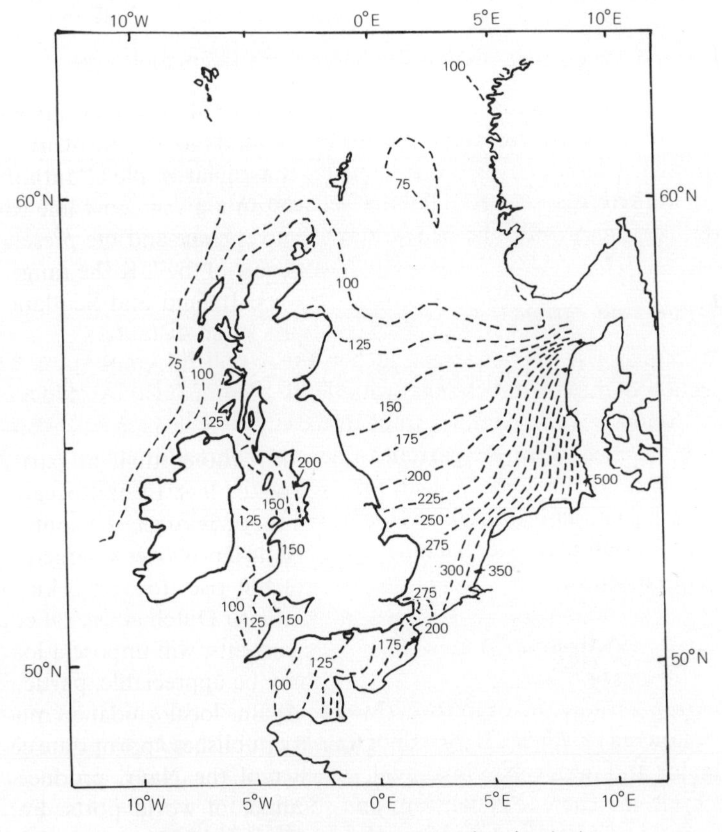

Figure 2.9 Fifty year storm surge elevation (cm)

(Heaps, 1979) (for a Storm Tide Warning Service) and also a chart for 50-year extreme levels in the North Sea, reproduced in Figure 2.9 (Crease, 1983). (Note that the surges also yield currents comparable with the strongest tidal flows although not necessarily in the same direction.) For important structures, or for anything other than preliminary design, it is again clearly essential to obtain as reliable data as possible and so it is wise to consult appropriate authorities such as the national meteorological or oceanographic service/organisation, where available.

2.6 Currents

Several main classes of current may be considered to exist (Wiegel, 1964):

(a) wind-drift currents of relatively short duration;
(b) currents related to surface waves;
(c) tidal currents;
(d) major ocean currents;
(e) nearshore currents due to discharge from rivers and bays.

Analysis of most of these is fairly complex, to match the movements resulting from the various phenomena. Wind drift of long duration, as in the large trade wind fields, is one of the principal components of the major ocean currents so there is some cross-linking between categories. 'Short duration', however, corresponds to one or two days.

2.6.1 *Wind drift currents*

Wind drift gives rise to phenomena such as the so-called Ekman spiral where the direction of the current changes with depth accompanied by reduction in velocity. Analysis is too complex to be dealt with here in such brief compass but, since the effect is for the current to be in a different direction to that of the wind, it results in iceberg drift, for example, not corresponding to the wind. It varies also with latitude and is affected by boundaries such as shores.

Waves are dealt with later (Chapter 2.7), and the current or mass transport effect is discussed there.

2.6.2 *Tidal currents*

Tidal currents rotate their direction through the tidal cycle, with the velocity vector following an ellipse. If the minor axis is zero, the current is alternating; if the major and minor axes are equal, the current is circular; that is during a tidal cycle the current is constant and its direction rotates through 360° like the hand of a clock. As in the other cases, velocities are relatively low in the main (although tidal streams can reach 10 knots) but the masses are large. Global force on a structure can therefore be appreciable. At amphidromic points currents are strong (Evans-Roberts, 1979) so level is not the sole consideration. Atlases or charts of tidal currents are published by various national authorities.

2.6.3 *Ocean currents*

The major circulation systems, as exemplified by the major ocean currents, are complex, as illustration of the surface indicates: Figure 2.10 is one graphic example (Hickling and Brown, 1973). Many of these systems form circulation cells—gyres—but it has to be remembered that the overall system is, in fact, three-dimensional and the figure therefore takes no account of bottom currents, upwelling and downwelling, convergence, and divergence. Deeper currents result mainly from thermal and densimetric differences but, in the present context, we are concerned primarily with the surface layers (in an oceanographic sense—say down to 1000 m or so). Governing factors are wind effects, variations in atmospheric pressure, and density and tidal forces (Vine, 1969). Various numerical models have been constructed and modified for these systems (Pond and Pickard, 1978) the details of which need not concern us here. One rule of thumb is that in trade-wind areas ocean current velocities are around 2–3% of the average wind velocity, say 1/2–1 knot (Vine, 1969).

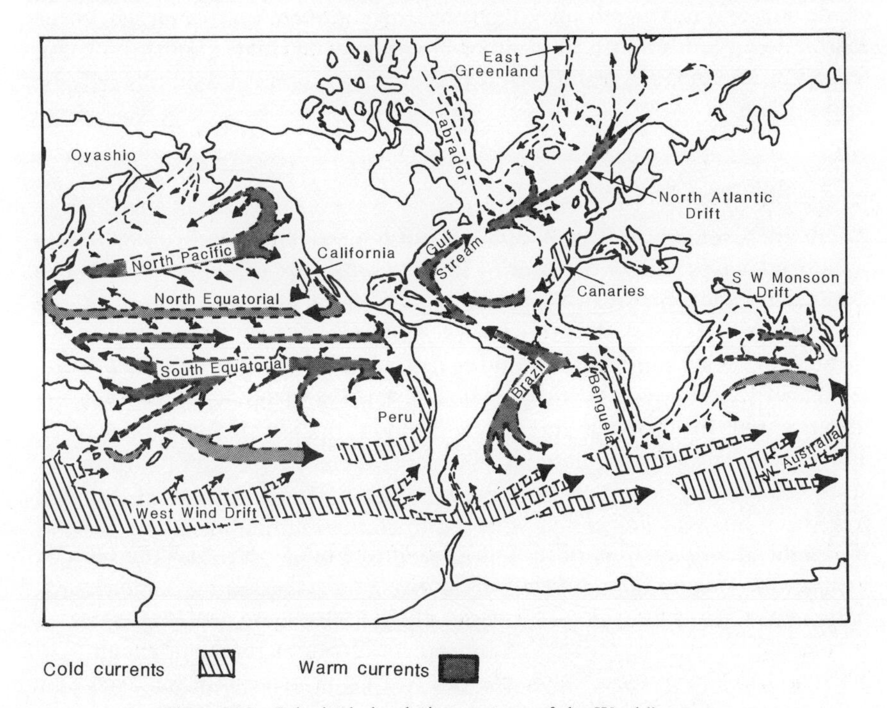

Figure 2.10 Principal circulation systems of the World's oceans

However, major currents such as the Gulf Stream can reach 5 knots and local effects can be significant. Nearshore currents are by their nature local and therefore necessarily site-specific.

Clearly currents may be significant and must be taken into account for loading, scouring, and route-planning for installation tow-out, especially for large structures. It is essential therefore to obtain detailed advice and data for each site. Direction as well as speed must be known and the possibility of wave steepening and breaking or acceleration has to be considered.

2.7 Waves

...waves usually account for a major proportion of the total environmental loading of large installations' (Dept. of Energy, 1980). It is obvious, therefore, that waves are important and their effects must be assessed properly. Unfortunately sea waves and their motions can be very difficult to quantify, and a very considerable effort has been and is being made to produce reliable methods of assessment. For these, mathematical representation is essential but it should always be remembered that it is not absolute, for all the apparent precision of the formulae and expressions used.

Two main approaches must be adopted, the deterministic and the stochastic, and it is worth spending some time on them since computation of load follows, , (dealt with in Chapter 3) and is dependent upon them. First, though, some general remarks and reference to empirical methods may be appropriate.

2.7.1 *Empirical relationships*

There are several main classes of wave but here we are concerned primarily with wind waves and swell, resulting from the effects of wind on the air/water interface (swell comprises wind waves which have travelled out of the wave-generating area—large distances may be involved). Forced waves are those in the generating area, becoming free waves, influenced only by gravity, as they travel beyond it. Generation is dependent on wind velocity, its duration and the distance over which it blows (the fetch). Table 2.2 (Myers, 1969) gives a useful compilation of the general characteristics of various sea states, illustrating how wave height is wind-dependent. A number of attempts have been made to link the various parameters, resulting in the production of graphical aids such as those reproduced in Figure 2.11 (Darbyshire and Draper, 1963) and Figure 2.12 (Ippen, 1966). As will become clear later, wave height *per se* is of limited value: a period must also be ascribed to it.

2.7.1.1 *Significant wave* Significant wave height is a term often used to describe a particular sea state. Its observational definition is the average of

Table 2.2 Wind and sea scale for fully arisen sea

		Wind				Sea								
						Wave height (m)								
Sea state	Description	Beaufort wind force	Description	Range. (knots)	Wind velocity (knots)	Average	Significant	Average highest (1/10)	Significant range of periods. (sec)	Period of maximum energy of spectrum	Average period (sec)	Average wavelength (m)	Minimum fetch. (nmi)	Minimum duration. (hr)
	Sea like a mirror	0	Calm	Less than 1	0	0	0	0						
0	Ripples with the appearance of scales are formed, but without foam crests.	1	Light airs	1–3	2	0.015	0.024	0.03	Up to 1.2 sec	0.7	0.5	0.25	5	18 min
1	Small wavelets, still short but more pronounced: crests have a glassy appearance, but do not break.	2	Light breeze	4–6	5	0.055	0.088	0.113	0.4–2.8	2.0	1.4	2.04	8	39 min
2	Large wavelets, crests begin to break. Foam of glassy appearance. Perhaps scattered white horses.	3	Gentle breeze	7–10	8.5	0.183	0.305	0.366	0.8–5.0	3.4	2.4	6.10	9.8	1.7 hr
					10	0.268	0.427	0.549	1.0–6.0	4	2.9	8.23	10	2.4
3	Small waves, becoming larger; fairly frequent white horses.	4	Moderate breeze	11–16	12	0.427	0.671	0.853	1.0–7.0	4.8	3.4	12.2	18	3.8
					13.5	0.549	0.384	1.13	1.4–7.6	5.4	3.9	15.8	24	4.8
					14	0.610	1.01	1.28	1.5–7.8	5.6	4.0	18.0	28	5.2
					16	0.884	1.40	1.77	2.0–8.8	6.5	4.6	21.6	40	6.6
4	Moderate waves, taking a more pronounced long form; many white horses are formed (chance of some spray).	5	Fresh breeze	17–21										
5					18	1.16	1.86	2.38	2.5–10.0	7.2	5.1	27.4	55	8.3
					19	1.31	2.10	2.65	2.8–10.5	7.7	5.4	30.2	65	9.2
					20	1.52	2.44	3.05	3.0–11.1	8.1	5.7	33.8	75	10

Table 2.2 (*Continued*)

Sea state	Description	Wind: Beaufort wind force	Wind: Description	Wind: Range. (knots)	Wind: Wind velocity (knots)	Sea: Wave height (m): Average	Sea: Wave height (m): Significant	Sea: Wave height (m): Average highest (1/10)	Sea: Significant range of periods. (sec)	Sea: Period of maximum energy of spectrum	Sea: Average period (sec)	Sea: Average wave-length (m)	Sea: Minimum fetch. (nmi)	Sea: Minimum duration. (hr)
5	Large waves begin to form; the white foam crests are more extensive everywhere (probably some spray).	6	Strong breeze	22–27	22	1.95	3.05	3.96	3.4–12.2	8.9	6.3	40.8	100	12
					24	2.41	3.66	4.88	3.7–13.5	9.7	6.8	48.8	130	14
6					24.5	2.50	3.96	5.18	3.8–13.6	9.9	7.0	50.0	140	15
					26	2.93	4.57	6.10	4.0–14.5	10.5	7.4	57.3	180	17
	Sea heaps up and white foam from breaking waves begins to be blown in streaks along the direction of the wind (spindrift begins to be seen).	7	Moderate gale	28–33	28	3.35	5.49	7.01	4.5–15.5	11.3	7.9	64.6	230	20
7					30	4.27	6.71	8.53	4.7–16.7	12.1	8.6	76.2	280	23
					30.5	4.27	7.01	8.84	4.8–17.0	12.4	8.7	78.6	290	24
					32	4.88	7.92	10.1	5.0–17.5	12.9	9.1	86.9	340	27
	Moderately high waves of greater length: edges of crests break into spindrift. The foam is blown in well-marked streaks along the direction of the wind. Spray affects visibility.	8	Fresh gale	34–40	34	5.79	9.14	11.6	5.6–18.5	13.6	9.7	98.1	420	30
					36	6.40	10.7	13.4	5.3–19.7	10.3	10.3	111	500	34
					37	7.01	11.3	14.2	6–20.5	14.9	10.5	115	530	37
					38	7.62	12.2	15.2	6.2–20.8	15.4	10.7	119	600	38
8					40	8.53	13.7	17.7	6.5–21.7	16.1	11.4	135	710	42
	High waves. Dense streaks of foam along the direction of the wind. Sea begins to roll. Visibility affected.	9	Strong gale	41–47	42	9.45	15.2	19.5	7–23	17.0	12.0	150	830	47
					44	11.0	17.7	22.3	7–24.2	17.7	12.5	163	960	52
					46	12.2	19.5	24.7	7–25	18.6	13.1	180	1110	57

9	Very high waves with long overhanging crests. The resulting foam is in great patches and is blown in dense white streaks along the direection of the wind. On the whole, the surface of the sea takes a white appearance. The rolling of the sea becomes heavy and shocklike. Visibility is affected.	10	Whole gale*	48–55	48	13.4	21.6	27.4	7.5–26	19.4	13.8	198	1250	63
					50	14.9	23.8	30.2	7.5–27	20.2	14.3	213	1420	69
					51.5	15.8	25.3	32.3	8–28.2	20.8	14.7	224	1560	73
					52	16.5	26.5	33.5	8–28.5	21.0	14.8	229	1610	75
					54	18.0	29.0	36.9	8–29.5	21.8	15.4	247	1800	81
	Exceptionally high waves (small and medium-sized ships might for a long time be lost to view behind the waves). The sea is completely covered with long white patches of foam lying along the direction of the wind. Everywhere the edges of the wave crests are blown into froth. Visibility affected.	11	Storm*	56–63	56	19.5	31.4	39.6	8.5–31	22.6	16.3	277	2100	88
					59.5	22.3	35.4	45.1	10–32	24	17.0	300	2500	101
	Air filled with foam and spray. Sea completely white with driving spray; visibility very seriously affected	12	Hurricane*	64–71	>64	>24.4‡	>39.0‡	>50‡	10–(35)	(26)	(18)	~	~	~

* For hurricane winds (and often whole gale and storm winds) required durations and fetches are rarely attained. **Seas are therefore not fully arisen.**

† A heavy box around this value means that the values tabulated are at the centre of the Beaufort range.

‡ For such high winds, the seas are confused. The wave crests blow off, and the water and the air mix.

Note: The apparent precision of wave heights and wave lengths is due to conversion from fps units to metric and minimising any further rounding off.

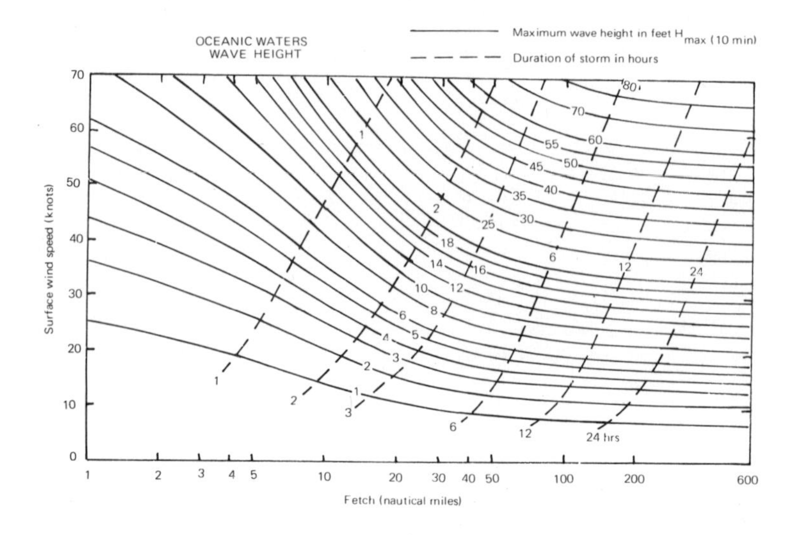

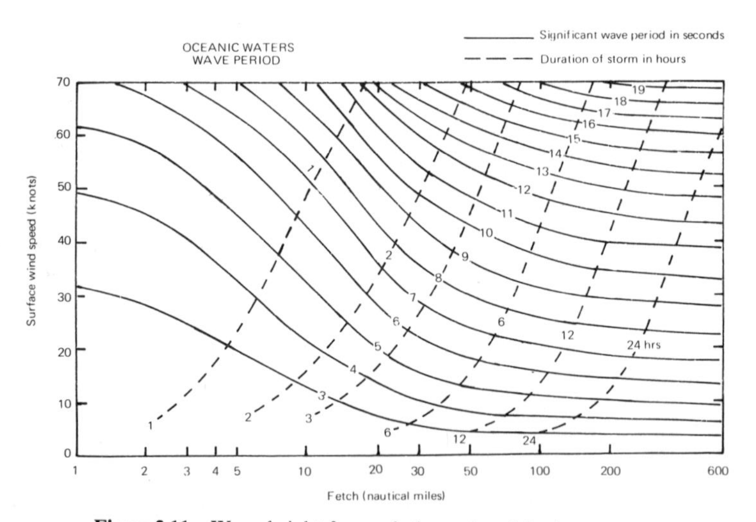

Figure 2.11 Wave height from wind speed and fetch or duration

the highest one-third of the waves. For a narrow band distribution of heights this is in good agreement with the 'mathematical' definition of $4\sqrt{m_0}$ where m_0 is the variance of the surface elevation (this is dealt with in a little more detail later in this chapter). The relationship between significant and maximum wave height will depend on the statistical distribution of heights.

2.7.1.2 *Design wave* The deterministic approach to wave loading may be epitomised by the *design wave* which can be defined loosely as the single

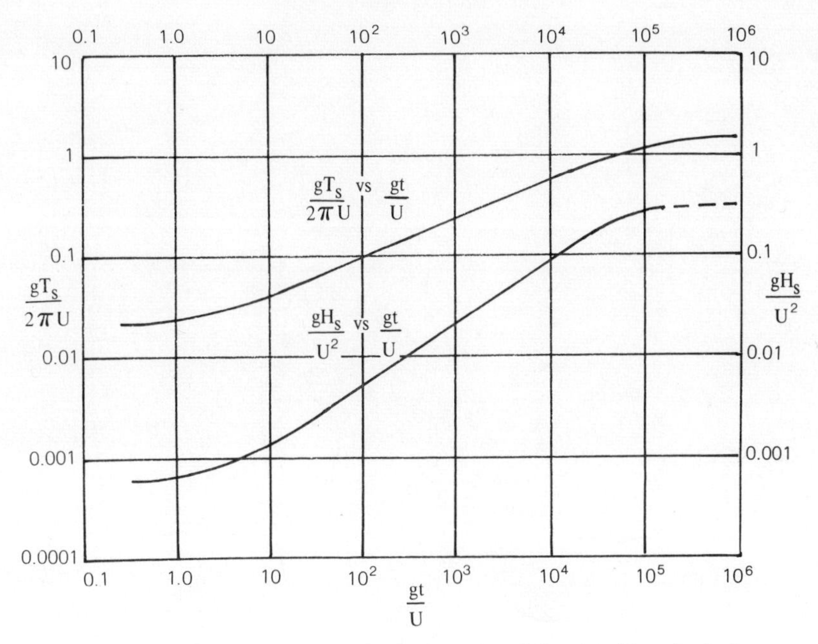

Figure 2.12 Wave height from wind speed and duration (short fetches)

wave against which the whole, or part,of a structure must be designed, to avoid collapse. Such an approach is not entirely deterministic, of course, because the wave to be selected must have a certain probability of occurrence within a given time span, generally for offshore structures, of 50 or 100 years. Figure 2.13 (Dept. of Energy, 1977) shows the UK Department of Energy 50-year wave heights and periods for the North Sea. Once wave height and period have been established, however, the process can be considered deterministic.

2.7.2 *Wave theories*

Computation of wave behaviour and loads depends on having a suitable theory available. St Denis (1969) gives a short historical review of the development of such theories, but it is also worth quoting a more general cautionary comment he makes. '[Deterministic waves are] waves that have been idealised and regularised to such an extent that the knowledge of one implies the knowledge of all. Such waves show up in textbooks and in the laboratory when properly coaxed, but never in the open sea. The waves that do appear in nature present a semblance of disorder—hence, unpredictability. Knowledge of one may, and then again may not, suggest what the next one might look like. When a sea is angered by a storm, there is disarray and confusion. Occasionally, for a brief interval, a relatively small part of the sea

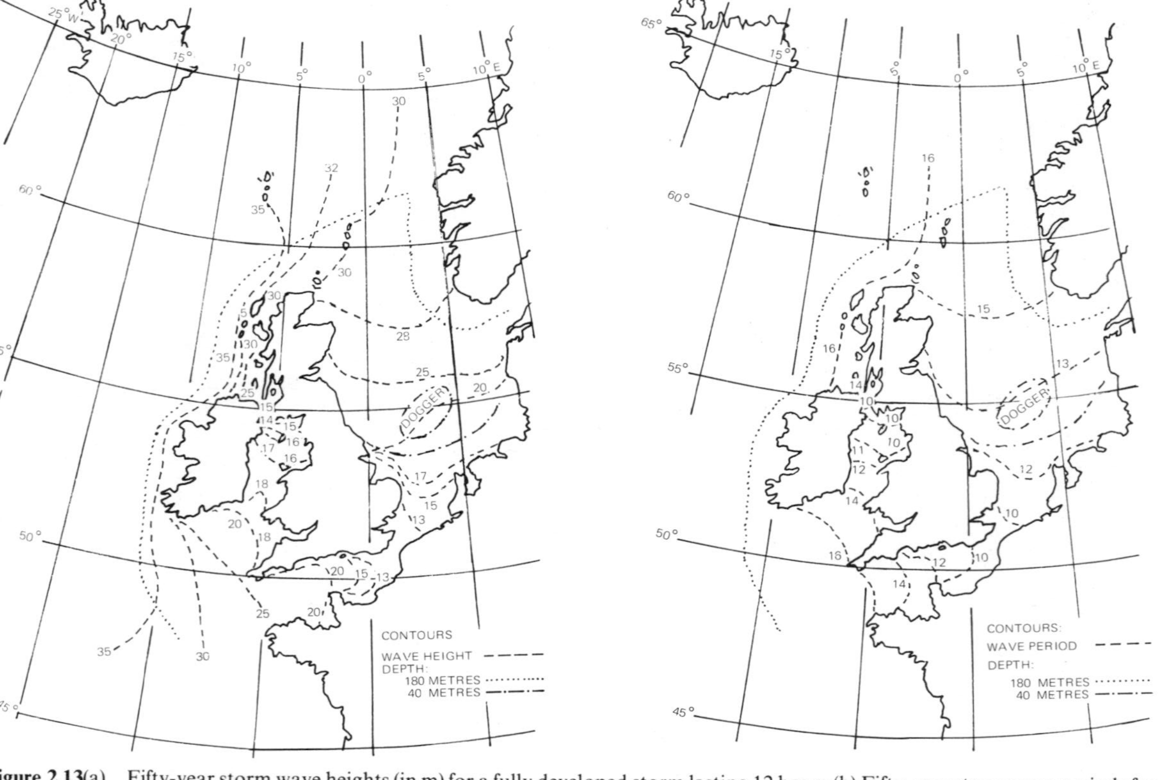

Figure 2.13(a) Fifty-year storm wave heights (in m) for a fully developed storm lasting 12 hours (b) Fifty-year storm wave periods for a fully developed storm lasting 12 hours

may be quiet, but the general appearance will be altogether chaotic. When designing structures to be exposed to the sea or when carrying out operations at sea, one *must* take this confusion into account...'

2.7.2.1 *Significance of water depth* As is perhaps to be expected, water depth has a significant influence on waves. However, in this context the word 'deep' has specific connotations. 'Deep water' is taken to be water whose depth is at least half a wavelength. Thus a puddle might have deep water waves while the sea might not. Clearly too, for some wave heights a given area may be deep, while for other heights it is not. Different theories have been developed, therefore, and Figure 2.14 indicates their range of applicability (Sarpkaya and Isaacson, 1981). The depth ranges are defined by:

shallow water waves	$\frac{1}{20} > \frac{d}{L}$	$0.0025 > \frac{d}{gT^2}$
intermediate depth waves	$\frac{1}{20} < \frac{d}{L} < \frac{1}{2}$	$0.0025 < \frac{d}{gT^2} < 0.08$
deep water waves	$\frac{d}{L} > \frac{1}{2}$	$\frac{d}{gT^2} > 0.08$

where d is water depth, L wave length and T wave period

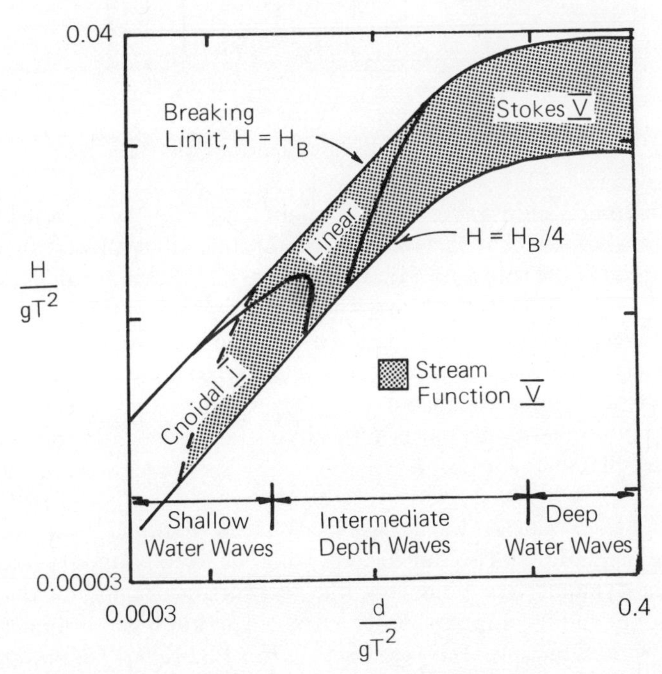

Figure 2.14 Suggested range of application of various wave theories (after Dean)

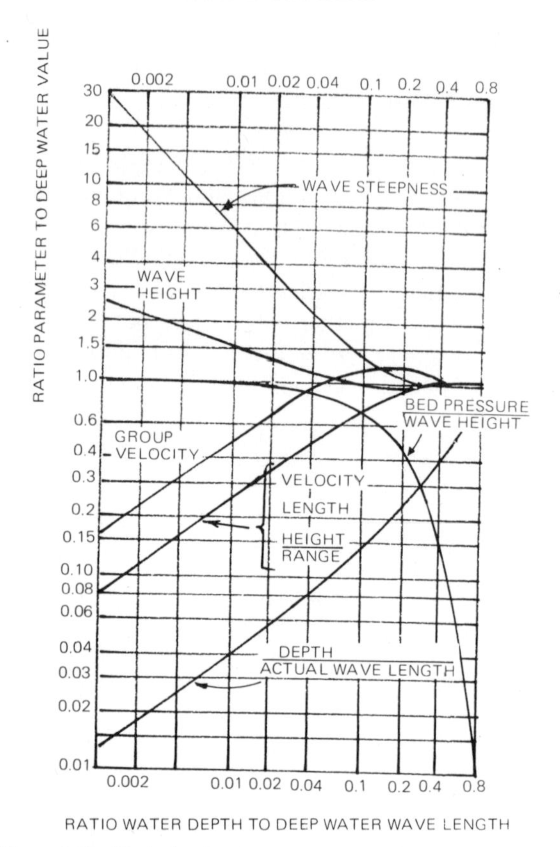

Figure 2.15 Variation in wave parameters with water depth

Furthermore, since waves travel, it might reasonably be expected that their characteristics will change as they move through water of varying depth; as they approach the shore, for example. Figure 2.15 (Barber and Tucker, 1962) shows how wave velocity, wave-length, and group velocity decrease as a wave travels from deep water (on the right) to shallow water (on the left). Conversely, wave height and steepness increase. (It is assumed that the waves approach a straight coast and that they remain sinusoidal.) Barber and Tucker (1962) also give a useful brief physical description of waves and their changing characteristics.

2.7.2.2 *Linear (Airy) wave theory* What is required is a solution to Laplace's equation in two dimensions, with one fixed boundary (the sea bed) and one dynamic free boundary (the sea surface), assuming that flow is irrotational and incompressible. The detailed solution will not be given here since it is in numerous texts (see Lamb, 1962; Lighthill, 1978; Crapper, 1984 for example). (It is interesting to note in passing that Lamb's book was first

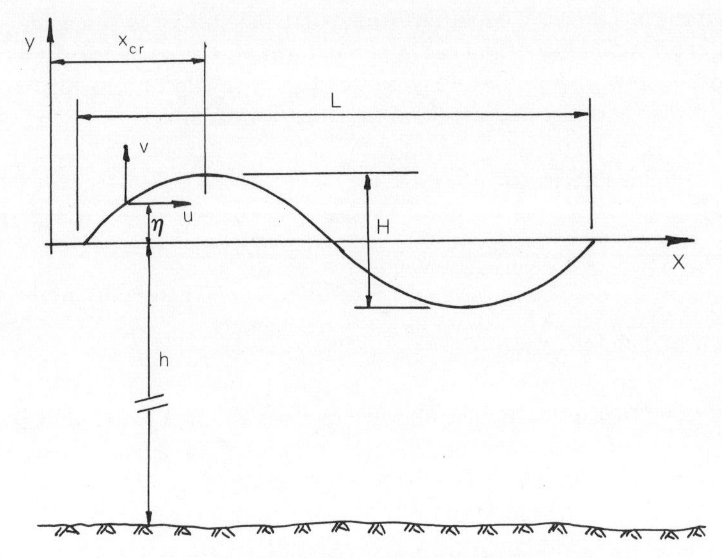

Figure 2.16 Wave form nomenclature

published in 1879.) However, it should be borne in mind that the principal requirements are for the surface profile, wave velocity (celerity), and water particle velocities and accelerations within the range (or depth) of action of the waves. Emphasis is given here to classical, small amplitude deep-water wave theory because of its ease of reproduction and its illustration of certain key points.

Figure 2.16 defines the various symbols used but before giving an outline of the results, it is as well to draw attention to one particular and potentially important point. Theory shows that the water surface is sinusoidal. Such a shape may be defined by a sine or a cosine wave and, in fact, some authors use one and some the other. The engineering advantage of using the cosine wave is that the origin lies below a wave crest and that is convenient when considering the effects of the wave advancing 'through' a structure having started at its leading edge. Generally, engineering texts seem to favour the cosine wave but there is a genuine risk of confusion in working from more than one text unless their starting point is made clear. The texts should be in phase, in other words.

The Laplace equation is

$$\frac{\partial^2 \phi}{\partial x^2} + \frac{\partial^2 \phi}{\partial y^2} = 0 \tag{2.1}$$

where ϕ is the velocity potential. The horizontal velocity of a water particle is $u = -\partial\phi/\partial x$ and the vertical velocity is $v = -\partial\phi/\partial y$.

At a fixed boundary, such as the bed, $\partial\phi/\partial n = 0$ where n is normal to the boundary.

Solution of Laplace's equation shows that the surface is sinusoidal and yields expressions for ϕ and wave speed through the so-called dispersion relation. Knowing ϕ it is an easy matter to derive expressions for particle velocity, displacement and acceleration since, for example:

$$\text{horizontal displacement } u = -\frac{\partial \phi}{\partial x} = \frac{\partial x}{\partial t}$$

$$\text{and so horizontal displacement } x = \int -\frac{\partial \phi}{\partial x} \cdot dt$$

$$\text{and horizontal acceleration} = \frac{\partial u}{\partial t}$$

One or two other terms are useful: wave number $k = 2\pi/L$, wave velocity (or celerity) $c = L/T$ and wave angular (or rotational or circular) frequency $\omega = 2\pi f = 2\pi/T = kc$, where T is the wave period for normal wave frequency f, and L is the wave length. The velocity potential ϕ is given by

$$\left.\begin{aligned} \phi &= -\frac{gH}{2\omega} \cdot \frac{\cosh k(h+y)}{\cosh kh} \cdot \sin(kx - \omega t) \\ &= -\frac{\pi H}{kT} \cdot \frac{\cosh k(h+y)}{\sinh kh} \cdot \sin(kx - \omega t) \end{aligned}\right\} \quad (2.2)$$

where H is the wave height.

Other expressions follow from this:

$$u = \frac{\pi H}{T} \cdot \frac{\cosh k(h+y)}{\sinh kh} \cdot \cos(kx - \omega t) \quad (2.3)$$

$$v = -\frac{\pi H}{T} \cdot \frac{\sinh k(h+y)}{\sinh kh} \cdot \sin(kx - \omega t) \quad (2.4)$$

$$\frac{\partial u}{\partial t} = \frac{2\pi^2 H}{T^2} \cdot \frac{\cosh k(h+y)}{\sinh kh} \cdot \sin(kx - \omega t) \quad (2.5)$$

$$\frac{\partial v}{\partial t} = \frac{2\pi^2 H}{T^2} \cdot \frac{\sinh k(h+y)}{\sinh kh} \cdot \cos(kx - \omega t) \quad (2.6)$$

$$\eta = \frac{H}{2} \cos(kx - \omega t) \quad (2.7)$$

and

$$c^2 = \frac{\omega^2}{k^2} = \frac{g}{k} \cdot \tanh kh \quad (2.8)$$

Equation (2.8) is known as the dispersion relation from which it can be seen that, in deep water (i.e. $h/L \geqslant 0.5$) $\tanh kh = 1$, so $c = \sqrt{g/k} = \sqrt{gL/2\pi}$ i.e.

wave velocity depends on wave length. On the other hand, in shallow water $\tanh kh \approx kh$ and so $c = \sqrt{gh}$ i.e. wave velocity depends on depth. Note also that

$$\theta = kx - \omega t = k(x - ct) = k(x - x_{cr})$$

where x_{cr} is the horizontal distance travelled by the crest in time t.

Convenient approximations can be made for velocities and accelerations in deep and shallow water:

	Deep water	Shallow water
u	$\frac{\pi H}{T} \cdot e^{ky} \cos\theta$	$\frac{\pi H}{T(kh)} \cdot \cos\theta$
v	$-\frac{\pi H}{T} \cdot e^{ky} \sin\theta$	$-\frac{\pi H}{T}\left(1 + \frac{y}{h}\right) \sin\theta$
$\frac{\partial u}{\partial t}$	$\frac{2\pi^2 H}{T^2} \cdot e^{ky} \sin\theta$	$\frac{2\pi^2 H}{T^2(kh)} \cdot \sin\theta$
$\frac{\partial v}{\partial t}$	$\frac{2\pi^2 H}{T^2} \cdot e^{ky} \cos\theta$	$\frac{2\pi^2 H}{T^2}\left(1 + \frac{y}{h}\right) \cos\theta$

It will be noted from the dispersion relation, eq. (2.8) that

$$c^2 = \frac{L^2}{T^2} = \frac{g}{k} \tanh kh = \frac{g}{k} \text{ in deep water}$$

whence

$$L = \frac{g}{2\pi} \cdot T^2$$

$$= 1.56\,T^2 \quad \text{for } L(m), T(s), g(\text{ms}^{-2}) \tag{2.9}$$

and

$$c = \sqrt{\frac{gL}{2\pi}} = 1.25\sqrt{L}\ \text{ms}^{-1} \tag{2.10}$$

Equation (2.9) is useful for speedy assessment of whether particular conditions are 'deep water' or not.

2.7.2.3 *Wave energy* Clearly waves possess energy and it is shown in most texts on waves that E, the total average energy per unit surface area is given by

$$E = \tfrac{1}{8}\rho g H^2 \tag{2.11}$$

i.e. it is dependent only on wave height and is independent of wave length and water depth. The kinetic and potential energies are equal. It can also be shown that waves travelling in groups (as they tend to) travel with a group velocity of half the wave velocity and this is the rate at which energy is transferred (in small amplitude waves there is no mass transfer but there is energy movement).

One further point may be worth remembering: the water particles in deep water, small amplitude waves move in circular orbits which become smaller with depth below the surface. However, as the water becomes shallower so the orbits become flatter and elliptical.

2.7.2.4 *Non-linear wave theories* While linear wave theory has its merits, ease of calculation being one, it has practical limitations, notably its restriction to small amplitude, deep-water waves. For higher, steeper waves and for shallow-water waves, other theories must be adopted. Indeed there may be some doubt as to the accuracy with which linear theory represents actual conditions. Clearly, therefore, it is desirable to adopt a procedure which comes closest to yielding 'true' velocities, accelerations, and profiles.

The application of perturbation theory to the solution of Laplace's equation is outlined by, for example, Stoker (1957) who deals with a number of general cases. The assumption is made that the velocity potential ϕ and surface elevation can be represented by power series expansions in terms of a parameter ε so that

$$\phi = \varepsilon\phi_1 + \varepsilon^2\phi_2 + \varepsilon^3\phi_3 + \cdots \tag{2.12}$$

and $$\eta(x, y, t) = \eta_0(x, y, t) + \varepsilon\eta_1(x, y, t) + \varepsilon^2\eta_2(x, y, t) + \varepsilon^3\eta_3(x, y, t) + \cdots \tag{2.13}$$

In the main, however, Stoker confines his text to theoretical development of the first term of the series which is, in fact, linear theory.

Representative of the extension of the approach are the so-called Stokes fifth-order theory and Dean's stream function theory. The first of these is worth outlining briefly because of its apparent wide use in practice and the second because of its wide range of application. Of course, the free application of these and other higher order theories has been facilitated by the ease and low cost of bulk computation, the procedures involved demanding considerable number crunching, particularly when applied to 'real' structures.

2.7.2.5 *Fifth-order theory* The Stokes fifth-order solution is given by Skjelbreia and Hendrickson (1960). ϕ, the velocity potential is obtained from

$$\frac{k\phi}{\bar{C}} = \sum_{n=1}^{5} A_n \cosh nk(h + y) \sin n\,\theta \tag{2.14}$$

with $$\bar{C}^2 = C_0^2(1 + \lambda^2 C_1 + \lambda^4 C_2)$$

A_n, C_1 and C_2 are constants, $\bar{C}$ is the wave celerity and $C_0\,(=\sqrt{(g/k)\tanh kh})$ is the linear wave celerity. λ is a factor dependent on the wave length and the various constants A_n depend on λ:

$$A_1 = \lambda A_{11} + \lambda^3 A_{13} + \lambda^5 A_{15}$$
$$A_2 = \lambda^2 A_{22} + \lambda^4 A_{24}$$
$$A_3 = \lambda^3 A_{33} + \lambda^5 A_{35}$$
$$A_4 = \lambda^4 A_{44}$$
$$A_5 = \lambda^5 A_{55}$$

A_{11}, A_{13}, A_{15}, etc. are hyperbolic-dependent coefficients of varying unwieldiness. For example, $A_{11} = 1/s$ and $A_{15} = -(1184c^{10} - 1440c^8 - 1992c^6 + 2641c^4 - 249c^2 + 18)/1536s^{11}$ where $s = \sinh kh$ and $c = \cosh \mathrm{kh}$ (and consequently depend upon h/L). A complete listing is given in the Appendix.

The surface profile is given by

$$ky = \sum_{n=1}^{5} B_n \cos n\theta \tag{2.15}$$

with $B_1 = 1, B_2 = (\lambda^2 B_{22} + \lambda^4 B_{24}), B_3 = (\lambda^3 B_{33} + \lambda^5 B_{35}),$
$B_4 = \lambda^4 B_{44}, B_5 = \lambda^5 B_{55}.$

B_{22}, B_{24} etc. are hyperbolic-dependent coefficients (see Appendix).

Computation of y and ϕ (and hence velocity and acceleration) is straightforward, if more than a little laborious, eased by the fact that Skjelbreia and Hendrickson provide tabulated values of A_{mn}, B_{mn} for various ratios of h/L. However, this presupposes that h/L and λ are known in any particular case, whereas normally the available or specified data are more likely to comprise the wave height (H) and period (T) and water depth (h), leaving the wave length L and the parameter λ unknown.

Two simultaneous equations must then be solved to obtain h/L and λ:

$$\frac{\pi H}{h} = \frac{1}{h/L}[\lambda + \lambda^3 B_{33} + \lambda^5 (B_{35} + B_{55})] \tag{2.16}$$

and

$$\frac{h}{L_0} = \frac{h}{L} \tanh kh \; [1 + \lambda^2 C_1 + \lambda^4 C_2] \tag{2.17}$$

The B and C constants are again hyperbolic coefficients and L_0 is the linear theory wave length ($L_0 = (g/2\pi)T^2$). While it remains true that these equations are difficult to solve and require use of a computer, it has been shown (Marshall, 1987) that their intractability can be reduced.

If eq. (2.17) is rearranged as a quadratic in λ^2 and the positive root taken (since we are dealing with real numbers) then

$$\lambda = \left[\frac{-C_1 + \left\{ C_1^2 - 4C_2 \left(1 - \dfrac{h/L_0}{h/L \tanh kh} \right) \right\}^{1/2}}{2C_2} \right]^{1/2} \tag{2.18}$$

h/L lies within a range defined by $h/L_0 = h/L \tanh kh$ and $h/L_0 = h/L$ so the search can be narrowed significantly. In fact, it can be reduced to four or five iterations, depending on the degree of accuracy required, by selecting initially two assumed values for h/L: the quarter points of the range defined above will suffice. Substituting these assumed values of h/L (let us call them h'_{a1} and h'_{a2} respectively) in eq. (2.18) gives values of λ (say λ_1 and λ_2) which, when substituted in eq. (2.16) give corresponding values of h/L, say h'_{c1} and h'_{c2}. The values of the various hyperbolic coefficients are either calculated from the appropriate formulae or from tables if available—these values correspond to h'_{a1} and h'_{a2}.

By making certain assumptions (Marshall, 1987) the next assumed value of h/L, h'_{a3} should be obtained from

$$h'_{a3} = \frac{h'_{c1}h'_{a2} - h'_{c2}h'_{a1}}{(h'_{c2} - h'_{c1}) - (h'_{a2} - h'_{a1})} \tag{2.19}$$

This then enables a further calculated value of λ, λ_3, to be produced and hence h'_{c3}. By substituting (h'_{c3}, h'_{a3}) and (h'_{c2}, h'_{a2}) for (h'_{c2}, h'_{a2}) and (h'_{c1}, h'_{a1}) in eq. (2.19), h'_{a4} can be obtained. Repeating the process until $h'_{an} = h'_{cn}$ produces the desired solution. The example quoted in the reference (Marshall, 1987) might make this clearer.

A location in the North Sea has a water depth of 100 m and a 50-year design wave height of 30.3 m, period 22.2s. Hence $L_0 = 768.83$ m and $h/L_0 = 0.13007$. The other limit of h/L $[=(h/L_0)/\tanh kh]$ is found to be 0.16662. The quarter points of the range then are 0.13921 and 0.15748. Table 2.3 gives numerical values corresponding to successive steps in the procedure.

In deep water the various coefficients are much simplified (see Appendix) and since $\tanh kh = 1$, there is only one limiting value for h/L. It is recommended therefore that $h'_{a1} = 1.05\, h/L_0$ and $h'_{a2} = 1.10\, h/L_0$ be used (or similar small multiples of h/L_0).

To complete this outline, the horizontal particle velocity u follows from partial differentiation of eq. (2.14) (note the signs in this particular

Table 2.3 Iterative determination of ratio water depth:wave length (h/L)

Step	h'_a	λ	h'_c	how h'_a obtained
1.	0.13921	0.25156	0.33759	'quarter point'
2.	0.15748	0.17862	0.20204	'three-quarter point'
3.	0.16277	0.12335	0.13346	eq. (2.19) applied to lines 1 and 2
4.	0.16067	0.14952	0.16458	eq. (2.19) applied to lines 2 and 3
5.	0.16092	0.14678	0.16123	eq. (2.19) applied to lines 3 and 4

development)

$$u = \frac{\partial \phi}{\partial x} = \bar{C} \sum_{n=1}^{5} nA_n \cosh nk(h + y) \cos n\theta \tag{2.20}$$

and hence the acceleration

$$\frac{\partial u}{\partial t} = \omega \bar{C} \sum_{n=1}^{5} n^2 A_n \cosh nk(h + y) \sin n\theta \tag{2.21}$$

2.7.2.6 *Stream function theory* Stream function theory, as postulated by Dean (1965, 1966, 1984), adopted a somewhat different approach since initially it proceeds from the matching of the theoretical wave profile to the measured profile. That is, it solves the Laplace equation by matching *all* the boundary conditions whereas linear and Stokes theory do not match the Kinematic free surface boundary condition over the full wave profile. Where the profile is unknown (for example when only the design wave height and period are specified) the procedure can be adapted accordingly. Of course crest level may be the most important point in the profile where structures are concerned, and it could be misleading to assume that crest and trough are equidistant from the mean surface as in the linear theory. It is therefore important to use a procedure which is likely to yield accurate estimates. Indeed, one particular merit claimed for the stream function theory is that it can be carried to higher orders without the same burden as Stokes-based theories. However, the procedure is not entirely straightforward and is difficult to summarise.

If there is a current, velocity U, present and ψ represents the stream function with η_p the free surface (which is a streamline) then it is assumed that (Dean, 1965)

$$\psi(x, y) = \left(\frac{L}{T} - U\right)y + \sum_{n=4,6,8}^{N-1} \sinh(n-2)\frac{\pi}{L}(h + y) \cdot \left[X(n)\cos(n-2)\frac{\pi}{L}x + X(n+1)\sin(n-2)\frac{\pi}{L}x\right] \tag{2.22}$$

$X(n)$ are the series coefficients which are unknown.

The analogy with potential theory is more clearly seen if this is put in slightly simpler form (Dean and Dalrymple, 1984) with no current present:

$$\psi(x, y) = Cy + \sum_{n=1}^{N} X(n) \sinh [nk(h + y)] \cos nkx \tag{2.23}$$

(compare eq. (2.14) for example).

Note that, in stream function theory, the coordinate reference system is presumed to travel with the wave celerity C so that, in keeping with previous terminology, $\theta = kx$ i.e. the system is rendered steady in relation to time.

The free surface follows from eq. (2.22) by setting $y=\eta_p$ to give a relationship for η_p. The procedure then goes on to calculate the various coefficients in given circumstances by following an iteration procedure to minimise the errors between the calculated and the actual profile (where one is available), or the 'theoretical' wave height and a balance of water quantity about the mean surface. Dean (1966) provides a set of graphs to enable the theory to be applied simply and refers (1984) to a set of tables which also may be used without resort to a computer (1974).

On the evidence presented, the stream function theory appears to be more accurate over a much wider range of application than linear or Stokes theory. For example, it may be used for relatively shallower water waves and for higher waves although Holmes et al (1983) point to the increasing inaccuracy in all these theories as waves approach their limiting height. Despite such limitations, it is questionable if the more recent developments (Cokelet, 1977; Chaplin, 1980; Chaplin and Anastasiou, 1980) have yet reached the stage of wide 'engineering applicability'. Nevertheless, one might speculate on whether or not the design wave is more, rather than less, likely to be steep, although it has been stated (Carter et al, 1986) that open water waves have steepness $< 1/16$, with limited fetch resulting in considerably higher values.

Limiting height, incidentally, is defined as the height at which the maximum particle velocity equals the wave velocity. If it is accepted that the limiting steepness of waves without breaking is $H/L=1/7$ then the inadequacies of linear theory when applied to high or steep waves can clearly be seen from the easily established deep water relation:

$$\frac{u_{\max}}{C}=\frac{\pi H}{L}e^{\pi H/L}$$

For $H/L=1/7$, the maximum value of $u_{\max}/C$ is 0.70—some way short of the limiting height definition.

2.7.2.7 *Comparing theories* A fairly simple method of testing the validity of any particular theory in given circumstances is suggested by Ebbesmeyer (1974). Profiles can be examined for monotonicity: 'a monotonic wave profile is one that continually decreases with increasing distance from the crest'. If the theory has been pushed beyond its range, bumps (or secondary crests) appear in the trough: no bumps, no problem! Dean is cited as noting that no fifth-order Stokes solution was obtainable for certain waves in the range $0.1 < h/T^2 < 0.5$ (fps units). This corresponds to $0.055 < h/L < 0.138$.

2.7.2.8 *Cnoidal waves* Shallow water, then, poses problems. An acceptable theory is that of cnoidal waves, reviewed by Wiegel (1964). The wave profile is formulated in terms of the Jacobian elliptic function cn (leading to the term cnoidal, analogous to sinusoidal). Elliptic functions are doubly periodic, i.e. they are periodic in **two** directions in the $x-y$ plane as distinct from simply periodic functions, such as the sine wave, which are periodic in one direction (Kyrala, 1972). However, for certain relevant conditions the function cn is singly periodic. The theory is too elaborate to be dealt with in a few lines and the reader is referred to Wiegel (1964) for the details.

2.7.3 *Wave refraction*

Waves are affected by a number of factors. They tend to radiate from the generation area thereby losing energy. In any case wave movement is a dispersive process—the longer waves travel faster so there is a spread of energy in the direction of travel. Currents and wind affect them, tending to make them steeper and eventually break when in opposition. Reduction in depth has the same effect when they become subject to bottom friction. Diffraction will be dealt with later in the chapter on loading—it occurs when the presence of an obstacle causes changes in direction and wave characteristics. One particularly significant process for coastal structures is refraction, the manner in which wave crests bend as they approach the coast. There is a tendency for crests to align with the bed contours with the result that energy focuses in some situations and disperses in others, as Figure 2.17 illustrates. Figure 2.18 shows what happens more clearly perhaps.

Bearing in mind that linear theory indicates that wave celerity $\propto h^{1/2}$, it will be clear that the waves will tend to slow down with reducing depth. Consequently, along line v_1v_2 for example, the wave will be slower the nearer it is to the coastline. Hence there will be a tendency for the crest to bend in such shoaling water. In addition, of course, the shallower the water, the more the wave will steepen until it breaks. Returning to Figure 2.17, on line v_1v_2 the crest will be faster in bay B than at headland A. Furthermore, since the frequency remains the same, where the waves slow so the wave lengths must shorten i.e. the crests will be closer together ($L = CT$ i.e. $L \propto C$ for constant

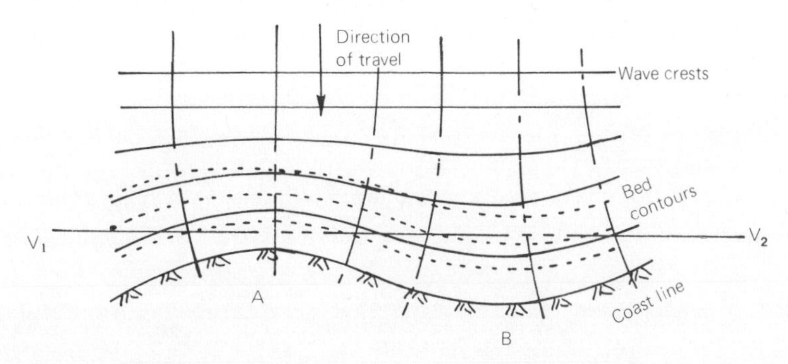

Figure 2.17 Wave refraction: the orthogonals illustrate how wave energy focuses on headland A and disperses in bay B

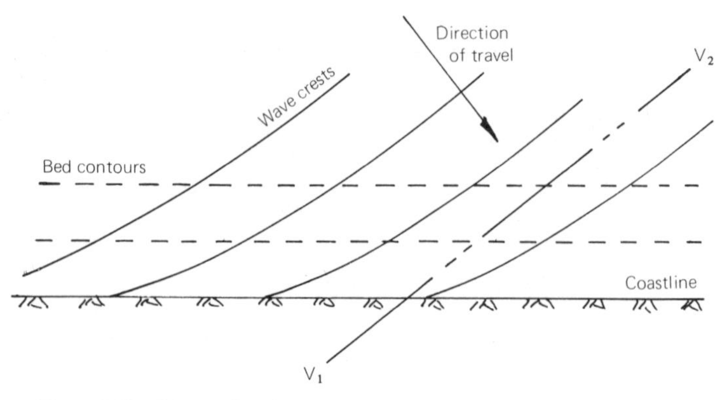

Figure 2.18 Wave refraction: waves approaching a coastline obliquely

T.) Combining such notions with Snell's law of wave refraction has led to the development of methods for calculating shore-line wave heights. More complete accounts are given elsewhere (Wiegel, 1964; Dean and Dalrymple, 1984, among others) so just a brief outline will be presented here.

In what follows it is assumed that wave energy remains constant between orthogonals. The simplest case is taken i.e. a straight shoreline with parallel contours, and tidal effects are ignored. The construction of ray models has been used for many years (Wiegel, 1964; Dean and Dalrymple, 1984) and seems most appropriate to modern computer solution as computational cost has fallen (Townend and Savell, 1983). This generally involves the graphical projection of orthogonals into shallower water, although computer methods seem to produce more accurate results by backtracking from a known point inshore rather than forward tracking from offshore. Forward tracking adopts waves from a fixed direction towards areas of interest while backtracking works from a fixed point to several directions of origin.

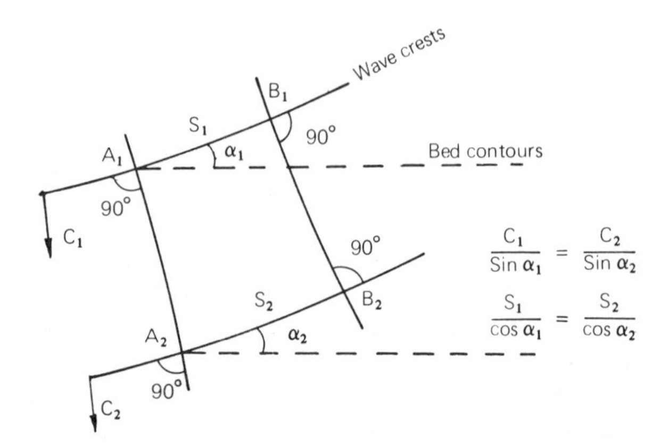

Figure 2.19 Wave refraction: oblique passage over bed contours

In Figure 2.19 A_1B_1 and A_2B_2 are sections of successive wave crests propagating towards a beach with parallel contours. Their velocities are C_1 and C_2 and they make angles α_1 and α_2 as shown. A_1A_2 and B_1B_2 are orthogonals (or rays) and the energy is assumed to be constant between them. (Note that A_1B_1 and A_2B_2 may alternatively be considered as successive positions of a wave crest after period T has elapsed.)

Then Snell's Law gives

$$\frac{C_1}{\sin\alpha_1} = \frac{C_2}{\sin\alpha_2} \tag{2.24}$$

and trigonometry indicates

$$\frac{S_1}{\cos\alpha_1} = \frac{S_2}{\cos\alpha_2} \tag{2.25}$$

assuming that the projections of A_1B_1 and A_2B_2 onto the beach contours are equal.

Now denote the deep water wave characteristics by the suffix 0 and the group velocity by C_g, which is the rate at which energy is transferred (the energy flux).

Since energy $E = (1/8)\rho g H^2$

then $$\tfrac{1}{8}\rho g H_0^2 L_{g0} S_0 = \tfrac{1}{8}\rho g H_1^2 L_{g1} S_1 = \tfrac{1}{8}\rho g H_2^2 L_{g2} S_2$$

But generally $L_g = C_g T$ and T is constant

hence $$H_0^2 C_{g0} S_0 = H_1^2 C_{g1} S_1 = H_2^2 C_{g2} S_2$$

and $$H_2 = H_1 \sqrt{\frac{C_{g1}}{C_{g2}}}\sqrt{\frac{S_1}{S_2}} \tag{2.26}$$

and $$H_2 = H_0 \sqrt{\frac{C_0}{2C_{g2}}}\sqrt{\frac{S_0}{S_2}}$$

$$= H_0 K_s K_r \tag{2.27}$$

where K_s is the shoaling coefficient and K_r is the refraction coefficient. From eq. (2.25) it is easy to see that $K_r = \sqrt{\cos\theta_0/\cos\theta_2}$.

To trace rays by hand the procedure is to start at the depth at which $h = 0.5L_0$ (the wave celerity being calculated for each depth contour). Equation (2.24) then gives the value of α at the next depth contour which enables the subsequent value of α to be determined and so on. Equation (2.27) yields the wave height at any given point. While they are easily calculated, values of K_s and K_r for a range of conditions are given in several sources—for example Dean and Dalrymple (1984) reproduce sets of graphs from the Shore Protection Manual (1975). It should be noted that, when the orthogonals intersect, i.e $S_2 \rightarrow 0$ in eq. (2.26) the wave height $H_2 \rightarrow \infty$ which is obviously unrealistic as diffraction effects then become significant (Holmes, 1983).

By following such a procedure, the wave height and direction can be estimated inshore for given conditions offshore; they are normally designed to work inshore from depths of 40 m or less (Carter et al, 1986). The wave height may also be limited by breaking: the limiting crest-to-trough height is 0.78 times the depth but it may be less when the bed slope is small (< 0.01).

2.7.4 *Design wave*

Implicit in all that has been written so far in this chapter is that a structure will be designed for a wave of certain specified characteristics to which the selected wave theory may then be applied. How then is this 'design' wave to be estimated?

The most satisfactory method of arriving at wave height appears to be the analysis of extreme values (Carter and Challenor, 1981). The wave of a certain small probability of occurrence within a suitable return period (usually 50 or 100 years) has to be found from the record which must have at least 5 years of data—10–15 years for confidence limits of reasonable proportions. This requirement creates some difficulty since data are more than likely sparse, other than for 'popular' or busy waters. Moreover, there are limitations inherent in the selection of a return period (Carter, 1983)—for the 50 years specified for UK waters and a structure with a working life of 25 years, there is a 5% probability that it will encounter the 500-year wave and a 0.25% probability of encountering the 10000 year wave (which is sometimes used in designing critical members of offshore platforms).

To overcome the problem of deficiency of data the Institute of Oceanographic Sciences (IOS) have adopted a method combining long-term and short-term statistics (Carter and Challenor, 1981). This entails recording 15–20 minute wave traces at three hourly intervals, giving an estimate of the significant wave height (H_s) and the most likely highest wave in a three-hour period ($H_{\max,3\,\mathrm{hr}}$). Having obtained these values over a period of at least a year, N-year return values are obtained by fitting a distribution to the data and then extrapolating. If, say, M values of $H_{\max,3\,\mathrm{hr}}$ are obtained from a population with a cumulative probability distribution function $P(h)$ so that

$$\mathrm{Prob}(H < h) = P(h)$$

then the N-year return value h_N is given by calculating

$$P(h_N) = 1 - \frac{1}{MN} \tag{2.28}$$

and finding the height corresponding to $P(h_N)$ from the assumed distribution.

H_{max} may be obtained from

$$\frac{H_{max}}{H_s} = \left[-0.5 \ln(1 - P^{1/N}) \right]^{1/2} \tag{2.29}$$

This is a very brief and indicative summary only of a procedure which can be very sophisticated, although the engineer may not always be as concerned with refinement as the oceanographer. However, the reader is advised to consult references Carter et al, 1986; Carter and Challenor, 1981; Carter, 1983 before engaging in serious estimation.

2.7.5 *Wave variability*

That the sea surface is highly variable is beyond question. It is not enough to consider wave height only, a period must be assigned to the design wave. Moreover, ideally one must look to the variability of the combination of height and period. Again, the possibility of fatigue cannot be ignored and so the frequency of occurrence of loads smaller than that from the design wave has to be assessed. Direction also is important. This then leads naturally to the notion of the wave spectrum: the variation of energy with frequency as in Figure 2.20(a), or with time, as in Figure 2.20(b). The foundation for this is the notion that the surface can be represented by a Fourier series i.e. the sum of a large number of harmonic components.

2.7.5.1 *Wave spectrum* It will be recollected from eq. (2.11) that wave energy is proportional to the square of the wave height (H) and so it is customary to plot H^2 against frequency (f) as in Figure 2.20(a). This has the further advantage that certain wave characteristics are related to moments of the spectrum, according to the assumptions made about the nature of the distribution of wave height. Furthermore, since data are not always available to produce a spectrum, certain 'synthetic' spectra have been derived or developed to facilitate assessment.

General introductions to spectral analysis may be found in texts such as Newland, 1975; Yuen and Fraser, 1979. A structure-oriented approach is to be found in Yang (1986) while a comprehensive discussion on the analysis of the sea surface is provided in Ochi, 1982.

Variation of the sea-surface elevation is a normal Gaussian process with variations above and below zero i.e. water level. However, applying the same considerations to wave height is not possible, since we cannot have negative wave height. In fact, wave height follows the Rayleigh distribution, demonstrated by Longuet-Higgins (1957): earlier, Rice (1954) had developed probability distributions associated with the maxima of random noise current and the maxima of its envelope. Subsequently, Longuet-Higgins (1957) derived a number of two-dimensional distributions applied to the sea surface,

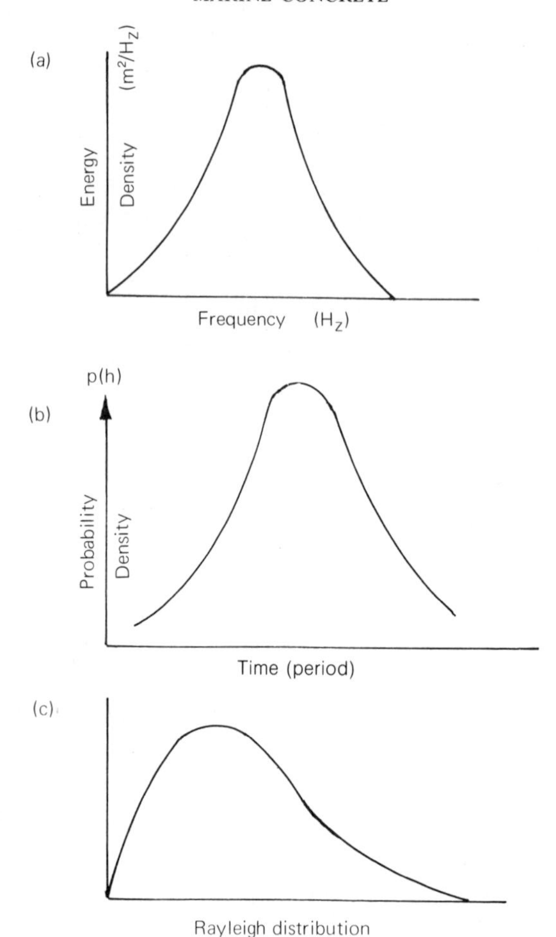

Figure 2.20 The wave spectrum. Variation of energy with (a) frequency and (b) time. (c) The Rayleigh distribution

followed by other developments over a period of years. These references seem to mark a turning point in sea-surface analysis.

The Rayleigh distribution is shown in Figure 2.20(c) and can be given variously as

$$\left.\begin{aligned} \bar{a}P(\eta) &= \mathrm{e}^{-\eta^2/\bar{a}^2}.\frac{2\eta}{\bar{a}} \quad \text{(Longuet-Higgins, 1952)} && \text{(i)} \\ P(\eta) &= \frac{1}{(2\pi)^{1/2} m_{00}^{1/2}} \cdot \exp[-\eta^2/2m_{00}] \quad \text{(L–H, 1957)} && \text{(ii)} \\ P(H) &= \frac{\pi}{2} \cdot \frac{H}{\bar{H}^2} \exp\left[-\frac{\pi}{4}\left(\frac{H}{\bar{H}}\right)^2\right] \quad \text{(Hallam et al, 1977)} && \text{(iii)} \end{aligned}\right\} \quad (2.30)$$

Equations (2.30) (i) and (ii) are for the one-dimensional spectrum, eq. (2.30) (ii) is for a two-dimensional spectrum. $\bar{a}^2$ is the variance, η the surface elevation (as H is the wave peak), m_{00} is the zeroth moment of the two-dimensional spectrum about the origin, $\bar{H}$ is the mean value of the peaks. Generally the theory relates to a narrow-banded distribution, applicable to sea conditions.

2.7.5.2 *Stochastic analysis* Figure 2.21, due to Ochi (1982) is a very effective way of summarising the procedures incorporated in stochastic analysis of sea waves. Recorded data in the time domain (i.e. measurement of wave height over a period of time) is analysed to provide the auto-correlation function. This then yields the wave spectrum which is in the frequency domain (the Wiener-Khintchine theorem shows that the autocorrelation function and

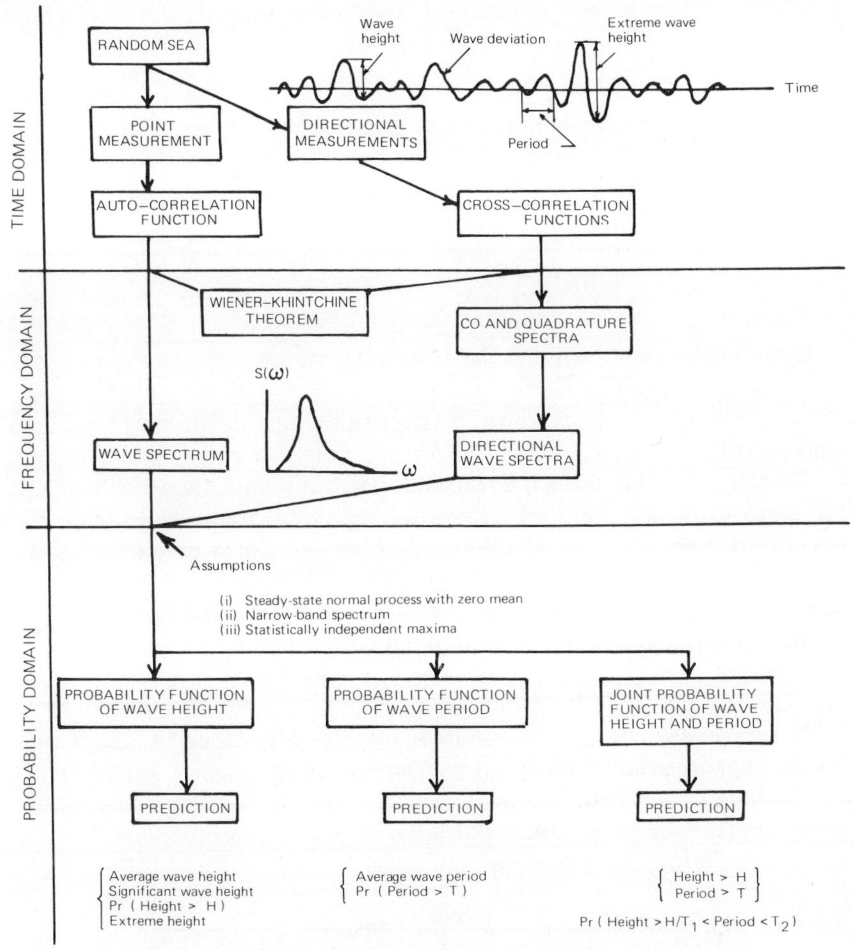

Figure 2.21 Principles and procedures of predicting random seas

the spectral density function—the wave spectrum—are a Fourier transform pair). From the spectrum can then be predicted various wave characteristics through the probability functions. There are three primary assumptions in this procedure (Ochi, 1982):

(a) waves are considered to be weakly steady-state ergodic, normal (Gaussian) random processes with zero mean;
(b) the wave spectral density function is narrow-banded;
(c) wave peaks (maxima) are statistically independent.

If the variation of wave height with time is represented by a function $x(t)$ within a time interval $-T$ to T and $x(t+\tau)$ represents the wave height after a time difference τ, the autocorrelation function $R(\tau)$ is then given by

$$R(\tau) = \lim_{T \to \infty} \frac{1}{2T} \int_{-T}^{T} x(t)x(t+\tau)\,\mathrm{d}t \qquad (2.31)$$

If the spectral density function is $S(\omega)$ where ω is the angular frequency as before, then the Wiener-Khintchine theorem states ($i = \sqrt{-1}$):

$$\left.\begin{aligned} S(\omega) &= \frac{1}{\pi} \int_{-\infty}^{\infty} R(\tau)\exp(-i\omega\tau)\,\mathrm{d}\tau \\ R(\tau) &= \frac{1}{2} \int_{-\infty}^{\infty} S(\omega)\exp(i\omega\tau)\,\mathrm{d}\tau \end{aligned}\right\} \qquad (2.32)$$

It can further be shown that the area under the wave spectrum represents the average wave energy and is equal to the variance of the wave displacement. This then enables the prediction of the statistical properties of wave amplitude and period.

Thus, by sampling the wave record at pairs of points τ apart to produce the autocorrelation function and hence the wave spectrum (strictly, the spectral density function), the door has been opened to prediction of the wave characteristics required for design. A brief description of the procedure adopted is given in Carter et al (1986) and the Fast Fourier Transform technique incorporated is described in publications such as Yuen and Fraser (1979).

2.7.5.3 *Wave spectra* Physical data in the ocean are difficult and expensive to obtain, consequently attention has been given to the generation of what we might term 'synthetic' spectra. Current practice favours the Bretschneider spectrum (Bretschneider, 1963) which for frequency takes the form

$$S_{\eta\eta}(f) = Af^{-a}\exp(Bf^{-b})$$

where f is frequency and A, B, a, b are constants.

Generally $a = 5$ and $b = 4$ in such as the Pierson–Moskowitz (P–M) spectrum (widely used) or its more recent 'fetch-limited seas' form, the JONSWAP spectrum (JONSWAP–Joint North Sea Wave Project).

P–M $$S(f) = \alpha_{PM} g^2 (2\pi)^{-4} f^{-5} \exp[-\beta(f/f_0)^{-4}]$$

α = Phillips constant = 0.0081

$\beta = 0.74$

$$f_0 = \frac{g}{2\pi U}$$

U = wind velocity at 19.5 m above surface

Examination of this shows that the spectrum is related to wind speed. For fetch-limited seas the situation is not so straightforward as JONSWAP shows. For example, α is not a constant but is in fact fetch-dependent (Ewing, 1974).

JONSWAP $$S(f) = \alpha g^2 (2\pi)^{-4} f^{-5} \exp\left[-\frac{5}{4}\left(\frac{f}{f_m}\right)^{-4}\right] \gamma^{\exp[-\{(f-f_m)^2\}/2\sigma^2 f_m^2]}$$

$\sigma = \sigma_a$ for $f \leqslant f_m$

σ_b for $f > f_m$

f_m = frequency of spectral peak

$\tilde{f}_m = 3.5\tilde{x}^{0.33}$ = non-dimensional frequency of spectral peak

x = fetch

$\tilde{x}$ = non-dimensional fetch

$\alpha = 0.076\tilde{x}^{-0.22}$

γ = ratio of maximum spectral energy to the maximum of the corresponding P–M spectrum

$\sigma_a = 0.07$

$\sigma_b = 0.09$

$\tilde{m}_0 = m_0 g^2/U_{10}^4 = 1.6 \times 10^{-7}\tilde{x}$ = non-dimensional variance

m_0 = variance (= zeroth moment of spectrum)

U_{10} = wind speed 10 m above the surface

For a wind speed of 20ms^{-1}, Table 2.4 (Ewing, 1974) gives an indication of the variation of significant wave height H_s, f_m and α with fetch.

Table 2.4 Variation in significant wave height (H_s), frequency of spectral peak (f_m) and α with fetch (x)

x *km*	f_m *Hz*	H_s *m*	α
500	0.076	7.2	9.6×10^{-3}
1000	0.061	10.2	8.2×10^{-3}
2000	0.049	14.5	7.1×10^{-3}

Clearly JONSWAP is more complex to apply than Pierson-Moskowitz and so certain applications of the spectrum are more easily demonstrated by Pierson–Moskowitz.

If m_n denotes the nth moment of the spectrum, then it can be shown (see Ochi, 1982, for example) that the significant wave height H_s is given by

$$H_s = 4\sqrt{m_0} \tag{2.33}$$

This corresponds to the physical definition of H_s as the mean of the highest one-third of the waves. m_0 is simply the area under the spectrum, of course.

The average of the highest $1/N^{\text{th}}$ wave heights $\bar{H}_{1/N}$ is given by (again, for the Rayleigh distribution)

$$\bar{H}_{1/N} = 2\sqrt{2 \ln N}\sqrt{m_0} \quad \text{for large } N \tag{2.34}$$

The 'mean' period T_1 is given by

$$T_1 = \frac{m_0}{m_1} \tag{2.35}$$

where 'mean' period is taken as the reciprocal of the mean frequency.

The mean period T_c taken as the mean time between successive crests follows from

$$T_c = \sqrt{\frac{m_2}{m_4}}$$

while the mean period T_z between successive zero up-crossings is $T_z = \sqrt{m_0/m_2}$.

It is stated (Carter et al, 1986) that the 'up-crossing' period is the one often used for engineering purposes although 'crest' period might be more appropriate for some fatigue problems. However m_4 is difficult to evaluate due to practical measurement problems (it is particularly affected by the tail of the distribution).

To return to the Pierson–Moskowitz spectrum as a representation of the Bretschneider spectrum:

$$m_n = \int_0^\infty f^n S(f)\,\mathrm{d}f \tag{2.36}$$

Substituting for $S(f)$ yields

$$m_n = \int_0^\infty f^n A f^{-5} \exp(B f^{-4})\,\mathrm{d}f$$

which, on integrating by substitution, results in

$$m_n = \frac{A}{4B^{1-n/4}} \int_0^\infty x^{-n/4} \mathrm{e}^{-x}\,\mathrm{d}x$$

and $\int_0^\infty x^{-n/4} e^{-x} \mathrm{d}x = \Gamma(1 - n/4)$ by definition of the gamma function (bear in mind too that $\Gamma(n+1) = n!$).

Hence
$$m_n = \frac{1}{4} A B^{(n/4-1)} \Gamma\left(1 - \frac{n}{4}\right) \text{ for } n < 4 \tag{2.37}$$

Consequently and
$$\left.\begin{aligned} H_s &= 2(A/B)^{1/2} = 0.0213 U_{19.5}^2 \\ T_1 &= \frac{1}{1.225 B^{1/4}} = 0.5635 U_{19.5} \end{aligned}\right\} \tag{2.38}$$

for the P–M spectrum.

It will be found that m_0 (and hence H_s) can also be derived directly from $m_0 = \int_0^\infty S(f)\mathrm{d}f$ on integrating by substitution but higher moments are not so tractable.

References

Barber, N.F. and Tucker, M.J. (1962) *Wind waves, The Sea*, ed. M.N. Hill, Vol. 1. Physical Oceanography, Interscience Publ.

Bretschneider, C.L. (1963) A one-dimensional gravity wave spectrum. *Proc. Conf. Ocean Wave Spectra*, Prentice-Hall.

Carter D.J.I. (1983) The probability of very high waves in deep water. *J. Soc. Underwater Tech.* Winter: 22–24.

Carter, D.J.I. and Challenor, P.G. (1981) Estimating return values of wave height. *Report no. 116, Inst. of Oceanographic Sciences.*

Carter, D.J.I., Challenor, P.G., Ewing, J.A., Pitt, E.G., Srokosz, M.A. Tucker, M.J. (1986) Estimating wave climate parameters for engineering applications. *Offshore Techy. Report OTH 86 228*, Dept. of Energy, HMSO, London.

Cartwright, P.E. (1978) Oceanic tides. *International Hydrographic Review, Monaco, LV(2)*, July.

Chaplin, J.R. (1980) Developments of stream function wave theory. *Coastal Engineering.* **3**: 179–205.

Chaplin, J.R. and Anastasiou, K. (1980) Some implications of recent advances in wave theories. *Proc. 17th Int. Conf., Coastal Engineering*: 31–49.

Coastal Engineering Research Center. (1975) *US Army, Shore Protection Manual.* US Govt. Printing Office, Washington DC.

Cokelet, E.D. (1977) Steep gravity waves in water of arbitrary uniform depth. *Phil. Trans. Royal Soc.* **A286**: 183–230.

Couper, A. (eds.) (1983) *The Times Atlas of the Oceans.* Times Books Ltd., London.

Crapper G.D. (1984) *Introduction to Water Waves.* Ellis Horwood, Chichester.

Crease, J. (1983) Extreme surge heights in the North Sea. *MIAS News Bull., No. 6, Inst. Oceanographic Sciences*, March: 8–9.

Darbyshire, M. and Draper, L. (1963) Forecasting wind-generated sea waves. *Engineering*, **195**, April: 482–484.

Dean, R.G. (1965) Stream function representation of non-linear ocean waves. *J. Geophys. Rsch.* **70(18)**, Sept: 4561–4572.

Dean, R.G. (1966) Stream function wave theory: validity and application. *Coastal Engineering, Proc. Santa Barbara Speciality Conf., ASCE.*

Dean R.G. (1974) Evaluation and development of water wave theories for engineering application, Vols. 1 and 2. *Sp. Report 1, US Army, Coastal Eng. Rsch. Center*, Fort Belvoir, Va.

Dean, R.G. and Dalrymple, R.A. (1984) *Water Wave Mechanics for Engineers & Scientists.* Prentice-Hall, Englewood Cliffs.

Directorate of Fisheries Research (1981) *Atlas of the Seas around the British Isles.* Min. of Agriculture, Fisheries & Food, Lowestoft.

Ebbesmeyer, C.C. (1974) Fifth Order Stokes wave profiles. *J. Waterways, Harbors & Coastal Eng. Div., Proc. ASCE*, **100**, WW3, Aug: 264–265.

Energy, Dept. of. (1977) *Offshore installations: Guidance on design and construction.* HMSO, London.

Energy, Dept. of. (1980) *Review of the fluid loading research programme, OTP7, CIRIA.* London.

Evans-Roberts, D.J. (1979) Tides in the Persian Gulf. *Consulting Engineer*, June: 46–48.

Ewing, J.A. (1974) Some results from the Joint North Sea Wave Project of interest to engineers. *Proc. Conf. Dynamics of Marine Vehicles & Structures in Waves, I Mech E.* London.

Garrison, C.J., Tørum, A., Iversen, C., Leivseth, S., Ebbesmeyer, C.C. (1974) Wave forces on large volume structures. *Proc. Ann. Offshore Tech. Conf.*, Houston.

Hallam, M.G., Heaf, N.J. and Wootton, L.R. (1977) *Dynamics of marine structures, CIRIA-UEG.* London.

Hanson, W. (1962) *Tides, The Sea*, (ed. M.N. Hill), Vol. 1., Physical Oceanography, Interscience Publ.

Heaps, N.S. (1979) Computer models for sea level prediction. *MIAS News Bull., No. 2, Inst. Oceanographic Sciences*, Aug: 10–13.

Hickling, C.F. and Brown, P.L. (1973) *The Seas & Oceans in Colour*, Blandford Press, London.

Holmes, P. (1983) Waves at shorelines. *Proc. Conf. Shoreline Protection, Inst. Civil Engs.* Thos. Telford, London.

Holmes, P., Chaplin, J.R. and Tickell, R.G. (1983) Wave loading and structure response. *Proc. Conf. Design in Offshore Structures, Inst. Civil Engrs.* Thos. Telford, London.

Ippen, A.T. (ed.) (1966) *Estuary and coastline hydromechanics.* Eng. Soc. Monograph, McGraw-Hill, New York, quoted in Hallam et al.

King, R.E. (1983) Exploration: the search is moving to new areas. *World Oil*, **197(1)**, July.

Kyrala, A. (1972) *Applied Functions of a Complex Variable.* Wiley-Interscience, New York.

Lamb, H. (1962) *Hydrodynamics 6th ed.* Cambridge Univ. Press.

Lighthill, J. (1978) *Waves in Fluids.* Cambridge Univ. Press.

Longuet-Higgins, M.S. (1952) On the statistical distribution of the heights of sea waves. *J. Marine Rsch.*, **X1(3)**: 245–266.

Longuet-Higgins, M.S. (1957) The statistical analysis of a random, moving surface. *Phil. Trans. Royal Soc., Series A*, **249(966)**, Feb: 321–387.

Marshall, A.L. (1987) On implementing Fifth Order gravity wave theory. *Underwater Technology*, **13(1)**: 2–4.

Marshall, A.L. (1988) Wave load estimation in deep water as affected by wave period errors. *Proc. 7th Int. Conf. Offshore Mechanics & Arctic Engineering, ASME*, Houston.

Myers, J.J. (ed.) (1969) *Handbook of Ocean and Underwater Engineering.* McGraw-Hill, New York.

Newland, D.E. (1975) *Random vibrations and spectral analysis.* Longman, London.

Ochi, M.K. (1982) Stochastic analysis and probabilistic prediction of random seas. *Advances in Hydroscience, Vol. 32*, Academic Press: 217–375.

Pond, S. and Pickard, G.L. (1978) *Introductory Dynamic Oceanography.* Pergamon Press, Oxford.

Pugh, D.T. (1980) Sea level variability and extremes. *MIAS News Bull. No. 3. Inst. of Oceanographic Sciences*, June: 6–10.

Rice, S.O. (1944–45, 1954) Mathematical analysis of random noise. *Bell System Tech. J. 23 & 24*, and *Selected Papers on Noise & Stochastic Processes*, ed. N. Wax, Dover, New York.

Sarpkaya, T. and Isaacson, M. (1981) *Mechanics of wave forces on offshore structures.* Van Nostrand Reinhold, New York.

Scottish Development Agency. (1985) *A review of international offshore oil and gas markets.* SDA, Aberdeen.

Skjelbreia, L. and Hendrickson, J.A. (1960) Fifth Order gravity wave theory. *Proc. 7th Coastal Eng. Conf.*, the Hague, Aug: 184–196.

St. Denis, M. (1969) On wind generated waves, propagation in the open sea. *Topics in Ocean Engineering* ed. C.I. Bretschneider, Gulf Publ. Co.

Stoker, J.J. (1957) *Water Waves.* Interscience Publ. New York.

Townend, I.H. & Savell, I.A. (1983) *The application of ray methods to wave refraction studies, Polymodel* 7. Conf. Mathematical Modelling, Sunderland Polytechnic.

Vine, A.C. (1969) Currents and circulation. *Handbook of Ocean & Underwater Engineering*, ed. J.J. Myers, McGraw-Hill, New York.

Wiegel, R.L. (1964) *Oceanographical Engineering*, Prentice-Hall, New Jersey.

Yang, C.Y. (1986) *Random Vibration of Structures.* Wiley, New York.

Yuen, C.K. and Fraser, D. (1979) *Digital Spectral Analysis.* Pitman, London & CSIRO, East Melbourne.

3 Actions/loads on structures

3.1 Introduction

In the preceding chapter we considered various aspects of the sea's behaviour. The problem for the engineer is how to translate these into quantified actions on structures so that fitness for purpose can be assured. Durability is considered elsewhere, here we consider static and dynamic loading. In the latter case, of course, response is an important factor so that load and effect form a system. Similarly, wave behaviour may be affected by the presence of the structure: through reflection from a wall, for example. Again, in dealing with a floating structure, the mooring system must be included as an integral part of the whole, and drift forces become important. Other forms of loading feature also: wave slam, wave slap, breaking waves, fatigue, to say nothing of the live and dead loads from equipment, stored liquids, surcharge, earthquake, ice, fire and tempest. The ancients' notions of the elements, earth, water, air and fire, are not far removed from the catalogue we need to examine, given one or two additions.

The engineer's duty is to assess and quantify the combination appropriate to the task in hand. Let us then first deal with the sea.

3.2 Pressure

The most obvious first consideration is the elementary one of hydrostatic pressure p at depth y below the surface:

$$p = -\rho g y \tag{3.1}$$

Allowance has to be made for variation in level due to the tide, atmospheric pressure possibly, and surge. There is, in addition, a dynamic fluctuation from the passage of a wave.

As has been shown (Dean and Dalrymple, 1984), below the surface of a wave

$$p = -\rho g y + \rho g \eta K_p(y) \tag{3.2}$$

where $\eta = (H/2)\cos\theta$ is the surface variation from mean water level and $K_p(y) = \cosh k(h + y)/\cosh kh$ is the pressure response factor

This follows from the Bernoulli equation, ignoring the velocity terms:

$$\frac{p}{\rho} = -gy + \frac{\partial \phi}{\partial t} \tag{3.3}$$

For pressure above the mean water level, it is shown (Dean and Dalrymple 1984) from linear theory that

$$p = \rho g(\eta - y) \tag{3.4}$$

i.e. the pressure variation is hydrostatic below the *actual* surface. Below mean water level eq. (3.2) should be used. It is clear that $K_p(y)$ will be less than unity and so the pressure will be less than the hydrostatic, measured from the actual surface level as distinct from the mean. Taking the pressure at the sea bed ($y = -h$) and maximum $\eta(=H/2)$, then

$$p_{\text{bed}} = \rho g\left[h + \frac{H}{2}\cdot K_p(-h)\right] \tag{3.5}$$

In deep water $K_p(-h) = 1/\cosh kh = 0.086$ for $h/L = 0.5$. That is

$$p_{\text{bed}} = \rho g[h + 0.043H]$$

$$= \rho gh\left[1 + 0.043\frac{H}{h}\right]$$

In shallow water $(kh < \pi/10)$, $K_p(-h) \approx 1$ for $h/L = 0.05$

That is $$p_{\text{bed}} = \rho g\left(h + \frac{H}{2}\right)$$

These two equations indicate that, in deep water, the passage of the wave makes little difference to *bottom* pressure (for the steepest unbroken wave $H/h = 0.71$ for $h/L = 0.5$). In shallow water the bottom pressure is the hydrostatic measured from the crest. Such simple calculations illustrate the point made by Dean and Dalrymple (1984) that bottom-mounted pressure-type devices for measuring wave frequency and height have definite limitations. For instance, their sensitivity in deep water is questionable, and they tend to filter out higher frequency waves (which have smaller heights).

3.3 Wave forces

Two broad categories of loading may be considered here, the first in which the waves are unaffected by the presence of the structure and the second in which they are. The usual dividing line is when $D \approx 0.2\,\text{L}$ (or $D \approx k^{-1}$) where D is the significant dimension of a structural member parallel to the wave crest and L is the wavelength. For $D > 0.2\,\text{L}$, the wave pattern is modified by

Table 3.1 Correlation of waveheight and limiting dimension D at various frequencies

Frequency	0.5	0.375	0.25	0.125	0.0625
Wave height (m)	0.45	0.79	1.78	7.13	28.53
D(m)	1.25	2.22	5.0	20.0	80.0

the structure. Henceforward, then, the terminology 'small diameter' and 'large diameter' implies 'in relation to this threshold value of one fifth of the wave length'. Generally too, that wavelength is for the wave adopted for design purposes: any given structure will interfere with waves of some sizes, as the table above demonstrates. If deep water linear theory conditions are assumed and a wave steepness of 1/14, say, then an indicative correlation of wave height H and limiting dimension D can be obtained for various frequencies; Table 3.1.

A structure 5m wide is 'large diameter' for wave frequencies greater than 0.25 and 'small diameter' for lower frequencies.

As a broad generalisation, offshore steel structures and the towers of typical concrete structures are 'small diameter' while the bases of concrete structures, breakwaters, and sea defence walls are 'large diameter' for design conditions in areas such as the North Sea.

Now let us consider loads affecting offshore concrete structures.

3.4 Wave loads on offshore structures

There are three main categories of wave load on a body: drag, inertia, and diffraction. Drag loads are the result of flow separation due to the velocity of fluid relative to the body—familiar to many through the performance of aircraft and automobiles. Inertial loads are the consequence of pressure gradients from the relative acceleration of the fluid. Diffraction loads result from wave scattering by the structure. Drag and inertia are considered together in the design of 'small' dimension structures and elements, while diffraction theory is employed in the design of 'large' dimension structures. Figure 3.1(a) (Hogben 1976) displays the predominant loading regime according to wave height and diameter. Clearly small diameter members in high waves are drag dominated, large diameter members in small waves are inertia or diffraction dominated, while very large members in all conditions are subject to diffraction.

Depth below the surface is also significant since waves are effectively 'reduced' in size and so the loading regimes might be expected to modify, as Figures 3.1(b) and (c) (Hogben, 1976) demonstrate (the wave heights indicated are, of course, the heights at the surface). It is easy, therefore, to visualise situations where different load regimes have to be applied to the same structure, as in a typical North Sea concrete oil production platform.

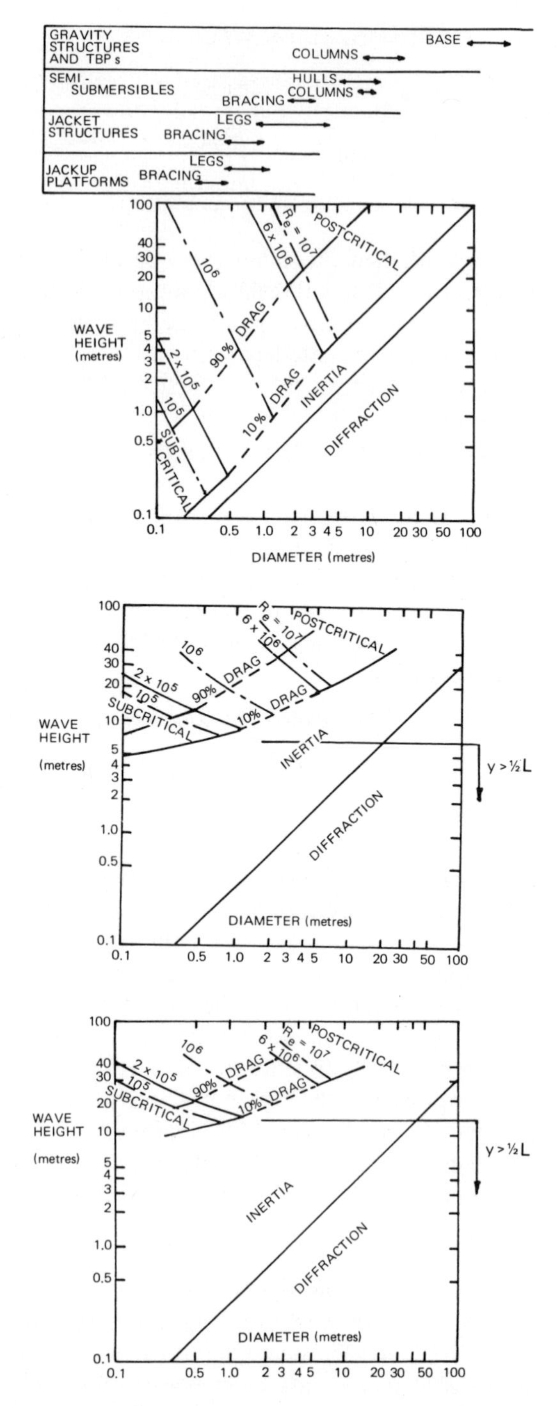

Figure 3.1 (a) Loading regimes at the sea surface (b) Loading regimes at a depth of 50 metres (c) Loading regimes at a depth of 100 metres

Clearly it is important to have a proper understanding of the phenomena involved in these regimes although that may be easier in aim than achievement since not all is known about the various processes. However, it is possible to make a broad assessment, with the general qualification that specific problems demand more detailed consideration than we can achieve here. Use of empirical co-efficients, effects of scale, variability, all invoke caution but, on the other hand, calculation might be impossible otherwise. This is not an unusual engineering dilemma, requiring intelligent application of the known and estimated rather than blind or slavish adherence to prescriptive rules. Nevertheless, in computing wave loads it is particularly important to remember that we are dealing with the idealisation of randomly varying uncertainties. This is especially true when employing software packages which may be extremely complex and costly.

3.4.1 Drag

Drag is defined as 'the force component, parallel to the relative velocity, exerted on the body by the moving fluid' (Streeter and Wylie, 1979) and the 'component of fluid dynamic force on a body in line with and in phase with the velocity of the undisturbed flow' (Hogben, 1974). It has a normal component known as 'lift'. There are in fact two categories of drag: skin friction and form drag (Vallentine, 1967) (Sarpkaya and Isaacson, 1981). As the names suggest, skin friction is due to friction between fluid and body (thus one would expect surface roughness to be significant, for example), while form drag is influenced by shape, which affects flow separation (and so a perfectly streamlined body would not have form drag while a bluff body would). Potential theory, of course, presupposes frictionless, irrotational flow which does not arise in nature, resulting in the familiar problem of accounting for reality. However, Prandtl's concept of the boundary layer provides the link between ideal and real fluid flow (Streeter and Wylie, 1979).

3.4.1.1 *Drag force* This is not the place to discuss boundary layer theory; it is sufficient to note the upshot which is that the drag force is a function of the square of the fluid velocity (Streeter and Wylie, 1979) (Vallentine, 1967) (Lamb, 1962), and may be calculated from the relation

$$F_D = C_D \rho u^2 A \tag{3.6}$$

where F_D is the drag force, C_D the drag coefficient, u the fluid velocity and A the surface area. For cylinders normal to the plane of wave travel it is customary to take the area as the projected area parallel to the wave crest; i.e. the width normal to the direction of flow multiplied by the immersed or unit length (in the latter case, a circular cylinder diameter D would result in $A = D$). C_D then incorporates any adjustment for cross-sectional shape e.g. square, elliptical, circular or whatever.

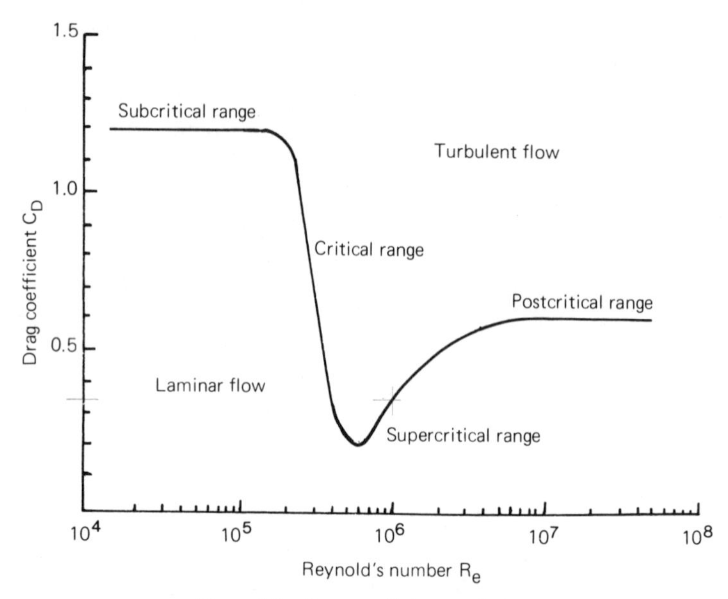

Figure 3.2 Drag coefficient variation

3.4.1.2 *Drag coefficient* Unfortunately, from a design point of view, C_D is not constant, and is in fact a function of Reynolds' Number, R_e (Lamb, 1962). Figure 3.2 (Dept. of Energy, 1980) indicates the pattern of variation to be expected. It is clear from this that laboratory wave measurement of C_D over the full range of uncertainty (corresponding to $R_e \approx 10^5$ to 10^7), and for anything other than very small diameter members, is quite impractical (the maximum horizontal particle velocity for a 2 m wave of 4 s period is 2.02 m/s).

In addition to dependence on Reynolds' Number, it has also been shown (Keulegan and Carpenter, 1958) that C_D varies with a dimensionless period parameter known generally as the Keulegan–Carpenter Number, $K = U_m T/D$ where U_m is the amplitude of the velocity oscillation, T the period, and D the diameter of the cylinder. The pattern of variation is shown in Figure 3.3 (Keulegan and Carpenter, 1958).

From this it seems that the drag coefficient is particularly uncertain for $K < 30$. Assuming that the results for such a model study can be translated to full scale wave flow, what implications does this have for structures?

For simplicity, let us use deep water linear theory. Then, below a depth of $-H/2$, the structure or member is always immersed and can be considered as subject to horizontal oscillatory flow (it is not clear what the effect is when the member/structure is in air at parts of the cycle, although it has been recommended (Sarpkaya, 1981) that $K = \pi H/D$ at the still water level be used). At a point x along the wave from the crest and at depth y, the horizontal velocity ${}_xu_y$ can be taken as a function of ${}_0u_y$ which will be the maximum

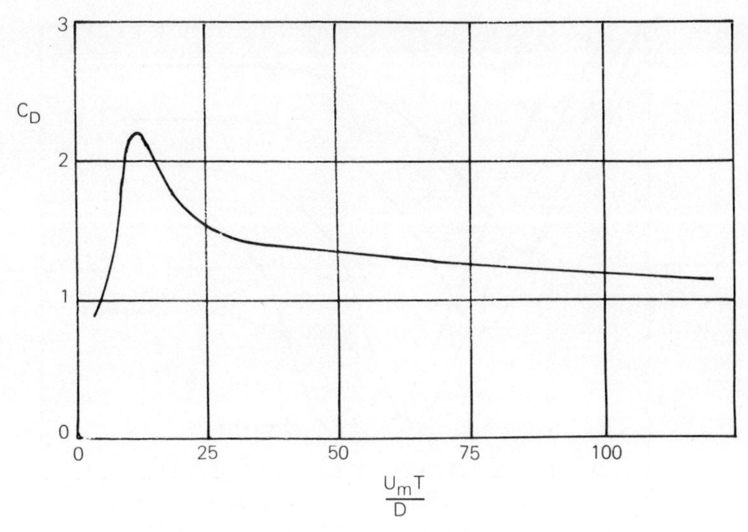

Figure 3.3 Variation of drag coefficient of cylinders

value of u_y, lying below the wave crest ($x = 0$). Hence ${}_xu_y = {}_0u_y \cos\theta$ and the value to be used in computing K is ${}_0u_y$. Clearly, therefore, K will vary with depth, reducing as the velocity reduces with depth.

Let us suppose waves of steepness 1/14. What diameters of structural members correspond to the 'critical' value of $K = 30$ indicated above at depths of $H/2$ below the mean water surface? The table below lists diameters for various wave heights for deep water linear theory.

Wave height (m)	10	15	20	25	30
'Critical' diameter (m)	0.837	1.255	1.673	2.092	2.510

This means, for example, that for a wave 25 m high the drag coefficient becomes particularly uncertain for members above 2 m diameter. Since the velocity reduces with depth, K becomes smaller than the 'critical' value and hence C_D is more problematical.

(Note: the 'critical' diameter is obtained from $D = U_m T/K$ with U_m being given by $U_m = {}_0u_{-H/2} = (\pi H/T)e^{-kH/2}$.)

Problems in estimating C_D do not end with R_e and K, unfortunately. It has already been mentioned that drag is affected by surface roughness. Marine growth is discussed later in this chapter but it will be obvious that hard growth in particular is going to affect the nature of the surface of 'real' structures. Figures 3.4(a) and (b) (Sarpkaya and Isaacson, 1981) show how C_D can vary on the laboratory scale.

3.4.1.3 *Lift* It was indicated above that drag is accompanied by a normal component known as lift. This is a complex phenomenon and, with its

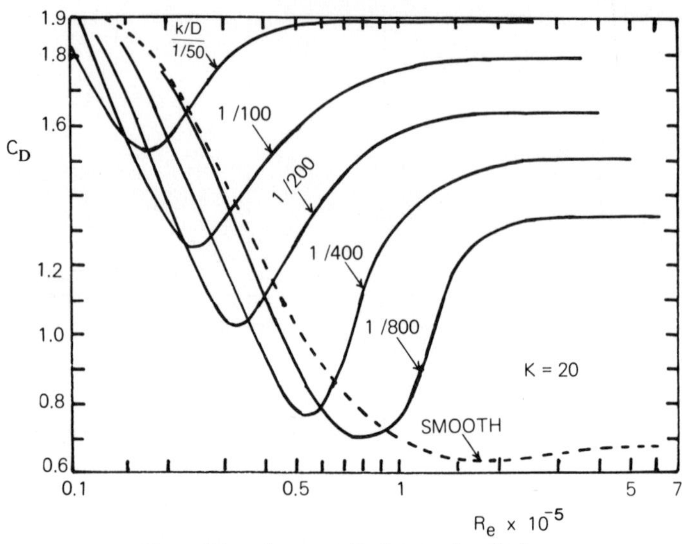

Figure 3.4a C_D versus R_e for rough cylinders

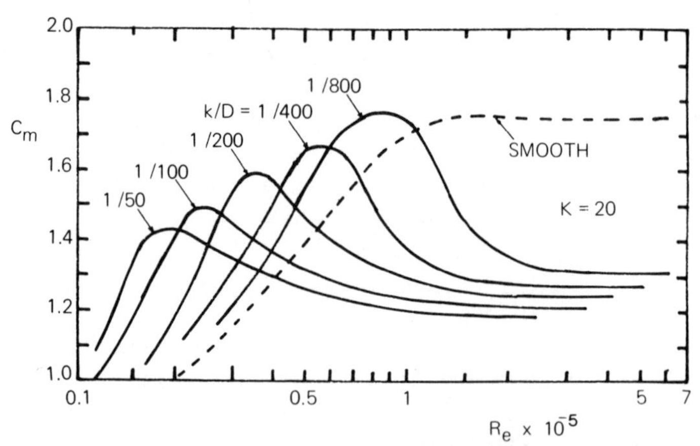

Figure 3.4b C_m versus R_e for rough cylinders

accompanying phenomenon of vortex shedding, is as yet imperfectly understood. In steady laminar flow, at the 'front' of a cylinder there is a 'stagnation point' which marks the start of the boundary layer. The layer follows the surface of the cylinder on both sides until it reaches a point at about 90° from the stagnation point where the flow becomes turbulent and breaks away from the surface, resulting in a wake immediately downstream of the cylinder (Lamb, 1962). When, however, the flow is unsteady and turbulent, the break-away point oscillates resulting in alternate vortex shedding from opposite sides of the cylinder. This turn gives rise to an oscillating transverse force—lift (Sarpkaya and Isaacson, 1981). (Lift is shape dependent so that for an aerofoil the breakaway point is in fact fixed.) Particularly for small-diameter flexible members such as risers, vortex shedding can develop resonant vibration with its concomitant problems. For steady flow about a smooth cylinder, shedding depends 'on the Reynolds number and the intensity and length-scale of the turbulence present in the ambient flow' (Sarpkaya and Isaacson, 1981). Oscillating flow, of course, produces in addition a fluctuating stagnation point.

Mention should be made also of the Strouhal number, $S_0 = f_0 D/V$, where f_0 is the vortex-shedding frequency, D the cylinder diameter and V the velocity of flow. Figure 3.5 shows the relationship between the Strouhal and Reynolds numbers as quoted by Sarpkaya and Isaacson (1981). The reader is directed there for a much more extended discussion of flow separation and its effects. Their significance is also considered elsewhere (Dean and Wootton, 1978).

The lift force F_L may be calculated in a similar fashion to that for drag force:

$$F_L = C_L \rho u^2 A \tag{3.7}$$

where C_L is the lift coefficient. The variation in this, however, is considerable up to Reynolds' numbers around 10^6, ranging from zero to 1.3 (Sarpkaya and Isaacson, 1981). Some other quoted evidence (Pearcey, 1979) appears to indicate that transverse forces are at least as great as in-line forces in real seas. For rigid members a lift coefficient of 0.3 to 0.4 might be used for $R_e > 10^6$ but, on the other hand, in time-varying flow there may be more need to think in terms of rms values.

3.4.2 *Inertia*

Certain aspects of inertia effects may be more easily visualised by considering the motion of a body through a fluid rather than the converse. It is then easy to see that the fluid will be disturbed throughout a region around the body. This in turn has led to theories built upon the premise that the body and the fluid region of interest be considered together as one dynamical system. Rather than consider the inertia of the body alone we then have to consider its virtual or total effective mass, which includes a hydrodynamic or added mass of fluid. This latter can be determined theoretically for a whole variety of shapes.

3.4.2.1 *Inertia force* The displaced fluid particles do not, in fact, return to their original positions so that there is a permanently displaced quantity, termed the drift, whose mass is the added mass (Darwin, 1953). Figure 3.6 shows typical particle trajectories, the numbers marking the times of travel to the respective points in multiples of the time for the cylinder to travel distance a (Darwin, 1953). The drift volume for unit length of a cylinder with

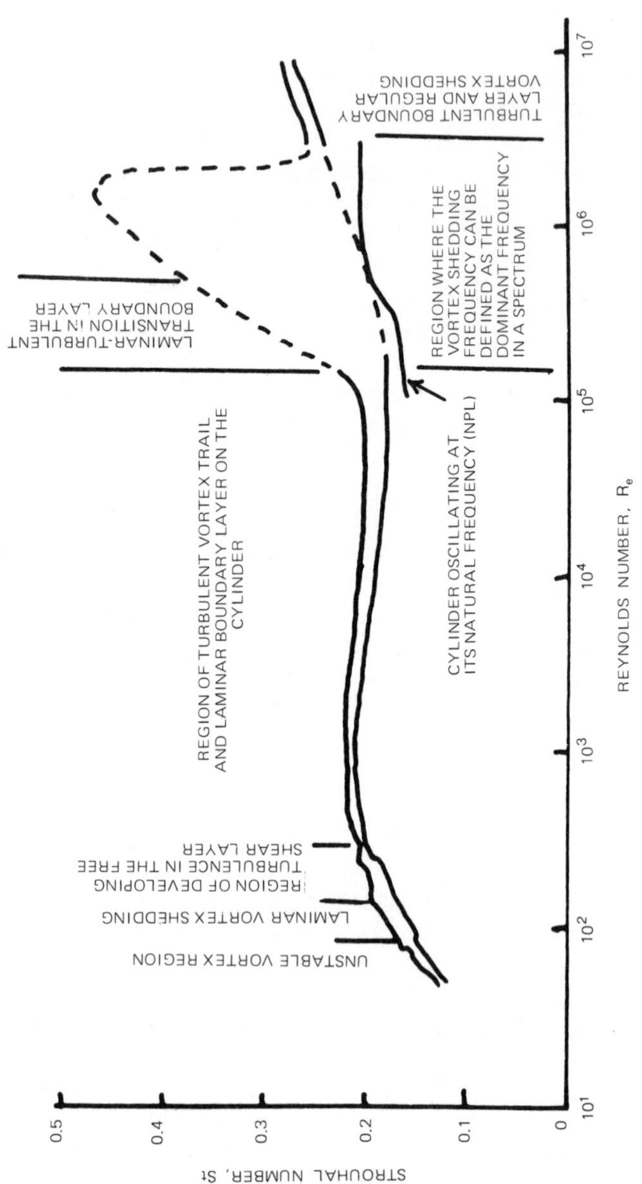

Figure 3.5 The Strouhal-Reynolds' number relationship for circular cylinders

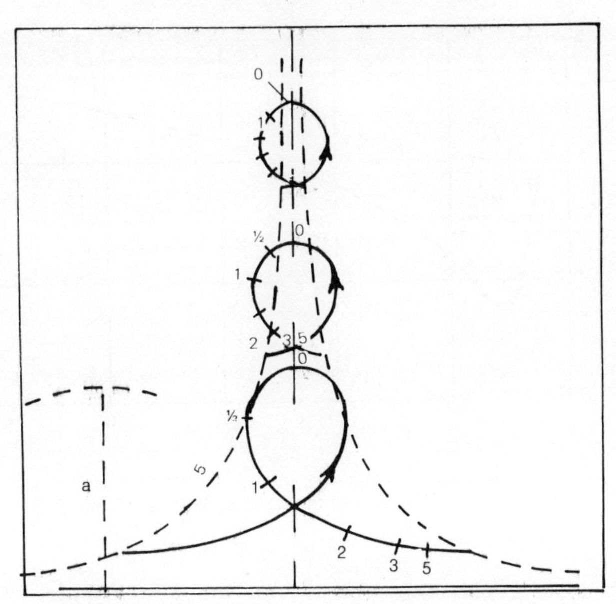

Figure 3.6 Typical particle trajectories. Broken line on left is upper half of wall of dye before passage of cylinder, that on right after passage. a is cylinder's radius. Firm lines show three trajectories. Numbers mark times of passage to these points in units in which cylinder moves distance a.

axis normal to the fluid surface surface is shown to be πa^2 i.e. the volume of the cylinder (Darwin, 1953). It has been shown (Lamb, 1962) that the force on the cylinder is its mass plus the added mass, multiplied by the acceleration, that is, the incremental inertial force dF_I is given by

$$\begin{aligned} dF_I &= (M + M')\frac{\partial u}{\partial t} \\ &= M(1 + k_M)\frac{\partial u}{\partial t} \\ &= C_M M\frac{\partial u}{\partial t} \end{aligned} \tag{3.8}$$

where M is the cylinder mass, M' the added mass, k_M the added mass coefficient, C_M the inertia coefficient, and u the velocity. As will be expected, for a cylinder (whose drift volume is πa^2), $M' = M$ i.e. $k = 1$ or $C_M = 2$ (the mass of a floating cylinder is of course the mass of the displaced volume of fluid).

Thus far we have been considering a cylinder moving through a fluid. In practice we are more often concerned with a fluid flowing past a fixed cylinder. Mathematically there is no difference between the two flow situations and so the same concepts apply. (Keulegan and Carpenter (1958) show that the inertia force is twice the displaced mass.)

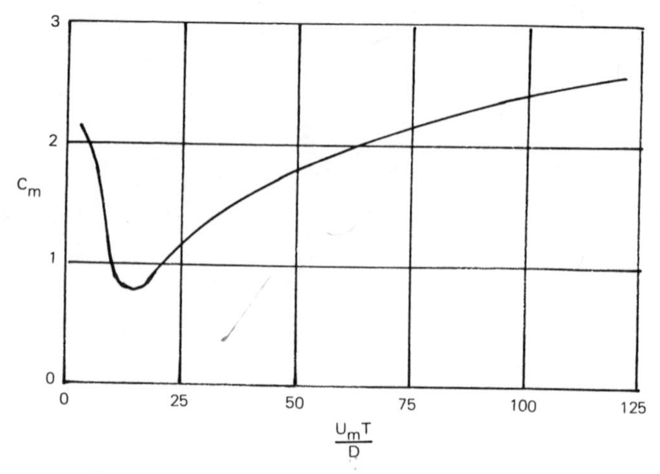

Figure 3.7 Variation of inertia coefficient of cylinders

$M \cdot \partial u/\partial t$ is sometimes known as the buoyancy or Froude–Krylov force, that which would exist in the fluid in the absence of the body (when, of course, M' would be zero). M' is a function of the kinetic energy of the fluid, $M' = 2T_f/u^2$ where T_f is the kinetic energy.

More commonly, perhaps, eq. (3.8) takes the form

$$\mathrm{d}F_I = C_M \rho \forall \frac{\partial u}{\partial t} \tag{3.8a}$$

where $\forall$ is the displaced volume. For a cylinder diameter D, $\forall = \pi D^2/4$ for unit length.

3.4.2.2 *Inertia coefficient* In real fluids M' (and thus C_M) is not constant. It is not, in fact, merely a function of body shape, as potential theory would suggest, but is a function of time in separated flow—C_M varies with wake geometry, for example. Figure 3.7 shows how it varies with Keulegan–Carpenter number (Keulegan and Carpenter, 1958).

3.4.3 *Total force*

3.4.3.1 *Morison equation* Calculation of the total force on a cylinder is based on the simple presumption of linear addition of the drag and inertia forces. That is, the incremental total horizontal force dF on unit height of a vertical cylinder is given by

$$\mathrm{d}F = \tfrac{1}{2} C_D \rho D u|u| + C_M \rho \frac{\pi D^2}{4} \frac{\partial u}{\partial t} \tag{3.9}$$

The modulus or absolute value $|u|$ of the velocity is used to ensure directional consistency. This is the so-called Morison equation, first verified some years ago in the laboratory (Morison et al, 1950). It is also sometimes referred to as the MOJS equation (Sarpkaya, 1986).

It should be noted that the calculation of total force and moment was based on integration of this expression between bed and *still* or mean water level, rather than the actual surface level at a given instant. This is a practice recommended for real situations (Dean and Dalrymple, 1984) in view of the uncertainty surrounding the selection of drag and inertia coefficients to suit the variations in velocity and acceleration (and hence Reynolds and Keulegan–Carpenter numbers) over the height of the cylinder.

The velocity according to linear wave theory is calculated from eq. (2.3)

$$u = \frac{\pi H}{T}\frac{\cosh k(h+y)}{\sinh kh}\cos\theta$$

and acceleration from eq. (2.5)

$$\frac{\partial u}{\partial t} = \frac{2\pi^2 H}{T^2}\frac{\cosh k(h+y)}{\sinh kh}\sin\theta$$

3.4.3.2 *Maximum force* Obviously the maximum horizontal velocity, and hence maximum horizontal drag force, will occur when $\cos\theta = 1$. At this point, of course, $\sin\theta = 0$ and so there will be no inertia force (according to linear theory and Morison's equation). Conversely there is no drag force when the inertia force is at its maximum. Maximum horizontal drag corresponds with the wave crest, and maximum inertia with the zero-crossing points of the wave. Since these maxima are out of phase it is not appropriate to calculate the maximum total force by simply adding them together. The exact location of the wave for maximum total force, however, will depend on the relative values of C_D and C_M, D and H, and H and L.*

It is shown (Dean and Dalrymple, 1984) that the ratio of the maxima is

$$\frac{(\mathrm{d}F_I)_{\max}}{(\mathrm{d}F_D)_{\max}} = \pi^2\frac{C_M}{C_D}\frac{1}{U_m T/D} \tag{3.10}$$

In other words the ratio is a function of the Keulegan–Carpenter number $K = U_M T/D$. Moreover, it is also shown that

$$\frac{(\mathrm{d}F_I)_{\max}}{(\mathrm{d}F_D)_{\max}} = \frac{\pi}{2}\frac{C_M}{C_D}\frac{1}{S/D} \tag{3.11}$$

*In deep water the appropriate value of θ can be found from the equation

$$\cos\theta = \frac{1}{\pi s}\ln\left[2\pi\cdot\frac{C_M}{C_D}\cdot\frac{D}{H}\cdot\frac{(\cos\theta - \pi s\sin^2\theta)}{(1+\pi s\cos\theta)\sin 2\theta}\right] \quad \text{where } s = H/L$$

where S is the maximum displacement of a water particle from its neutral position. S/D is a so-called displacement parameter. For small and large values of the period and displacement parameters, the inertia and drag components dominate respectively. The discussion (Dean and Dalrymple, 1984) considering earlier evidence, states that 'the inertia coefficient C_M is approximated quite well by the potential flow value of 2 for S/D values less than 0.5... For S/D values larger than 2.0, the drag and inertia coefficients oscillate with time'. The displacement parameter then, provides some guidance on the limits of validity of potential theory.

A singular merit of the Morison equation is its simplicity, and it might be considered, in view of the variability and uncertainty of the wave parameters and coefficients, that there is little justification for greater complexity, at least under 'real' marine conditions as distinct from the artificiality of the laboratory. Certainly, where the coefficients have been derived for, or measured on, test structures simulating reality, they should be used in like manner. Nevertheless, laboratory tests should not be dismissed too readily since they do much to improve understanding of the phenomena involved. Perhaps, then, improved understanding will lead to greater wisdom in application (although at times it might be easier to get the job done with a measure of ignorance!).

3.4.3.3 *Modifications to Morison's equation* There have been attempts of modify the Morison equation in consequence. It has been shown (Keulegan and Carpenter, 1958) for example, that the simplified form of the equation, strictly speaking, applies only to situations where C_D and C_M are constant. Where they vary with phase (as they do in waves) then an additional term or terms in required, known as the 'remainder' or 'residue'. Unfortunately this is not constant either and requires additional coefficient(s) of varying complexity: for instance a three-term equation (Sarpkaya and Isaacson, 1981) was subsequently improved by one of four-terms (Sarpkaya, 1981) and a further possible (different) three-term (Sarpkaya, 1986) version. This last attributes the residue, particularly in the drag-inertia dominated regime, to the inability of the unmodified equation to account for the in-line contribution of the transverse pressure gradient. It has been shown, in fact, that the lift coefficient, like the drag and inertia coefficients, varies with Reynolds number, Keulegan–Carpenter number, and relative roughness. This has been done (see Sarpkaya (1986) for the full references) through the introduction of a frequency parameter $\beta = R_e/K = D^2/\upsilon T$ (see also Sarpkaya (1976) for data relating C_D and C_M to β in oscillatory flow). General conclusions include (i) for smooth cylinders, C_D, C_L and C_M depend on both R_e and K; (ii) for rough cylinders, these coefficients are independent of R_e above a critical value and depend only on K and relative roughness, k/D; (iii) for both smooth and rough cylinders, the relationship between C_D and C_M is not unique and depends on K; (iv) the transverse force is a significant fraction of the total resistance for all R_e and must be considered in design.

Sometimes, in preference to using separate coefficients, a 'total force' coefficient may be used. Certainly where the waves are random in direction as well as amplitude, such an approach is potentially more useful. However, in examining data for possible use which might have been collected on a sizeable structure, care has to be taken to ensure that the force and velocity sensors which have been employed are close enough to be in phase—simultaneous measurements taken some distance apart are quite likely to be out of phase with each other. This is particularly the case where peak values are employed and has led to the use of rms values in preference. Obviously this has additional merit in randomised situations.

Such considerations lie behind the proposed modified Morison equation (Bishop, 1980) for the mean square value of wave force:

$$\bar{F}^2 = A^2\overline{u^4} + B^2\bar{\dot{u}}^2 \tag{3.12}$$

where

$$A = \tfrac{1}{2}C_D\rho DL$$

and

$$B = C_M\frac{\pi}{4}\rho D^2 L$$

u is the total vector value of horizontal velocity due to both waves and current and $\dot{u}$ is the horizontal acceleration. Re-arranging this equation slightly gives

$$\bar{F}^2 = (\tfrac{1}{2}\rho DL)^2\left[C_D^2\overline{u^4} + C_M^2\left(\frac{\pi}{2}D\right)^2\bar{\dot{u}}^2\right]$$

which leads to the definition of the total force coefficient C_{F*}

$$C_{F*} = \left[\frac{F^2}{(\tfrac{1}{2}\rho DL)^2\left[\overline{u^4} + \left(\frac{\pi D}{2}\right)^2\bar{\dot{u}}^2\right]}\right]^{1/2} \tag{3.13}$$

when drag is dominant $C_{F*} \approx C_D$ and when inertia is dominant $C_{F*} \approx C_M$.

Arising from this is a re-defined Keulegan–Carpenter number for irregular waves, K_*

$$K_* = \frac{2\pi}{0.866D}\left[\frac{\overline{u^4}}{\bar{\dot{u}}^2}\right]^{1/2} \tag{3.14}$$

The total force coefficient C_{F*} has been used extensively in the analysis of data for the Christchurch Bay tower which is a sea-based test structure. It seems likely that this is the sort of approach which will have to be adopted by the designer in applying results from such test structures as more information becomes generally available.

So far it has been assumed that the structures concerned in the discussion are rigid. However, if they are dynamically responsive, compliant, buoyant, or form parts of a floating body, then the dynamic form of the Morison equation must be used (Hallam et al, 1977):

$$F = \tfrac{1}{2}C_D\rho D|(u-u_b)|(u-u_b)\mathrm{d}l + C_M\rho A(\dot{u}-\dot{u}_b)\mathrm{d}l + (\rho A\mathrm{d}l - M)\dot{u}_b \quad (3.15)$$

where u_b and $\dot{u}_b$ are the velocity and acceleration of the incremental section of the structural member of mass M. For a floating body, or when the acceleration is zero, the third term is zero and the first two then represent the standard form of Morison's equation with vectorial addition of velocity and acceleration. For *parts* of a floating body, however, the full form will be needed since the displacement of the incremental section is not the same as its mass.

3.4.4 *Diffraction*

In general the preceding discussion has dealt with the situation where the structure does not interfere with the wave pattern. However, where a structure is sufficiently large, then the waves *are* affected and application of the theories so far outlined would tend to over-estimate loads (Sarpkaya and Isaacson, 1981). Diffraction theory should then be used.

A very crude distinction between the drag and inertia, and the diffraction regimes is to relate drag and inertia to space frame type structures and diffraction to monoliths. More specifically the 'border line' is set when the effective diameter of the body or member is greater than about one fifth of the wave length (Hogben, 1974). In seeking to define 'large', three relevant length scales in wave-body interaction have been suggested (Mei, 1983): the characteristic body dimension a ($=D$ for a cylindrical body), the wavelength $2\pi/k$ ($=L$), and the wave amplitude A ($=H/2$ for linear theory). The body is 'large' when $ka \geqslant O(1)$ i.e. when $D \geqslant L/2\pi$, which corresponds roughly to the rule of thumb already mentioned. Inviscid, linearised diffraction theory (with considerable experimental confirmation) is stated to be fairly well developed for this scale when $A/a \ll 1$ (i.e. for small amplitude waves when, of course, linearised wave theory is applicable). When $A/a \geqslant O(1)$ and $ka \ll 1$, pure theory cannot be applied but experimental work has been intensive. The most difficult and least explored area is reckoned to be the intermediate case of $A/a \geqslant O(1)$ and $ka = O(1)$. In cylindrical body, linear wave theory terms, that is when $H \geqslant 2D$ and $D/L \geqslant 1/2\pi$ i.e. when the waves can no longer be considered small amplitude and so non-linear diffraction is involved as well as flow separation.

As further, very crude assessment, one might say that space-frame type structures involve (relatively) simple hydrodynamic load estimation on complex structures while monolithic type structures involve complex load estimation on simple structures.

3.4.4.1 *General background* Diffraction theory and the detailed solution of consequent problems is too extensive and requires too specialised techniques to be considered in detail here. However, in view of its importance in the design of concrete offshore structures in particular, a general introduction cannot be omitted from a text such as this.

Other than for the simplest of shapes, analytical solution is not possible; consequently the requirement is for numerical integration, finite difference, finite element techniques and so on. As before, the fluid is assumed to be inviscid and incompressible and flow is irrotational. The majority of solutions are also based upon linear wave theory with its presumptions of small amplitude waves of moderate steepness (this has the effect that velocity terms can be neglected in applying Bernoulli's equation to the free surface boundary).

While this is not the place for detailed mathematical development (which can be sought in sources dealing with the hydrodynamics of the system), there are certain key elements, essential to load calculation, which are frequently referred to. Two or more wave systems are invoked. There is the incident (or incoming) wave system (potential Φ_I) determined by the wave climate and which is the source of the loads to be estimated. The structure is then assumed variously to scatter the waves (potential Φ_S) and/or to radiate waves (potential Φ_R). A simple surface-penetrating vertical cylinder may be assumed to radiate cylindrical waves rather as ripples radiate when a stone is dropped into a pond. Other vertical solids of revolution (otherwise known as axisymmetrical) may radiate in similar fashion. However, more complex shapes or elongated or non-axisymmetrical shapes cannot be dealt with in this way and are normally considered to be divided into surface elements, each of which has a pulsating source at its centre. A distinction made between scattering and radiation (Mei, 1978) separates the problems into two: the diffraction problem where all rigid boundaries are stationary, and the radiation problem where parts of the rigid boundaries oscillate (as on a floating structure). If then the total potential is Φ

$$\Phi = \Phi_I + \Phi_S + \Phi_R \tag{3.16}$$

Since Φ_S may represent a number of sources, then (Hogben, 1974)

$$\Phi_S = \sum_{n}^{N} m_n \Phi_n \tag{3.17}$$

where Φ_n is the potential of an individual source of strength m_n. Φ_R also may be represented as a sum (Standing, 1981) according to the number of degrees of freedom of the structure:

$$\Phi_R = \sum_{j=1}^{N} n_j \Phi_j \tag{3.18}$$

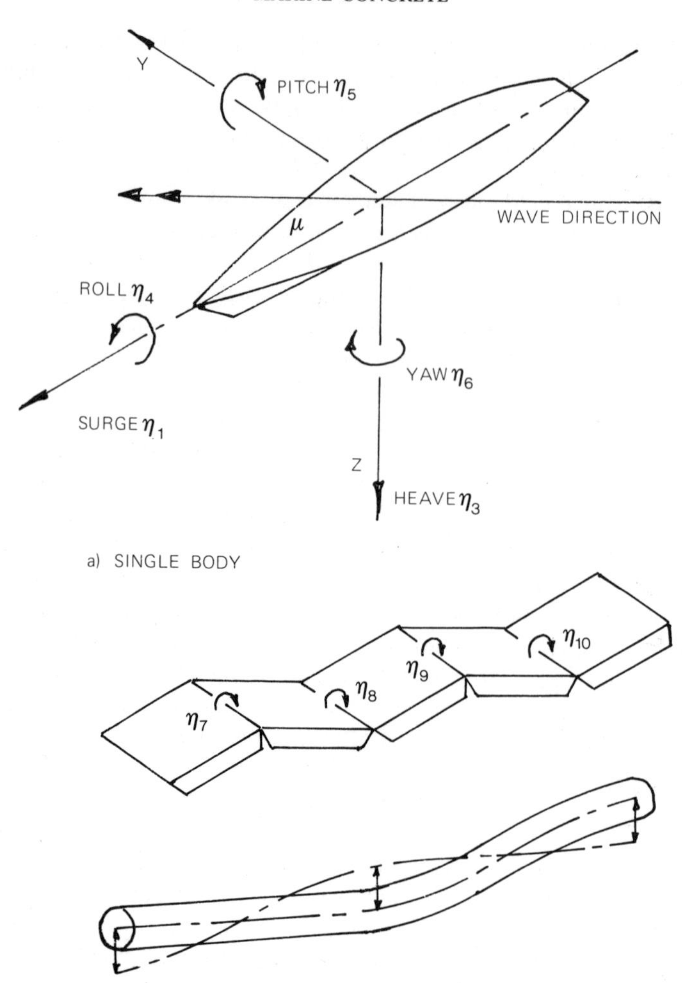

Figure 3.8 Coordinate system and motions for diffraction analysis

where n_j is the structure's response in its jth degree of freedom. The degrees of freedom are defined in Figure 3.8 (Standing, 1981) for a single rigid body, and for articulated and bending structures. Generally, for a fixed concrete structure, $\Phi_R = 0$. Where it is, say, buoyant or articulated, as in a tower or a wave-generating device, then Φ_R must be taken into account.

3.4.4.2 *Diffraction coefficient* In a manner analogous to eq. 3.16 forces have been equated (Hogben et al, 1977):

$$F = F_i + F_d \tag{3.19}$$

where F_i is the contribution due to the undisturbed incident wave and F_d is the scattering wave component plus the local disturbance corresponding to the added mass term in Morison's equation. F_i is generally known as the Froude–Krylov force (see Inertia, 3.4.2) in which case it may more conveniently be designated F_k. This results in the postulation of a *diffraction coefficient* C_h (Sarpkaya and Isaacson, 1981) defined by

$$C_h = F/F_k \tag{3.20}$$

When the diameter tends towards a small fraction of the wavelength, wave scattering is negligible and so (Hogben, 1976)

$$F = C_h F_k \approx C_M \rho \forall \frac{\partial u}{\partial t}$$

Hence diffraction theory can be used for calculating the inertia coefficient C_M.

3.4.4.3 *Effective inertia coefficient* As an alternative to the diffraction coefficient approach an *effective inertia* coefficient C_M can be adopted (Sarpkaya and Isaacson, 1981). This makes use of the inertia component of the Morison equation, allowing for the phase difference between it and the acceleration of the incident waves at a reference point. On unit length of the structure (cross-sectional area A), the incremental force will be

$$dF = \rho A C_M \frac{\partial u}{\partial t} \cos(\omega t - \delta) \tag{3.21}$$

$\partial u/\partial t$ is the maximum amplitude of the acceleration at the reference point, say the section centre.

In line with this approach Sarpakaya and Isaacson (1981) quote the results of some Canadian work giving appropriate values which may be used in the following equations for a vertical cylinder:

$$\left.\begin{aligned} dF &= \frac{\pi}{8}\rho g H k D^2 \frac{\cosh k(h+y)}{\cosh kh} C_M \cos(\omega t - \delta) && \text{(a)} \\ F &= \frac{\pi}{8}\rho g H D^2 \tanh kh C_M \cos(\omega t - \delta) && \text{(b)} \\ M &= \frac{\pi}{8}\rho g H \frac{D^2}{k}\left[\frac{kh \sinh(kh) + 1 - \cosh kh}{\cosh kh}\right] C_M \cos(\omega t - \delta) && \text{(c)} \end{aligned}\right\} \tag{3.22}$$

Values of C_h, C_M and δ for values of $ka (= \pi D/L)$ are given in Figure 3.9. Runup profiles and maximum runup are shown in Figures 3.10(a) and (b) (Sarpkaya and Isaacson, 1981). From these figures and eqs (3.22) it is a straightforward matter to calculate the total force F and maximum moment M for a given wave on a cylinder of pre-determined diameter in water of known depth.

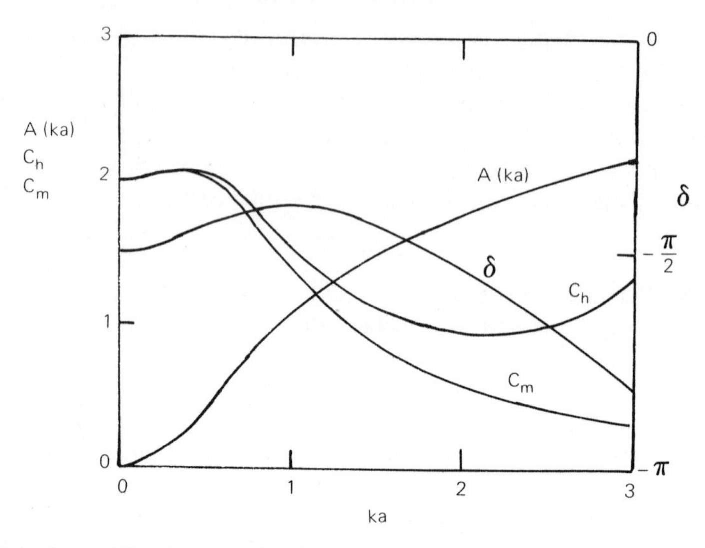

Figure 3.9 Wave diffraction around a circular cylinder: variations of A (ka), C_h, C_m, and δ with ka

3.4.4.4 *Vertical cylinder-analytical approach* The analytical approach to the vertical cylinder problem is set out by MacCamy and Fuchs (1954). The incremental force per unit height is given by:

$$dF = \frac{2\rho g H}{k}\frac{\cosh k(h+y)}{\cosh kh} A\left(\frac{D}{L}\right)\cos(\omega t - \delta) \tag{3.23}$$

where

$$\tan\delta = \frac{J'_1\left(\pi\frac{D}{L}\right)}{Y'_1\left(\pi\frac{D}{L}\right)}\left(=\frac{J'_1(ka)}{Y'_1(ka)}\right)$$

$$A\left(\frac{D}{L}\right) = \frac{1}{\sqrt{J'^2_1\left(\pi\frac{D}{L}\right) + Y'^2_1\left(\pi\frac{D}{L}\right)}}$$

$$= \frac{1}{\sqrt{J'^2_1(ka) + Y'^2_1(ka)}}$$

and the surface elevation is given by

$$\eta = \frac{H}{2}\sin(kx - \omega t)$$

J_1 and Y_1 are the Bessel Functions of the first and second kinds respectively,

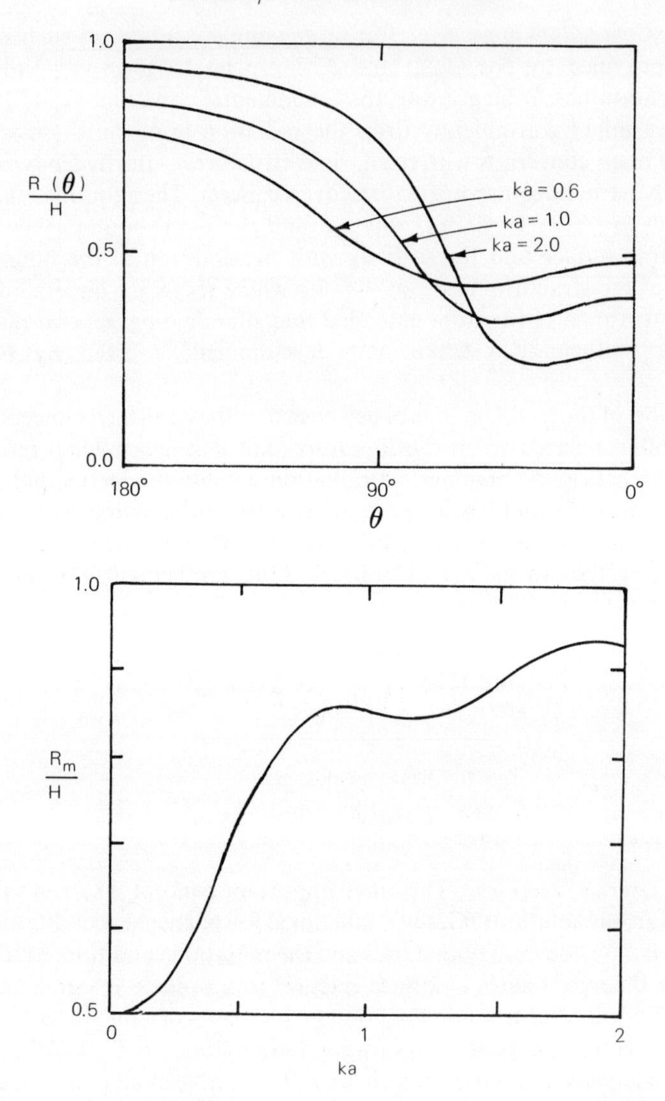

Figure 3.10 (a) Runup profiles $R(\theta)/H$ around a circular cylinder for various values of ka (b) Maximum dimensionless runup R_m/H for a circular cylinder as a function of ka

and the primes indicate differentiation. $A(ka)(= AD/L)$ is plotted on Figure 3.9 as is δ. Horizontal force is obtained by integrating eq. (3.23) over the appropriate depth range, and total force F is found easily to be given by

$$F = \frac{2\rho g H}{k^2} \tanh kh A(ka) \cos(\omega t - \delta) \tag{3.24}$$

Moments may be found similarly.

3.4.4.5 *Generalised approach* For more complex structures such solutions as are provided by eqs (3.22) and 3.23 are no longer valid and further consideration has to be given to the fundamental equation, Eq. (3.16), from which we might conveniently drop the radiation term (on the assumption that our main concern is with fixed, rigid structures—the hydrodynamics of compliant structures are not to be covered here). The requirement then is to find Φ to satisfy Laplace's equation and the usual boundary conditions (at the free surface and the seabed), with the addition of the boundary or surface of the structure. Here $\Phi_I = -\Phi_R$ where these are understood to be normal to the face. (A more extended and illuminating general discussion omitting mathematical detail and development is given by Newman (1972).)

Solution of the problem makes use of two well-established concepts which are so often referred to in the literature that it is worthwhile re-iterating them. The first is the Sommerfeld radiation condition which stipulates that radiating waves should behave at infinity like progressing waves moving away from the source of disturbance (Stoker, 1957). In the present case this means that the waves are cylindrical. One mathematical form for the condition is (Mei, 1978)

$$\lim_{kR\to\infty} (kR)^{1/2}\left(\frac{\partial}{\partial R} - ik\right)(\phi - \phi_I) = 0 \qquad (3.25)$$

where

$$R^2 = x^2 + y^2 + z^2 \quad \text{and} \quad i = \sqrt{-1}$$

3.4.4.6 *Green's functions* The other important concept is Green's theorem and its various solutions (Green's functions) for particular conditions which are, in this case, the fluid boundaries and the radiation condition. Mathematically the theorem relates a surface integral to a volume integral, that is, it relates physical conditions on the boundary surfaces of a region to conditions at points within the region. Following Lamb (1962), if U, V, W are three finite, single-valued functions which are differentiable at all points of a region completely bounded by surfaces S, it can be shown that

$$\iint (lU + mV + nW)\,\mathrm{d}S = -\iiint\left(\frac{\partial U}{\partial x} + \frac{\partial V}{\partial y} + \frac{\partial W}{\partial z}\right)\mathrm{d}x\,\mathrm{d}y\,\mathrm{d}z \qquad (3.26)$$

where l, m and n are the direction cosines of the normals to an element $\mathrm{d}S$ of one of the surfaces.

Setting U, V and W equal to $\phi\cdot(\partial\phi'/\partial x)$, $\phi\cdot(\partial\phi'/\partial y)$, $\phi\cdot(\partial\phi'/\partial z)$ so that $lU + mV + nW = \phi\cdot(\partial\phi'/\partial n)$ where ϕ and ϕ' are two functions, the formal statement of Green's theorem then is

$$\iint \phi \frac{\partial \phi'}{\partial n} \mathrm{d}S = - \iiint \left(\frac{\partial \phi}{\partial x} \frac{\partial \phi'}{\partial x} + \frac{\partial \phi}{\partial y} \frac{\partial \phi'}{\partial y} + \frac{\partial \phi}{\partial z} \frac{\partial \phi'}{\partial z} \right) \mathrm{d}x \, \mathrm{d}y \, \mathrm{d}z$$
$$- \iiint \phi \nabla^2 \phi' \mathrm{d}x \, \mathrm{d}y \, \mathrm{d}z \tag{3.27}$$

together with a similar expression interchanging ϕ and ϕ'.

If the region is in a fluid and ϕ and ϕ' are irrotational potential functions where Laplace's equation applies i.e. $\nabla^2 \phi = \nabla^2 \phi' = 0$, then it follows that

$$\iint \phi \frac{\partial \phi'}{\partial n} \partial S = \iint \phi' \frac{\partial \phi}{\partial n} \partial S \tag{3.28}$$

From this basis it is then shown (Lamb, 1962) that any acyclic, irrotational motion of a liquid mass may be regarded as due to a distribution of simple and/or double sources over the boundary. Cyclic irrotational motion can be represented by distributed double sources over the boundary. Extending this idea, (Wehausen and Laitone, 1960) a Green's function can in fact be derived representing a source of pulsating strength in three dimensions. Furthermore, it has also been shown (John, 1950) that Green's formula enables an arbitrary wave function to be separated into two components: an everywhere regular wave function, and a wave function which satisfies the radiation condition.

This, of course, is the situation represented by eq. (3.16) ($\Phi = \Phi_I + \Phi_R + \Phi_S$). Hence it is clear that an appropriate Green's function should provide a solution to the problem, a feasible route being in terms of a suitable source distribution over the boundary formed by the surface of the body (the structure to be designed, in other words), since the other boundaries are the free surface and the seabed, the horizontal extent being assumed infinite. Mathematically, the aim is to represent $\Phi_S (= Re \phi_S \mathrm{e}^{-i\omega t})$ by a distribution of sources over the surfaces S of the body obstructing flow (the structure) where ϕ_S is given by

$$\phi_S = \frac{1}{4\pi} \iint_S f(\xi, \eta, \zeta) G(x, y, z; \xi, \eta, \zeta) \, \mathrm{d}S \tag{3.29}$$

$f(\xi, \eta, \zeta)$ is the (unknown) source distribution function and $G(\ )$ is the 'appropriate Green's function' with (ξ, η, ζ) a point on the surface of the body. The form of the equation follows that for spherical radiation from a simple source (Lamb, 1962). $G(\ )$ for a point wave source of unit strength (the pulsating source referred to above) is given by

$$G(\ ) = \left[\frac{1}{r} + \frac{1}{r_2} + PV \int_0^\infty \frac{2(k+v)\mathrm{e}^{-kh} \cosh k(\eta + h) \cosh k(y+h)}{k \sinh kh - v \cosh kh} J_0(kR) \mathrm{d}k \right] \cos \omega t$$
$$+ \frac{2\pi(m_0^2 - v^2) \cosh m_0(\eta + h) \cosh m_0(y+h)}{m_0^2 h - v^2 h + v} J_0(mR) \sin \omega t \tag{3.30}$$

where $m_0 \tanh m_0 v - v = 0$; $v = \omega^2/g$; $r^2 = (x-\xi)^2 + (y-\eta)^2 + (z-\zeta)^2$; $r_2^2 = (x-\xi)^2 + (y+2h+\eta)^2 + (z-\zeta)^2$; $R^2 = (x-\xi)^2 + (z-\zeta)^2$ (for the radiation condition—eq. (3.25)); PV denotes the principal value of the integral; J_0 is the Bessel function of the first kind, order zero.

A series form for $G(\ )$, attributed to John (1950) is given by Garrison (1978)

$$G(\ ) = \frac{2\pi(v^2 - k^2)}{k^2 h - v^2 h + v} \cosh[k(y+h)] \cosh[k(\eta+h)][Y_0(kR - iJ_0(kR))]$$

$$+ 4 \sum_{k=1}^{\infty} \frac{(\mu_k^2 + v^2)}{\mu_k^2 h + v^2 h - v} \cos[\mu_k(y+h)] \cos[\mu_k(\eta+h)] K_0(\mu_k R) \quad (3.31)$$

where μ_k are the real positive roots of the equation

$$\mu_k \tan(\mu_k h) + v = 0$$

and J_0 and Y_0 are Bessel functions of the first and second kind of order zero and K_0 is the modified Bessel function of the second kind of order zero.

This is valid for large R and, in fact, in the numerical procedure devised by Garrison (1978) which follows, the series form for $G(\ )$ is employed except for the region close to the body where the integral version has to be used (as $kR \to 0$ so $K_0(\mu_k R) \to \infty$).

The boundary condition on the surface of the body S has to be established before use can be made of the Green's function given above to arrive at the source distribution function and, in due course, the pressure distribution on the body. The problem here is that we are dealing with singularities (the sources), whose velocity tends to infinity at their centres. Any general integration must therefore avoid such points.

From such considerations, the boundary condition on the body surface is (Wehausen and Laitone, 1960)

$$0 = \frac{\partial \phi_I}{\partial n} + \frac{\partial \phi_S}{\partial n} = \frac{\partial \phi_I}{\partial n} - \frac{1}{2} f(x, y, z)$$

$$+ \frac{1}{4\pi} \iint_S f(\xi, \eta, \zeta) \frac{\partial}{\partial n} G(x, y, z; \xi, \eta, \zeta)\, dS \quad (3.32)$$

or

$$f(x, y, z) = 2 \cdot \frac{\partial \phi_I}{\partial n} + \frac{1}{2\pi} \iint_S f(\xi, \eta, \zeta) \frac{\partial G}{\partial n}\, dS \quad (3.33)$$

$f(x, y, z)$ is the source distribution function giving the source strength vector (which is unknown). $f(\xi, \eta, \zeta)$ is the source strength distribution function representing the effect at a point (ξ, η, ζ) on the body of all the other sources. The justification for this may be found in Stoker (1957) and Mei (1983).

3.4.4.7 *Numerical solution procedure* The numerical procedure followed by Garrison (1978) is to divide the body surface into N quadrilateral patches or panels (or facets) of area $\Delta S_j (j = 1, 2, \ldots N)$ with the centre of each panel being the location of a source. It is assumed that the source strength is distributed uniformly over the panel. Equation (3.33) is then replaced by the N equations

$$-f_i + \alpha_{ij} f_j = 2v_{ni} \quad i, j = 1, 2, \ldots, N \tag{3.34}$$

where $f_i = f(x, y, z)$; $v_{n_i} = -\dfrac{\partial \phi_I}{\partial n}$; $f_j = f(\xi, \eta, \zeta)$;

$$\alpha_{ij} = \frac{1}{2\pi} \iint_{\Delta S_j} \frac{\partial G}{\partial n}(x_i, y_i, z_i; \xi, \eta, \zeta)\, \mathrm{d}S$$

α_{ij} is the velocity induced at panel i by a unit source on panel j.

Since v_{n_i} is due to the incoming wave, it is known. α_{ij} is also known since the Green's function can be calculated from eqs. (3.30) and (3.31). Consequently the unknown f_j can be found using matrix techniques.

Using similar notation in eq. (3.29), it becomes

$$\phi_i = \beta_{ij} f_j \tag{3.35}$$

where $$\beta_{ij} = \frac{1}{4\pi} \iint_S G(x_i, y_i, z_i; \xi, \eta, \zeta)\, \mathrm{d}S$$

Hence ϕ_i can be found since the remaining terms are already known.

Velocities can be found from $\partial\phi/\partial x, y, z = \alpha_{xj,yj,zj} f_j$ where

$$\alpha_{xj} = \frac{1}{4\pi} \iint_{\Delta S_j} \frac{\partial G}{\partial x}(x, y, z; \xi, \eta, \zeta)\, \mathrm{d}S, \text{ etc.}$$

Forces and moments are of prime importance, of course, and they follow from the velocity potential. Thus

$$\mathbf{P} = \rho\omega Re[i\phi e^{-i\omega t}] - \frac{1}{2}\rho Re[(\phi_x^2 + \phi_y^2 + \phi_z^2) e^{-i2\omega t}]$$

$$-\frac{1}{2}\rho[|\phi_x|^2 + |\phi_y|^2 + |\phi_z|^2] - \rho g y \tag{3.36}$$

with the $(\phi_{x,y,z})^2$ terms normally being neglected. The force and moment vectors are obtained from the first term

$$\mathbf{F} = -\rho\sigma \iint_S \mathrm{Re}\,[i\phi \mathrm{e}^{-i\omega t}]\mathbf{n}\,\mathrm{d}S \tag{3.37}$$

$$\mathbf{M} = -\rho\sigma \iint_S \mathrm{Re}\,[i\phi \mathrm{e}^{-i\omega t}](\mathbf{r}' \times \mathbf{n})\,\mathrm{d}S$$

where $\mathbf{r}'$ is the position vector extending from the point where the forces and moment are resolved, to the local point on the surface.

There are various refinements to be considered in preparing and applying computer programs to solve problems. Factors include facet shape, size, aspect ratio and curvature, the proximity of the ith to the jth element, what happens as (x, y, z) tends to (ξ, η, ζ), and so on. These and other points are dealt with in various sources including Hogben and Standing (1974) and Garrison (1974 and 1978) to which the reader is advised to refer.

From this outline of the Green's function approach it will be clear that it is by no means simple or straightforward. The derivation and development of techniques is difficult and necessitates computer solution which can be demanding in machine time. Commercial considerations are therefore significant and there will be a cost commensurate with the complexity, sophistication, and versatility of an appropriate software package, especially if it has been physically verified against models or prototypes. In other words, development of a comprehensive package is expensive.

3.4.4.8 *Variational principles and finite element techniques* It requires little step of the imagination to anticipate the application of finite element techniques to diffraction problems. Since this appears to have been most successful through the adoption of the variational approach, some of the terminology should be introduced.

In some respects variational or extremum principles may be more familiar or easier to visualise in considering the displacements of a loaded beam, dealt with (*inter alia*) in some detail by Richards (1977). At the risk of gross oversimplification by paraphrasing, we search for a function to represent the deformed shape of the beam by a technique of minimising the differences between the real (and unknown) beam and some arbitrary or assumed shape. Thus in Figure 3.11 (Richards, 1977), $y = f(x)$ is the (unknown) function being sought and the assumed function is $y = f(x) + \varepsilon\eta(x)$ lying very close to $f(x)$ with ε being an infinitesimal parameter; $\eta(x)$ is an arbitrary function which coincides with $f(x)$ at $x = a$ and b. The function $\varepsilon\eta(x)$ is called a *variation* in y, the conventional notation being $\delta y (= \varepsilon\eta(x))$. The integral $I(y)$ is called a *functional* of y and obviously a variation in y will result in a variation in I. The conditions when the integral I is stationary are expected to yield y, or

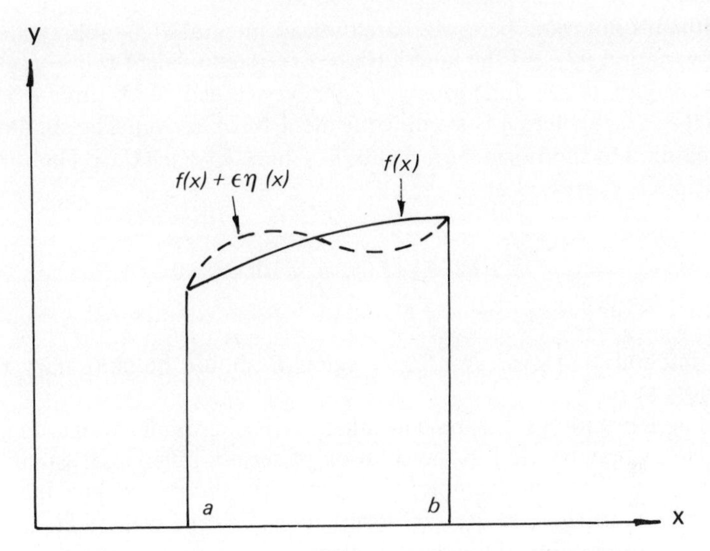

Figure 3.11 Graphical representation of the variational principle

at least that is the objective behind the technique. For a beam, conditions at the ends or at extrema (*local* maxima and minima i.e. humps or hollows) should enable solution. For the sea, conditions at the boundaries (such as the free surface, the seabed, the surface of a body) should do likewise.

To add to the definitions, an example quoted by Richards (1977) is useful.

If
$$I(y) = \int_a^b F(x, y, y')\mathrm{d}x$$

then
$$I(y + \varepsilon\eta) = \int_a^b F(x, (y + \varepsilon\eta), (y' + \varepsilon\eta'))\,\mathrm{d}x$$

and
$$\Delta I = I(y + \varepsilon\eta) - I(y) = \int_a^b [F(x, (y + \varepsilon\eta), (y' + \varepsilon\eta')) - F(x, y, y')]\,\mathrm{d}x$$

Expanding the right hand side by Taylor's series yields

$$\Delta I = (\delta I)\varepsilon + \frac{1}{2!}(\delta^2 I)\varepsilon^2 + \ldots$$

δI is the first variation in I, $\delta^2 I$ the second variation and so on. (Note: $y' = \mathrm{d}y/\mathrm{d}x; \eta' = \mathrm{d}\eta/\mathrm{d}x$.)

The application of variational principles to waves is discussed by Whitham (1974). Here, of course, we are dealing with a continuum and, additionally, time. It is appropriate therefore to use the relevant conventional notation where, for example $\phi_x = \partial\phi/\partial x, \phi_t = \partial\phi/\partial t, \phi_{xx} = \partial^2\phi/\partial x^2$ and so on. Using

Whitham's notation, here we search for an integral $J(\phi)$ (abbreviated to J), first variation δJ, and the function $\phi = \phi(\mathbf{x}, t)$ (equivalent to $y = f(x)$ above). The neighbouring functions we are concerned with are $\phi(\mathbf{x}, t)$ and $\phi(\mathbf{x}, t) + h(\mathbf{x}, t)$ where h is small (equivalent to $\varepsilon\eta$ above). The smallness of h is measured by the 'norm' $\| h \| = \max|h| + \max|h_t| + \max|x_i|$. The variational principle is expressed as

$$\delta J = \delta \iint_R L(\phi_t, \phi_x, \phi)\, \mathrm{d}t\, \mathrm{d}x = 0 \tag{3.38}$$

The integral $J(\phi)$ over any finite region R should be stationary to small changes of ϕ.

Some terminology can be confusing. As Richards illustrates, *Lagrangian multipliers* may be used in the solution of certain differential equations (see Bajpai, et al, 1977 for example). However, they should not be confused with the term *Lagrangian* as used by Whitham (1974) and others. The properties of any conservative dynamical system can be expressed by Lagrange's equations (Lighthill, 1979) entirely in terms of the Lagrangian function L, defined as the kinetic energy of the system minus its potential energy. This (the Lagrangian) is regarded as a function of the generalised coordinates of the system and their time derivatives (which therefore are the generalised velocities). 'Hamilton's principle states that Lagrange's equations are equivalent to a variational statement; namely, that a time-integral $\int_{t_1}^{t_2}, L\mathrm{d}t$ of the Lagrangian is stationary in any small variation of the generalised coordinates as functions of time with zero variation at the end-points t_1 and t_2' (Lighthill, 1979). An alternative version of Hamilton's principle (Mitchell and Wait, 1977) states that 'the motion of a system from time t_0 to time t_1 is such that the time integral of the difference between the kinetic and potential energies is stationary for the true path.'

If T and V denote the kinetic and potential energies then $L = T - V$ and $I = \int_{t_0}^{t_1} L\mathrm{d}t = \int_{t_0}^{t_1} (T - V)\mathrm{d}t$. Associated with this is the so-called Euler or Euler-Lagrange equation(s) for a system with n generalised coordinates

$$\frac{\mathrm{d}}{\mathrm{d}t}\left(\frac{\partial T}{\partial \dot{q}_r}\right) - \frac{\partial}{\partial q_r}(T - V) = 0 \quad (r = 1, 2, \ldots \mathrm{n}) \tag{3.39}$$

$q_r = \partial q_r / \partial t$ in this context.

In essence the argument is that the Lagrangian is the variational principle with the route to solution being found through the Euler-Lagrange equations. It might be noted that the term Hamiltonian is often used for the *total* energy of a system—kinetic *plus* potential—expressed as a function of generalised coordinates and associated generalized momenta (Lighthill, 1979).)

These concepts may be expressed in a perhaps more familiar form

(Zienkiewicz, 1970): 'The Euler theorem states that if the integral

$$I(u) = \iiint f\left(x, y, z, u, \frac{\partial u}{\partial x}, \frac{\partial u}{\partial y}, \frac{\partial u}{\partial z}\right) \mathrm{d}x\,\mathrm{d}y\,\mathrm{d}z$$

is to be minimised, then the necessary and sufficient condition for this to be reached is that the unknown function $u(x, y, z)$ should satisfy the following differential equation

$$\frac{\partial}{\partial x}\left[\frac{\partial f}{\partial(\partial u/\partial x)}\right] + \frac{\partial}{\partial y}\left[\frac{\partial f}{\partial(\partial u/\partial y)}\right] + \frac{\partial}{\partial z}\left[\frac{\partial f}{\partial(\partial u/\partial z)}\right] - \frac{\partial f}{\partial u} = 0$$

within the same region, provided u satisfies the same boundary conditions in both cases'.

Referring to the 'quasi-harmonic' equation for the behaviour of an unknown quantity ϕ as

$$\frac{\partial}{\partial x}\left(k_x \frac{\partial \phi}{\partial x}\right) + \frac{\partial}{\partial y}\left(k_y \frac{\partial \phi}{\partial y}\right) + \frac{\partial}{\partial z}\left(k_z \frac{\partial \phi}{\partial z}\right) + Q = 0$$

where k_x, k_y, k_z and Q are known, specified, functions of x, y and z, Zienkiewicz (1970) shows that the minimising requirement of Euler's theorem is for the volume integral χ (the functional) to be given by

$$\chi = \iiint \left[\frac{1}{2}\left\{k_x\left(\frac{\partial \phi}{\partial x}\right)^2 + k_y\left(\frac{\partial \phi}{\partial y}\right)^2 + k_z\left(\frac{\partial \phi}{\partial z}\right)^2\right\} - Q\phi\right] \mathrm{d}x\,\mathrm{d}y\,\mathrm{d}z$$

or, in two dimensions.

$$\chi = \iint \left[\frac{1}{2}\left\{k_x\left(\frac{\partial \phi}{\partial x}\right)^2 + k_y\left(\frac{\partial \phi}{\partial y}\right)^2\right\} - Q\phi\right] dx\,dy \qquad (3.40)$$

For a wave velocity c and group velocity c_g, this becomes (Zienkiewicz et al, 1978)

$$\Pi = \iint_{\Omega} \frac{1}{2}\left[cc_g(\nabla\phi)^2 - \frac{\omega^2 c_g}{c}\phi^2\right] \mathrm{d}x\,\mathrm{d}y - \int_{\Gamma} \phi\left(\frac{\partial \phi}{\partial n}\right)_P \mathrm{d}\Gamma \qquad (3.41)$$

where Γ is the surface or boundary as indicated on Figure 3.12.

Figure 3.12 in fact distinguishes among several different boundaries. Γ_4 is the boundary (or surface) of the structure(s) or object(s) on or at which the load is to be found. At Γ_B the wave field is divided into the (known) incoming wave and the reflected wave. Γ_C is the limit of the finite element model. Γ_D is the far-field boundary. Γ_B can range from coincidence with Γ_A to coincidence with Γ_C. Inside Γ_B the potential is that of the total wave ϕ, outside it is ϕ_0 (known from the incoming wave) and ϕ_S, that of the reflected

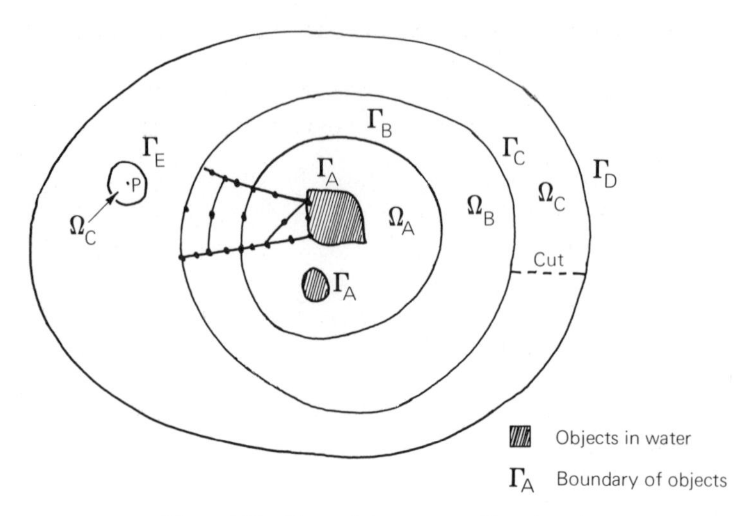

Figure 3.12 Plane of the wave problem domain

wave. On substituting $\phi_0 + \phi_S$ for ϕ, eq. (3.41) becomes

$$\Pi = \iint_{\Omega_B + \Omega_C} \frac{1}{2}\left[cc_g(\nabla\phi_S)^2 - \frac{\omega^2 c_g}{c}\phi_S^2 \right] \mathrm{d}x\,\mathrm{d}y + \int_{\Gamma_B} cc_g\left[\frac{\partial\phi_0}{\partial x}\phi_S\,\mathrm{d}y - \frac{\partial\phi_0}{\partial y}\phi_S\,\mathrm{d}x \right] \tag{3.42}$$

Having expressed the boundary-value problem as a variational principle, the procedure then commonly follows these steps (Mei, 1983):

1. The region is discretised into finite elements (triangles, quadrilaterals, etc. for two dimensions);
2. the solution inside the elements is approximated by linear, quadratic, etc. interpolating functions which have unknown coefficients;
3. differentiation and integration required by the functional is carried out for each element; the functional is quadratic for linear problems and is expressible as a bilinear form in the unknown coefficients;
4. these element bilinear forms are assembled to produce a global bilinear form for the total functional;
5. the functional is extremised with respect to each unknown coefficient to produce a set of linear algebraic equations for the coefficients;
6. the equations are solved;
7. the desired quantities are computed.

According to Mei (1983), experience shows that elements should be less than 10% of the incident wave length and, where there is sharp curvature, they should be smaller than the radius of curvature. The exterior boundary should also be several wavelengths distant from the bodies.

3.4.4.9 *Other techniques* The simplest and most direct approach (Sarpkaya and Isaacson, 1981) is to adopt 'radiation' boundaries at some reasonably large (but finite) distance from the body and apply the radiation condition directly. The *hybrid element* method on the other hand, has an interior region around the body and an exterior region (extending to infinity) for which there is an analytical solution. An alternative is to have a *boundary integral* solution for the exterior region. This, with its distribution of singularities over the matching boundary of the two regions, permits a more general shape for it than does the analytical solution. A fourth method is to use *infinite elements* where the outermost elements extend to infinity and possess exponentially decaying interpolation functions which ensure satisfaction of the radiation condition.

Frequently adopted interpolation functions have a quadratic variation in velocity *potential* across an element, thus providing a linear variation in velocity. At element boundaries the potential is then continuous but velocity is not (Sarpkaya and Isaacson, 1981).

Since the size of element within a grid is governed by the wave length being considered, it is difficult to cover a wide range of wavelengths (Mei, 1983). Dampers are then used in some techniques. Further references on these are given in Zienkiewicz et al (1978) and Hara et al (1979), in the first of which may be found an outline of the formulations required for elements and for solution of the general problem. These aspects will not be dealt with here any further since they fall outside the present remit.

3.4.5 *Stochastic approaches*

3.4.5.1 *Spectral density of force* Whatever the wave theory adopted, the deterministic methods outlined for the calculation of wave loads all have in common the presumption that the waves are regular, long-crested, generally small amplitude, and uni-directional. In reality of course, as the previous chapter indicated, they are generally highly variable and random and the spectral methods mentioned in the previous chapter have been developed to take account of that fact. It is therefore natural to look to the development of similar approaches to load calculation, not only in relation to fatigue but also for the longer-term critical loads.

Using the notation of eqs (2.2) to (2.8), it is easy to see that the spectral densities of velocity, $S_{vv}(f)$, and acceleration, $S_{aa}(f)$, will be related to the spectral density of the sea surface, $S_{\eta\eta}(f)$, by the expressions

$$S_{vv}(f) = \omega^2 \cdot \frac{\cosh^2 k(h+y)}{\sinh^2 kh} \cdot S_{\eta\eta}(f) \tag{3.43}$$

and

$$S_{aa}(f) = \omega^4 \cdot \frac{\cosh^2 k(h+y)}{\sinh^2 kh} \cdot S_{\eta\eta}(f) \tag{3.44}$$

The spectral density of force, on the other hand, is rather more difficult to quantify where drag forces predominate, since the force is dependent on the *square* of the velocity. Consequently the drag term in Morison's equation has to be linearised to maintain a straightforward relationship to the surface, although where inertia (and then diffraction) effects become predominant, the problem diminishes and ultimately disappears.

The employment of stochastic methods thus resolves into two alternatives:

(a) linearisation (or elimination) of the drag contribution;
(b) development of a non-linear method.

3.4.5.2 *Linear approach* Borgman (1967) showed that the total force spectral density could be represented in series form by

$$S_{FF}(f) = \frac{A^2\sigma^4}{\pi}\left[\frac{8S_{vv}(f)}{\sigma^2} + \frac{4\{S_{vv}(f)\}^{*3}}{3\sigma^6} + \ldots\right] + B^2 S_{aa}(f) \qquad (3.45)$$

where $A = C_D \cdot \frac{\gamma}{2g} \cdot D$ and $B = C_M \cdot \frac{\gamma}{2g} \cdot \forall$ in Morison's equation (3.9)

$$\sigma^2 = 2\int_0^\infty S_W(F)\,\mathrm{d}F$$

The symbol $*3$ here represents 'convolution', defined by

$$[S(f)]^{*n} = \int_{-\infty}^{\infty} [S(g)]^{*(n-1)} S(f-g)\,\mathrm{d}g$$

Where drag is not the dominant component, the first term only of the series in eq. (3.45) may be used, yielding the linearised spectral form of Morison's equation

$$S_{FF}(f) \approx \frac{8A^2\sigma^2}{\pi} S_{vv}(f) + B^2 S_{aa}(f) \qquad (3.46)$$

Some authors express this equation in terms of the variances of velocity and acceleration, which of course are the areas under the respective spectral density curves. As used by Holmes et al (1978) for example, eq. (3.46) becomes

$$\sigma_f^2 = \frac{8}{\pi} A^2 \sigma_u^4 + B^2 \sigma_{\dot{u}}^2 \qquad (3.47)$$

where σ_f^2, σ_u^2 and $\sigma_{\dot{u}}^2$ are the variances of force, velocity and acceleration respectively.

Borgman (1967) goes on to show that the total force $F(t)$ at time t is given by

$$F(t) = C_D \frac{\gamma}{2g} D v_{rms} \sqrt{\frac{8}{\pi}} \cdot V(t) + C_M \frac{\gamma}{g} \frac{\pi D^2}{4} A(t) \tag{3.48}$$

where $V(t)$ and $A(t)$ represent the temporal variations of velocity and acceleration respectively.

From eqs (3.43), (3.44) and (3.46) it is clear that the total force is Gaussian in nature for the linearised case, since it is the sum of two Gaussian processes (since they in turn are dependent on surface elevation).

3.4.5.3 *Non-linear approach* When drag is significant the force is non-Gaussian. Its probability distribution function is then given by (Holmes et al, 1978), (Tickell et al, 1976)

$$p(F) = \frac{1}{2\pi A \sigma_u \sigma_{\dot{u}}} \int_{-\infty}^{\infty} \exp\left(- \left| \frac{u^2}{\sigma_u^2} + \frac{\dot{u}^2}{\sigma_{\dot{u}}^2} \right| \frac{1}{2} \right) \mathrm{d}u \tag{3.49}$$

whereas in the linear case it becomes

$$p(F) = \frac{1}{\sqrt{2\pi} \cdot \sigma_f} \exp\left(-\frac{1}{2} \left(\frac{F}{\sigma_F} \right)^2 \right) \tag{3.50}$$

These two equations lead in turn to probability distributions of peak force given by (Tickell et al, 1976)

$$P_{p1}(F) = 1.0 - \int_0^{\infty} \dot{F} p(F, \dot{F}) \, \mathrm{d}\dot{F} \Big/ \int_0^{\infty} \dot{F} p(F = 0, \dot{F}) \, \mathrm{d}\dot{F} \tag{3.51}$$

for the non-linear case and

$$P_{P2}(F) = 1.0 - p(F)/p(O) \tag{3.52}$$

for the linear case.

Equation (3.51) is stated (Holmes et al, 1978) to be extremely lengthy in evaluation for the non-Gaussian situation. (A numerical integration procedure for the solution of eq. (3.49) is outlined in the same reference.) Equation (3.51) derives from the assumption that positive peaks (i.e. wave crests) follow zero up-crossings: a general outline of such theoretical considerations may be found in Newland (1975), for example.

It is important to realise that eqs (3.49) to (3.52) relate to a specific sea state, represented by a specific spectrum or variance. Consequently for longer-term distributions or the evaluation of extreme conditions, it is necessary in turn to weight each such short-term state according to the probability of its occurrence. This may be done by representing each state by its significant wave $H_{1/3}$. As a result the long term force distribution function is (Holmes et al, 1978)

$$P_{LT}(F) = \sum_{i}^{\text{all}\, H_{1/3}} P(F/H_{1/3i}) * \text{Prob}(H_{1/3i}) \tag{3.53}$$

For the peak distribution $P_{PLT}(F)$ the equation becomes

$$P_{PLT}(F) = \frac{\sum_{i}^{\text{all}\, H_{1/3}} \sum_{j}^{\text{all}\, T_z} P_p(F/H_{1/3i}) * \text{Prob}(H_{1/3i}, T_{zj})/T_{zj}}{\overline{T_z^{-1}}} \tag{3.54}$$

where $\overline{T_z^{-1}} = \sum_i^{\text{all}\, H_{1/3}} \sum_j^{T_2} \text{Prob}\,(H_{1/3i}, T_{zj})/T_{zj}$ and is called the gross mean period factor. T_z is the mean zero crossing period calculated from the moments of the spectrum as shown in the preceding chapter subsequent to eq. (2.36) ($T_z = \sqrt{m_0/m_2}$).

For the purposes of calculation, or due to the insufficiency of data, it may be necessary to assume that the individual spectra are given by, say, the Pierson–Moskowitz spectrum. The use of $H_{1/3}$ and T_z has the merit of associating a particular period with a wave height. The probability of its occurrence can be assessed from a bivariate histogram when one is available, as indicated by Holmes et al (1978), from which Figure 3.13 serves as illustration. The significant wave can also be related to the maximum wave in a given period of time, of course, depending on the assumptions made about the distributions (see Longuet–Higgins, 1952 for example).

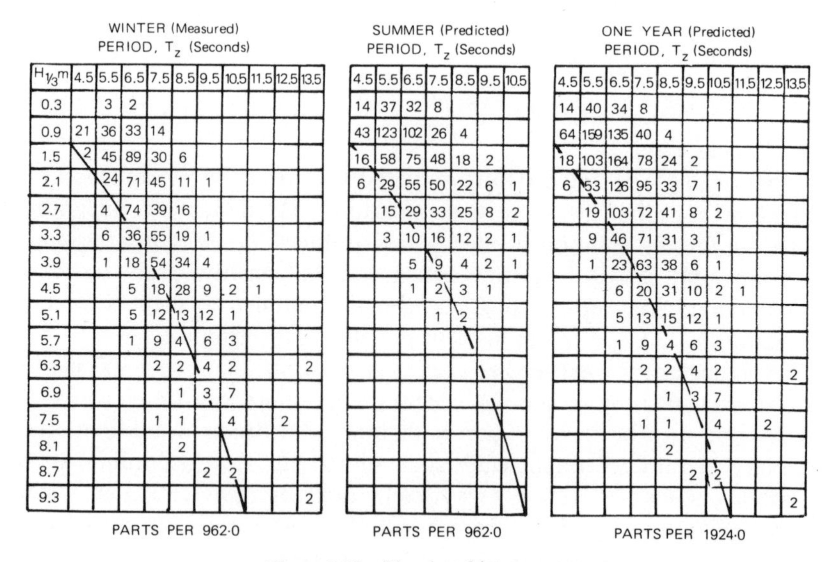

WINTER (Measured)
PERIOD, T_z (Seconds)

$H_{1/3}$ m	4.5	5.5	6.5	7.5	8.5	9.5	10.5	11.5	12.5	13.5
0.3		3	2							
0.9	21	36	33	14						
1.5	2	45	89	30	6					
2.1		24	71	45	11	1				
2.7		4	74	39	16					
3.3		6	36	55	19	1				
3.9		1	18	54	34	4				
4.5			5	18	28	9	2	1		
5.1			5	12	13	12	1			
5.7			1	9	4	6	3			
6.3				2	2	4	2			2
6.9					1	3	7			
7.5				1	1		4		2	
8.1					2					
8.7						2	2			
9.3										2

PARTS PER 962·0

SUMMER (Predicted)
PERIOD, T_z (Seconds)

$H_{1/3}$ m	4.5	5.5	6.5	7.5	8.5	9.5	10.5
0.3	14	37	32	8			
0.9	43	123	102	26	4		
1.5	16	58	75	48	18	2	
2.1	6	29	55	50	22	6	1
2.7		15	29	33	25	8	2
3.3		3	10	16	12	2	1
3.9			5	9	4	2	1
4.5			1	2	3	1	
5.1				1	2		
5.7							
6.3							
6.9							
7.5							
8.1							
8.7							
9.3							

PARTS PER 962·0

ONE YEAR (Predicted)
PERIOD, T_z (Seconds)

$H_{1/3}$ m	4.5	5.5	6.5	7.5	8.5	9.5	10.5	11.5	12.5	13.5
0.3	14	40	34	8						
0.9	64	159	135	40	4					
1.5	18	103	164	78	24	2				
2.1	6	53	126	95	33	7	1			
2.7		19	103	72	41	8	2			
3.3		9	46	71	31	3	1			
3.9		1	23	63	38	6	1			
4.5			6	20	31	10	2	1		
5.1			5	13	15	12	1			
5.7			1	9	4	6	3			
6.3				2	2	4	2			2
6.9					1	3	7			
7.5				1	1		4		2	
8.1					2					
8.7						2	2			
9.3										2

PARTS PER 1924·0

Figure 3.13 Bivariate histograms

3.4.5.4 *Rms force and moment* Another approach (HRS, 1981) for surface-piercing vertical cylinders where drag is not predominant, utilises the Pierson–Moskowitz spectrum in calculating the rms force F_{rms} and moment M_{rms} which are given by

$$m_0' = \int_0^\infty \frac{A}{\sigma^5 \alpha_m^2} \exp\left(-\frac{B}{\sigma^4}\right)\left[C_M\left(\frac{\beta}{2\gamma}\right)\tanh\beta\right]^2 \mathrm{d}\sigma \tag{3.55}$$

and
$$m_0'' = \int_0^\infty \frac{A}{\sigma^5 \alpha_m^2} \exp\left(-\frac{B}{\sigma^4}\right)\left(\frac{C_M(\beta/2\gamma)}{\beta}(\alpha - 1 + \operatorname{sech}\beta)\right)^2 \mathrm{d}\sigma \tag{3.56}$$

where
$$m_0' = \left(\frac{F_{\text{rms}}}{\rho g R^2 \pi h}\right)^2$$

and
$$m_0'' = \left(\frac{M_{\text{rms}}}{\rho g R^2 \pi h^2}\right)^2$$

$\alpha_m = 4\pi^2 f_m^2 h/g$ and $\gamma = h/2R$ and form the coordinate axes basis of Figures 3.14 and 3.15 which have been produced as design guides. In the drag-dominated regions alternative methods have to be employed. $\sigma = f/f_m$;

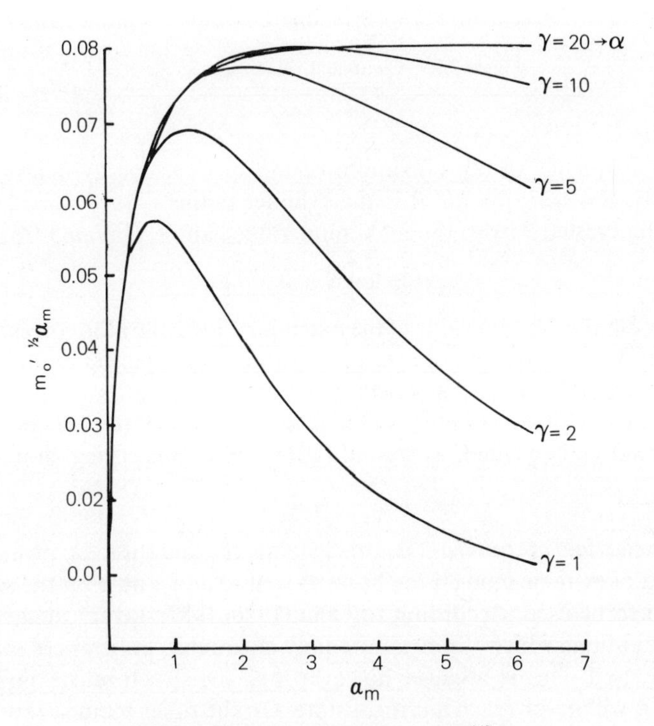

Figure 3.14 Computed values of $m_0'^{1/2}\alpha_m$

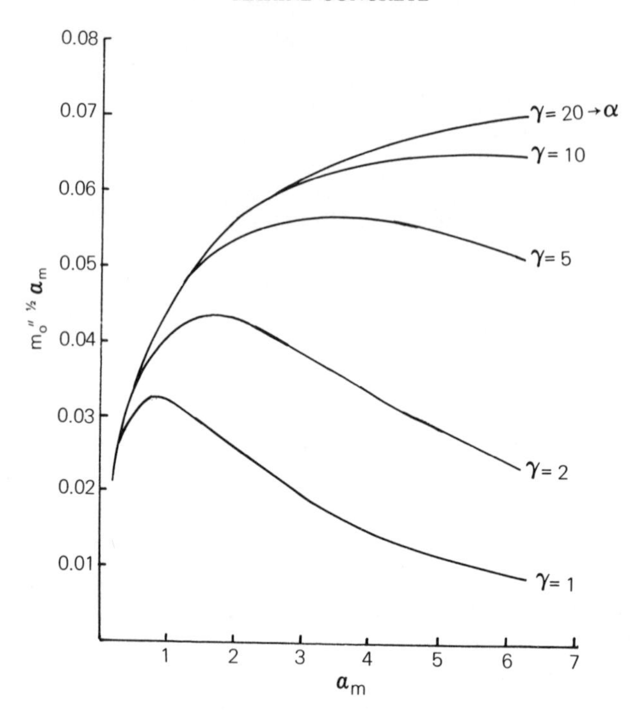

Figure 3.15 Computed values of $m_0''^{1/2}\alpha_m$

f_m is the frequency of the maximum of the spectral energy density curve; $A = 0.0081$; $B = 1.25$; $\beta = kh$; R is the cylinder radius.

For long crested waves, the maximum value can be obtained from

$$\tfrac{1}{2}\overline{\eta}^2 = \log N + 0.5772$$

where $\bar{\eta}$ is the expectation value of the maximum divided by the corresponding rms value, N is the number of zero up-crossings in the period of interest and the number 0.5772 is Euler's constant.

There are one or two limiting qualifications for which the reference (HRS, 1981) should be consulted, as should (HRS, 1979) concerning short-crested waves.

3.4.5.5 *Selection of spectra* Before leaving the question of probabilistic prediction of extreme loads it might be as well to add a note on the selection of spectra to be used. According to Ochi (1976, 1982), rather than consider *all* sea conditions which the structure may encounter, only severe sea states should be used. This is because, however long the structure is exposed to a mild sea, it will never reach a critical state. On the other hand, a severe state need only last a short time before critical responses are generated. This argument provides the basis for calculating the design extreme wave height $\hat{H}_n$:

$$\hat{H}_n = \left[2m_0 \ln\left(\frac{n}{\alpha}\right)\right]^{1/2}$$

$$= \left[2m_0 \ln\left\{\frac{(60)^2 T}{2\pi\alpha}\left(\frac{m_2}{m_0}\right)^{1/2}\right\}\right]^{1/2} \tag{3.57}$$

where n is the number of observations in T hours and α is the *risk parameter* and is a very small numer e.g. 0.005 or less.

3.4.5.6 *Fatigue* Finally mention ought to be made again of fatigue loading. This is considered by Holmes et al (1978) and emphasis is given to the need to consider the **range** of load rather than extremes. In other words, fatigue may be related, for example, to the number of occurrences in a given time (i.e. frequency) of stress reversals of a particular size, say, which of course will result from successive peaks and troughs of a pre-determined difference.

3.5 Wave slam

Slamming is familiar to anyone who has observed a ship or other vessel pitching in heavy seas when the bow emerges from the water and then re-enters. The resulting impulsive force produces a rapid deceleration—it is a hydrodynamic impact phenomenon (Bhattacharyya, 1978). Similar effects are produced in fixed and other bodies of large radius or curvature in waves, giving rise to forces which are generally much greater than those in fully immersed flow such as we have been considering.

Relevant situations in which slamming loads are significant are (Miller 1980):

1. horizontal bracing members at splash zone levels on fixed offshore structures subject to waves passing through;
2. bracing members on semi-submersibles due to waves and vessel motion;
3. towing and launch of jacket structures;
4. wave breaking on vertical members, particularly in inshore and coastal waters where the depth is much reduced.

Category 3 is least likely to apply to concrete structures although as subsea structures (as distinct from jackets) proliferate it may not be unknown.

Miller discusses the phenomenon, physical considerations, experimental work, and design approaches and the concerned reader is advised to consult the reference. Here only a short summary can be given.

Again, deterministic and stochastic approaches are necessary according to

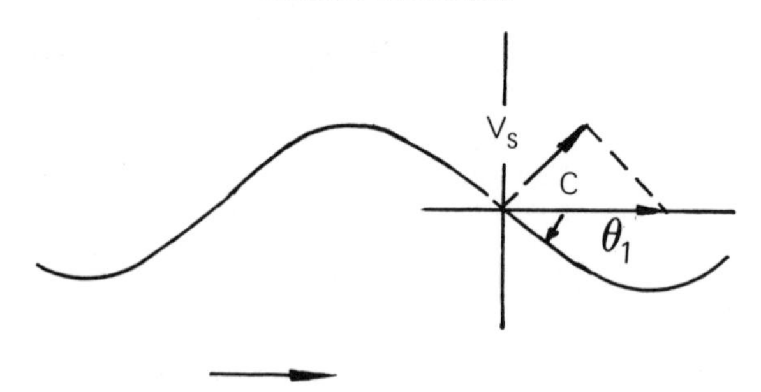

Figure 3.16 Definition of slamming velocity for waves of constant form

whether peak or probable maximum loads are required. Of course, for both fatigue estimates and motion responsive loads the stochastic approach is essential.

In the deterministic methods the maximum slamming velocity V (which is the velocity of the surface normal to itself at impact as in Figure 3.16) is estimated from an appropriate wave theory and eq. (3.58). For a fixed structure the design wave so used should be that which gives maximum velocity at the elevation of the particular member being designed.

$$V = \dot{\eta}(1 + \eta'^2)^{-1/2} \tag{3.58}$$

$$\eta' = \frac{\partial \eta}{\partial x}; \quad \dot{\eta} = \frac{\partial \eta}{\partial t} = c\eta'$$

where η is the surface elevation and c is the phase velocity. This assumes a wave of constant shape but an alternative, exact for any wave theory, is to use eq. (3.59) (Figure 3.17)

$$V = (u^2 + v^2)^{1/2} \sin(\theta_2 - \theta_1) \tag{3.59}$$

$$\tan\theta_1 = \eta' \quad \text{and} \quad \tan\theta_2 = \frac{u}{v}$$

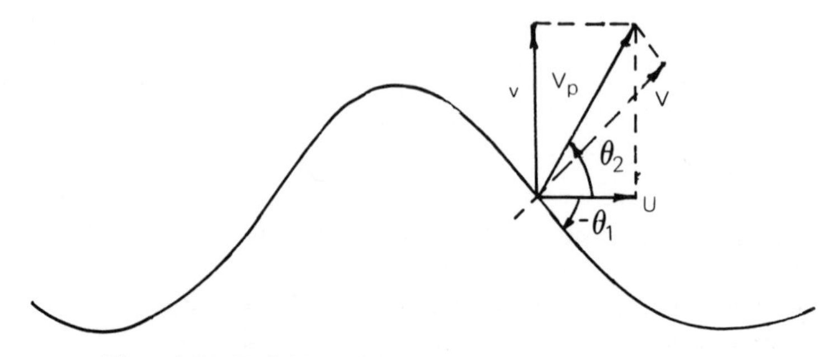

Figure 3.17 Definition of slamming velocity valid for all waves

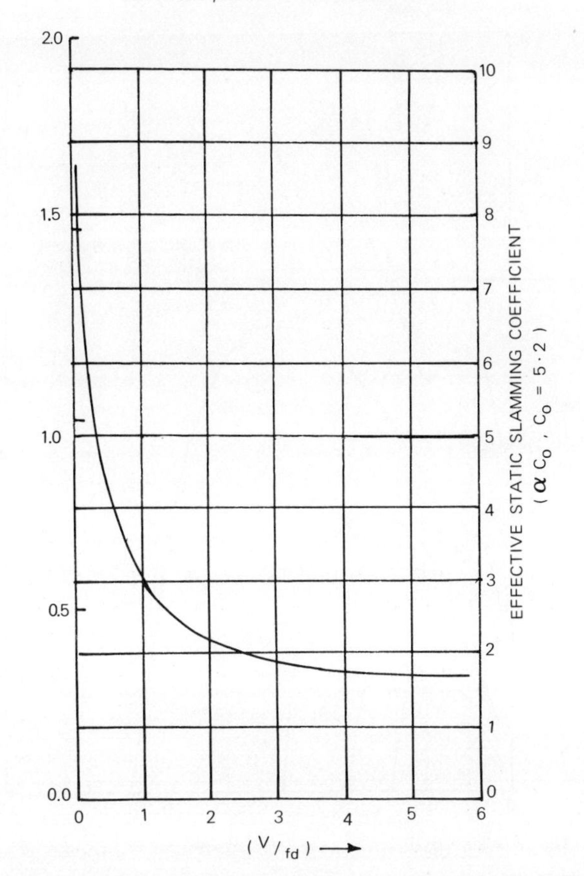

Figure 3.18 Dynamic amplification factor for zero impact angle

The range of angle of inclination θ of the member over which impact may occur must then be estimated as must its natural frequency f. These yield parameters V/fd and $\beta[(l/d)\sin\theta]$ where d is the diameter of the member and l its length over which slamming occurs (for horizontal members l would be the overall length; for inclined members some part may always be above the wave). The effective slamming coefficient for zero impact angle (αC_0) can then be obtained from Figure 3.18 and stresses computed from a static uniformly distributed load (udl), $w = (1/2)\rho V^2 \alpha C_0 d$.

Should the stresses be too high, Figure 3.19 must then be used to give $\bar{\alpha}$ (α_{mean} over the range of β) for V/fd. The udl, w is then computed from $w = (1/2\rho V^2(\bar{\alpha}C_0)d$ where $C_0 = 5.2$. If the stresses are still too high, the member must be redesigned.

Miller also suggests probabilistic and fatigue methods employing further equations and curves but they will not be given here, the reader being referred to the original paper for the details.

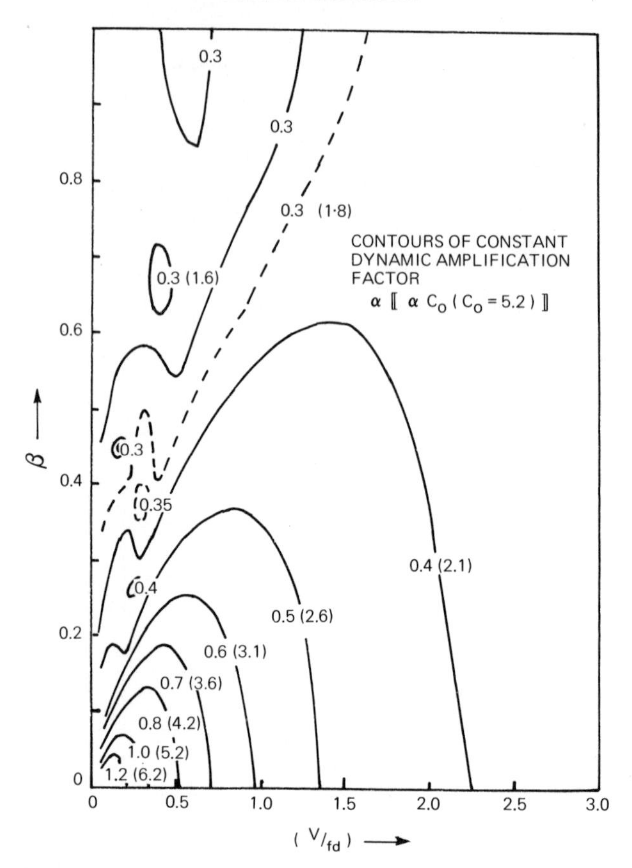

Figure 3.19 Sketch of variation of dynamic amplification factor with β and V/fd

3.6 Wave breaking

3.6.1 *Offshore structures*

Wave breaking is familiar to anyone who has walked along a beach. While it might therefore be expected to affect inshore and coastal structures, offshore structures are not immune from it since waves can often reach their limiting steepness under storm conditions (Hallam et al, 1977). Although design methods as yet are not fully developed, it has been shown (Cokelet, 1979) that forward particle velocities at the wave 'tip' can be more than 30% greater than the deep water wave speed, while accelerations can exceed g. Despite some reduction in force due to entrained air and foaming (which adds to the difficulties in load calculation), the forces can still be substantial, with an increase of up to 30% over long wave theory forces on horizontal cylinders at small depths of submergence (Vinje and Brevig, 1981). In the top 20% of the wave the force may be 3 to 4 times greater (Easson and Greated, 1985). Clearly then some account has to be taken of the phenomenon where it is likely to occur: there is evidence to suggest (Peregrine, 1979) that it may be the *combination* of high waves with high steepness rather than one *or* the other which gives rise to breaking. Vinje and Brevig (1981) refer to other recommendations of a design breaker of steepness 0.135 with a peak pressure of $p_0/\rho g H = 1.43$ and a corresponding horizontal force of $F_x/\rho g H^2 = 0.18$.

3.6.2 *Coastal structures*

While the whole breaking process is very similar in both deep and shallow water (Peregrine, 1979), appropriate design methods for coastal waters are more readily available. However, the availability of theories does not necessarily mean that they are universally applicable (or indeed in agreement), and the designer consequently has to exercise his judgement. Comparative calculations are required to test the sensitivity of a particular design to different theories (Clifford, 1984). The increase in forces due to breaking can

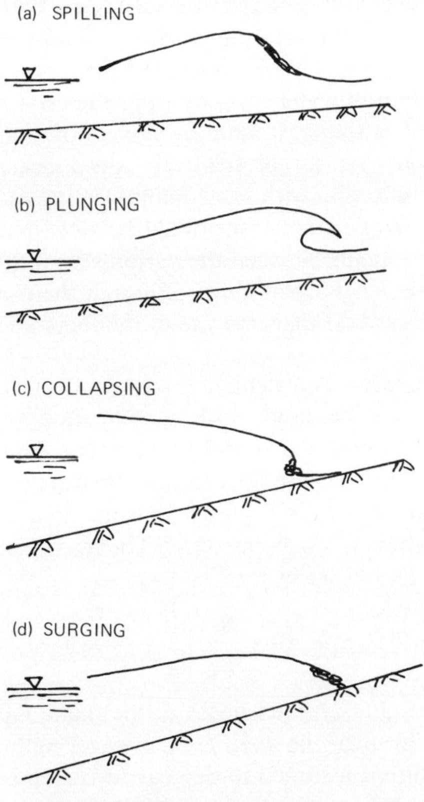

Figure 3.20 Different breaking wave profiles

Table 3.2 Iribarren number for different types of breaker

Breaker Type	N_{I1}	N_{I2}
Surging or Collapsing	>3.3	>2.0
Plunging	$3.3 > N_{I1} > 0.5$	$2.0 > N_{I2} > 0.4$
Spilling	<0.5	<0.4

N_I is the Iribarren Number defined as $N_I = \dfrac{\tan\theta}{(H/L)^{1/2}}$
where θ = beach slope angle
N_{I1}: $H/L = H_0/L_0$ where 0 denotes deep water values
N_{I2}: $H/L = H_b/L_0$ where b denotes breaking.

be considerable: particle velocities twice as great as those predicted from linear theory and forces up to five times as great (Barber, 1986) for coastal structures.

3.6.2.1 *Classification of breaking waves* The general classification of breaking waves is as **spilling, plunging, collapsing** or **surging**. They are illustrated in Figure 3.20 (Sarpkaya and Isaacson, 1981). Spilling breakers have foam spilling from the crest down over the forward face of the wave, and are typical of deep water waves or gentle beach slopes. Plunging breakers have a well defined jet of water forming from the crest and falling onto the water surface ahead of the crest, and are typical of moderate beach slopes. Surging breakers are typical of relatively steep beaches where there is considerable wave reflection with some foam forming near the surface of the beach. Collapsing breakers are transitional between plunging and surging. A distinction may be made between the various types through the Iribarren Number as in Table 3.2 (Barber, 1986) although the change from one type to another is more gradual than the use of numbers might imply.

3.6.2.2 *Forces on walls* To calculate forces on vertical walls the so-called Minikin Method may be used, with modification for walls on rubble foundations or of low height (Coastal Engineering Research Center, 1984), although caution is needed in use since force estimates can be extremely high.

The definition sketch is as in Figure 3.21 which indicates a dynamic force component to be added to the hydrostatic. The maximum dynamic pressure p_m assumed to act at still water level is

$$p_m = 101\gamma \cdot \frac{H_b}{L_D} \cdot \frac{h_w}{h_D}(h_D + h_w) \tag{3.60}$$

H_b is the breaker height, h_w water depth at the toe of the wall, h_D the depth one wavelength in front of the wall, L_D the wavelength in water depth h_D. The dynamic pressure is assumed to decrease parabolically from p_m to zero at $H_b/2$ above and below still water level. The total dynamic force (the

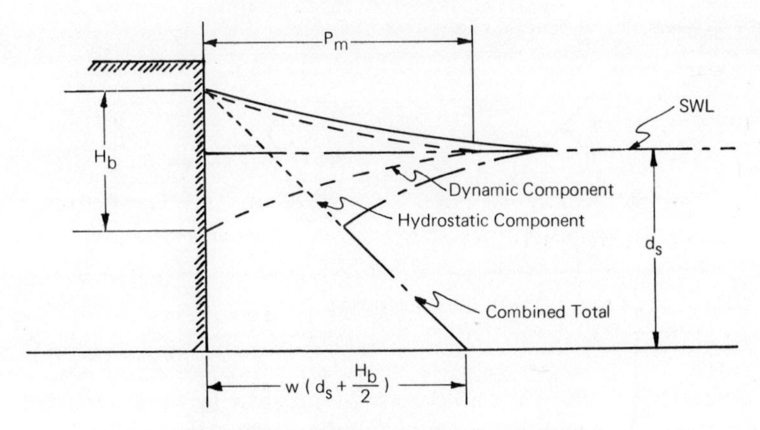

Figure 3.21 Minikin wave pressure diagram

area ‘under’ the dynamic pressure curve) is

$$R_m = \frac{p_m H_b}{3} \tag{3.61}$$

and the overturning moment about the toe is

$$M_m = R_m h_w = \frac{p_m H_b h_w}{3} \tag{3.62}$$

The force and moment due to the hydrostatic pressure distribution would have to be added to the values given by eqs (3.61) and (3.62) of course.

The reference also gives modifications for a wall on a rubble foundation and a wall of low height and also for the situation where a wave has already broken before it reaches the structure. In this case the crest height above still water level is taken as 0.78 H_b and the dynamic and hydrostatic pressures modified accordingly, with the maximum dynamic pressure p_m being given by $p_m = \gamma h_b/2$ where h_b is the breaking wave depth.

Where a wall is shoreward of the still-water line (i.e. the line at which still water level intersects the beach slope) the pressure will be reduced since the wave has to run up the shore to the structure. The reader is referred to the reference for the necessary modifications.

Some variations on wall shape are shown in Figure 3.22 (Coastal Engineering Research Center, 1984). In the case of a sloping wall 3.22(a), the force R should be modified to $R \sin^2 \theta$ and the moment adjusted accordingly. The remaining cases may be assumed to be vertical.

3.6.2.3 *Other design factors* Force is not the sole consideration in design, of course; wave run-up and overtopping have to be considered as well, for example. However, the topic merits a great deal more space than can be

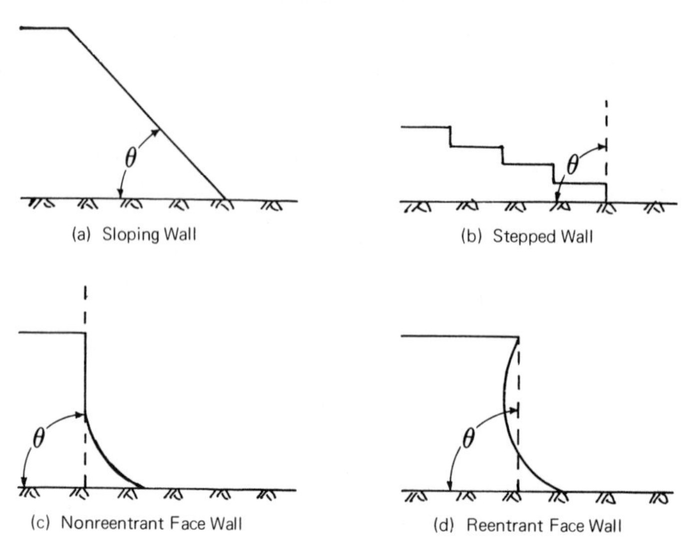

Figure 3.22 Wall shapes

allotted here: the reference gives a series of design charts for instance. The reader who needs more detailed guidance is therefore directed there as a much more comprehensive source.

Wave directionality and variability clearly must have an influence on design (although Barber (1986) remarks that 'wave conditions in shallow water appear to be better represented by a number of discrete uni-directional wave trains rather than continuous models of directional spread'). It is evident, moreover, that prior activity cannot be ignored—breaking waves in shallow water can hardly be considered progressive: reflection, lateral translation, downrush from preceding waves must all be influential. The general lesson is that all methods must be treated with some reserve and that there is no substitute for modelling if the expense can be justified in relation to the project cost. Fortunately, the future may lie more with mathematical than with physical modelling, and may therefore be becoming quicker and cheaper; that appears to be some way off yet for general application. Verification with real-time physical data is by no means always available. Nevertheless the ultimate model seems likely to remain the prototype, the design of which, however, is becoming more informed—a conclusion which is as much as one might reasonably seek, perhaps.

3.7 Variability and uncertainty in wave load calculation

It will be clear from what has gone before that there can be considerable variation in, and uncertainty about, the values of the parameters and coefficients to be used in wave load computation. What effect is this likely to have on the design and behaviour of structures? What guidance is available to the engineer on the selection of appropriate values?

The first question may be the easier to answer but it does little to ease the dilemma posed in seeking a reasonable compromise between, on the one hand, the conservatism inherent in safety and, on the other, the economy of increased risk. Such a statement of the problem, of course, embraces several implicit serious questions and increasing effort is being put towards answering them. Briefly, they require the definition and assessment of 'safety', 'conservative', 'economy', and 'risk', any one of which must be qualified by circumstances. For example, it is easy to insist that hazard to life cannot be tolerated, but not every risk is of loss of life or limb or necessarily directly involves people. There is an infinity of possible gradations even on a scale of nought to one.

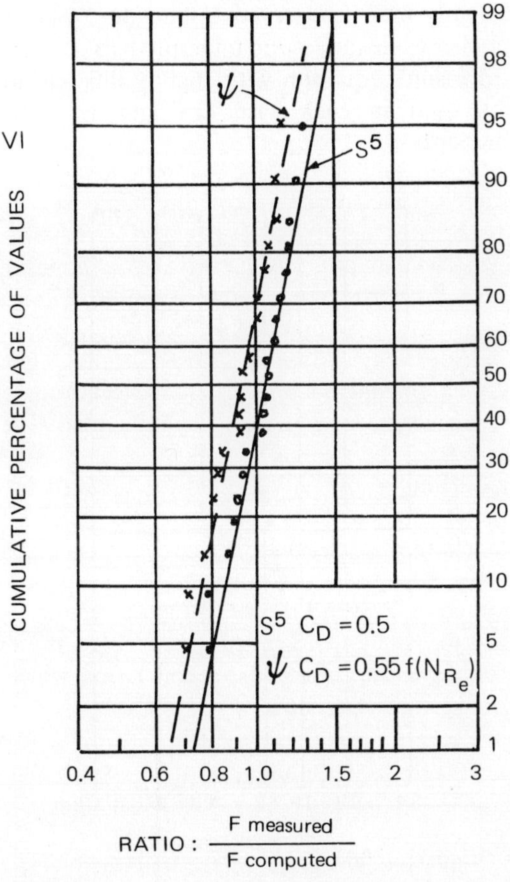

Figure 3.23 Twenty highest measured total wave forces

This is not a text on reliability and risk analysis and so detailed discussion will not be attempted here although the topic will be returned to in subsequent chapters. However, the extent of inherent uncertainty is revealed easily by looking at a small literature sample which, although relating primarily to steel structures, provides some useful indicators.

3.7.1 *Variations due to choice of wave theory*

Bea and Lai (1978) examined loads measured on a single pile test structure 0.92 m diameter (36 in.) in 31 m (100 ft.) water depth in the Gulf of Mexico. Peak measured and computed forces were plotted as ratios (ratios > 1.0 are conservative) on a relative frequency basis (the cumulative percentages of ratios equal to or less than a given ratio of measured to computed load): Figures 3.23 and 3.24. The computed values were based on stream function and Stokes fifth-order theories and it can be seen that the stream function values are consistently more conservative than the Stokes values, the first tending to overpredict while the second underpredicts. However, calculations were based on Morison's equation with slightly different drag and inertia coefficients being used in each case (0.5 and 1.5 – Stokes, 0.55 and 1.33 – stream function).

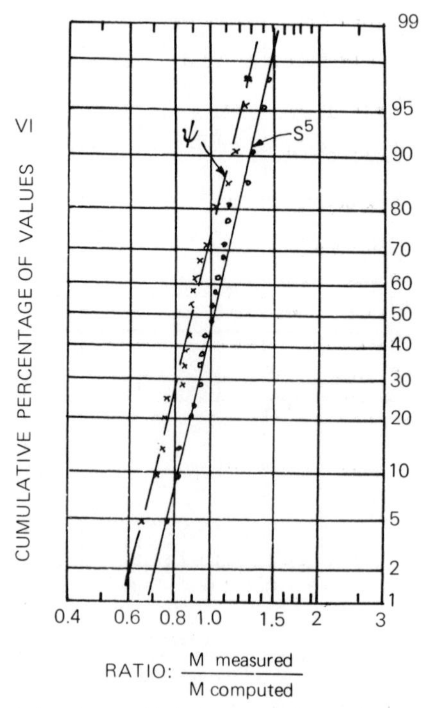

Figure 3.24 Twenty highest measured wave force moments

3.7.2 *Uncertainties in force coefficients*

The uncertainty in force coefficients (such as C_D and C_M) is obviously of some concern. However, it has been suggested (Nolte, 1978) that 'the increase in the maximum force due to the uncertainty in force coefficients is relatively small and is much less than that which would be determined from the mean value plus one standard deviation of the uncertainty'. (It is accepted that there is a random scatter from wave to wave in force coefficients calculated from field measurements.) Some qualification is needed to such a statement, however, judging by the results of another study (Abdelradi and Miller, 1986). Here it seems that the question is very much of whether or not the forces being considered are in-line, lateral or vertical.

The authors compared the recommended values for drag and inertia coefficients of three major classification societies and then proceeded to apply these to a hypothetical jacket structure (with four columns at 4.0 m diameter and bracings 1.5 and 2.0 m diameter). The C_M and C_D values are shown at Table 3.3 and the load calculations were made for three selected wave frequencies and for smooth and rough cylinders, taking account of the variations in the coefficients with Reynolds and Keulegan-Carpenter numbers and relative roughness. Adopting the Lloyd's Register values as base, the following conclusions were reached:

(a) *DnV* values for surge force ranged from −3.3% to +14.9%; *BV* values ranged from −3.0% to +5.3%
(b) *DnV* values for heave forces ranged from +36.1% to +50.9%; the *BV* range was from +21.2% to +42.7%
(c) *DnV* sway values ranged from +33.4% to +37.6%; the *BV* range was from +9.1% to +44.5%

There were also considerable variations in moments.

Linear wave theory was used but it is stated higher order theories substantially increase the velocities and accelerations and consequently the

Table 3.3 Comparison of recommended C_M and C_D values for three classification societies (Abdelradi and Miller, 1986)

Society	C_M		C_D
Lloyd's Register	$R < 1.75$m	1.5	0.5
	$R > 1.75$m	2.0	
Det norske Veritas	$2R/L > 1.0$	1.5	0.5 to 1.2 depending
	$2R/L < 0.1$	2.0	on R_e and roughness
	Intermediate values		>0.7 for $R_e > 3 \times 10^6$
	by interpolation		
Bureau Veritas	$R < 0.725$	1.5	0.75
	$R > 0.725$		
	$1 + (\log_{10} 20R - 0.8)^{1/2}$		

wave loads and stresses. To what extent this affects the 'discrepancies' is not indicated but it does seem that a major factor is whether or not roughness is allowed for.

3.7.3 *Load sensitivity to design parameters*

Preliminary results of a UK Department of Energy study have been published (DoEn, 1986) from which Figure 3.25 is taken. This summarises the static strength sensitivities of two Northern North Sea platforms and one Southern North Sea: one barge launch, one self floater and one shallow water. The greatest effect is from variation in the drag coefficient, particularly the use of rough drag values, which bears out the remarks above.

Other work (Marshall, 1986) suggests that sensitivities for individual members can be very different from those for the structure as a whole. For drag-dominated or inertia-dominated loading regimes it is straightforward to assess the effect of varying such factors as coefficient, wave height, member size, and so on. Wave period is more difficult but for deep water linear theory it has been shown that

$$F_{D2}/F_{D1} = (1+r)^{-2} f_{rD} f_{xD} \tag{3.63}$$

where $f_{yD} = \exp\{4\pi[(1+r)^{-2} - 1]y'\}$

$$f_{xD} = \frac{\cos^2[2\pi(1+r)^{-2}x']}{\cos^2 2\pi x'}$$

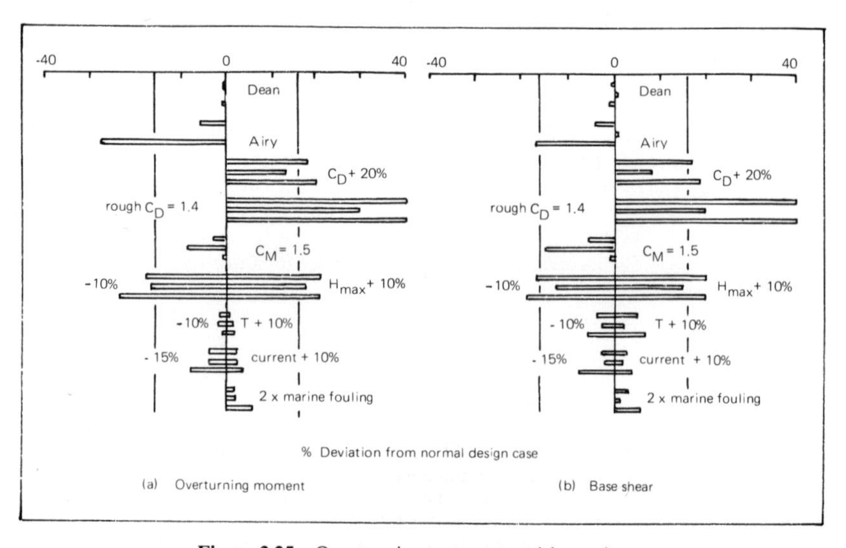

Figure 3.25 Overturning moment and base shear

where F_{D1} can be taken as the incremental horizontal drag force for period T_1
F_{D2} is the force for period $T_2 = (1+r)T_1$
$x' = x/L$ where x is the location of the member
$y' = y/L$ where y is the depth of the member

f_{yD} and f_{xD} are known as depth and length factors. Clearly it is easy to compute the effect on force of varying the period by a given proportion r. The comparable inertia equation is

$$\frac{F_{I2}}{F_{I1}} = (1+r)^{-2} f_{yI} f_{xI} \tag{3.64}$$

where $f_{yI} = \exp\{2\pi[(1+r)^{-2} - 1]y'\}$

$$f_{xI} = \frac{\sin[2\pi(1+r)^{-2}x']}{\sin 2\pi x'}$$

Equivalent expressions have been derived for deep-water Stokes fifth-order theory (Marshall, 1988).

In general it appears that variations in period can have a considerable effect on local wave force computation and a significant effect on individual vertical members. In the inertia dominated case $\pm 10\%$ variation in period gives a variation of -4.5 to -6.5% in horizontal load on a vertical column while $\pm 20\%$ period variation increases the load variation to -14 to -37%. Fifth-order theory yields similar results if a little smaller.

In the words of Abdelradi and Miller (1986), 'This is not to suggest that the structures which are produced are unsafe. The very good record of jacket behaviour in the North Sea is proof of the adequacy of the safety factors which must have been applied to the design loads...' As has already been

Table 3.4 Uncertainty measures of extreme hydrodynamic loading on a vertical pile (Guedes Soares and Moan, 1983)

Main Variables	Total *cov*	H	T	h	c	kt	k	C_D	C_M	$\tilde{v}$
$H = 30$m $h = 80$m $D = 4.0$m	0.37	0.67	0.02	0.00	0.15	0.01	0.03	0.08	0.00	0.04
$H = 30$ $h = 120$ $D = 4.0$	0.40	0.66	0.04	0.00	0.16	0.01	0.03	0.06	0.00	0.04
$H = 30$ $h = 80$ $D = 0.5$	0.34	0.44	0.03	0.00	0.11	0.30	0.01	0.09	0.00	0.03
$H = 22$ $h = 80$ $D = 4.0$	0.45	0.68	0.04	0.00	0.16	0.00	0.03	0.05	0.00	0.04

cov coefficient of variation; c wave speed; t fouling thickness; k roughness; $\tilde{v}$ wave kinematics.

mentioned at the beginning of this section, the designer has to reconcile economy and safety. Informed judgement is essential and, as ever, one must beware of assertions of absolute certitude.

Finally, reference may be made to a study (Guedes Soares and Moan, 1983) of the relative importance of the different factors affecting variation in extreme hydrodynamic loading on a vertical pile. Table 3.4 shows that variation in wave height is by far the most important although reduced somewhat by depth. Next comes variation in wave speed, although roughness is a significant contributor for small diameters. Thus uncertainty does not necessarily correspond with sensitivity.

As a footnote, it might be added that increasing attention is being given to risk analysis for offshore structures. A useful brief introduction is given by Bea (1975) who sets out to establish a general framework for evaluating risk, defining a tolerable level for it and developing environmental criteria based on this level. The discussion refers primarily to steel platforms but it is more generally relevant despite the slant of this particular statement which is worth quoting. 'In the meantime, engineering has at its disposal the knowledge to install reliable platforms. This reliability is achieved with additional steel (increasing the design loadings) rather than by reducing the uncertainties in loadings or resistances'. Nonetheless, in view of some of the results quoted above, even increasing the design loadings is still imbued with some uncertainty. (It should also be noted that, despite the emphasis on offshore platforms, the same techniques can be applied to the design of coastal structures (Dover and Bea, 1979).)

3.8 Wind loading

While wind loading is generally considered to be minor in relation to wave loading for the main structure in offshore applications in particular, it cannot be ignored. It is of course especially significant for the superstructure, parts of which may be 100 m or so above the mean sea surface. This also has implications for the deck support structure. Furthermore, in some locations, onshore and inshore as well as offshore, winds may reach hurricane force.

There are two main considerations, given the existence of appropriate theories for load transfer: first, the velocity profile and second, variability and the prediction of extremes.

Following the description in the Shore Protection Manual (Coastal Engineering Research Center, 1984), the wind can be considered to be driven by large-scale, near steady-state pressure gradients in the atmosphere, resulting in a profile as in Figure 3.26. From about 1000 m above the surface the driving force is the geostrophic balance between Coriolis and local pressure forces. Nearer the surface frictional effects become increasingly significant with surface roughness of obvious importance and temperature

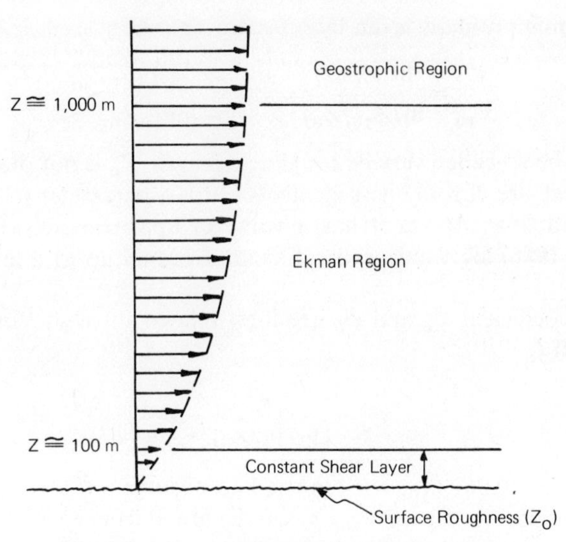

Figure 3.26 Atmospheric boundary layer over waves

variations, horizontally and vertically, having an influence. For our present purpose, however, it is the constant shear layer which is the predominant consideration and here alternative theories are available.

3.8.1 *Velocity distribution*

Within this atmospheric boundary layer it is customary to relate the wind velocity distribution to the velocity at a fixed level, generally 10 m above mean water level. Apart from a full surface layer similarity approach (which applies dimensional analysis to empirical data), it is normal to adopt either a power law distribution or a logarithmic profile. However, despite the popularity of the power law approach, recent advice (Shearman, 1985) is against it. For extremes particularly, the logarithmic profile is to be preferred as being more justifiable and scientifically acceptable (while the similarity approach is advisable when greater accuracy is required). This is borne out by a major wind measurement project in the North Sea (Wills et al, 1986). Nevertheless, both the power law and logarithmic profile are given here for completeness.

The power law is

$$\frac{V}{V_{10}} = \left(\frac{Z}{Z_{10}}\right)^p \tag{3.65}$$

where V is the windspeed at height Z
V_{10} is the windspeed at 10 m (Z_{10})
p is the power, generally 0.125 over the sea.

The logarithmic profile is given by

$$\frac{V}{V_{10}} = \frac{\ln(Z/Z_0)}{\ln(Z_{10}/Z_0)} = \frac{\log_{10}Z - \log_{10}Z_0}{1 - \log_{10}Z_0} \tag{3.66}$$

where Z_0 is the so-called surface roughness length. Z_0 is not directly related to the physical size of roughness elements but is a parameter related to their effect on fluid flow. At sea it has a value of approximately 1.5×10^{-4} m (Wills et al., 1986) for windspeeds of around 10 m/s up to a level of about 80 m.

The drag coefficient C_d and Z_0 are inter-related through Von Karman's constant k_a (0.4):

$$C_d = \frac{k_a^2}{[\ln(10/Z_0)]^2} \tag{3.67}$$

while $C_d = 0.75 + 0.067\,V_{10}$, $\quad 4\,\text{m/s} < V_{10} < 21\,\text{m/s}$

For a particular windspeed V_{10}, Z_0 can be found from eqs (3.66) and (3.67) and hence the scaling rule of eq. (3.65) can be applied i.e. the velocity distribution can be found. Table 3.5 details the factors to be applied (Shearman, 1985). From this follows the wind load on various parts of the superstructure by using the drag and lift equations as for fluid flow

$$F = \tfrac{1}{2} C_{D,L} \rho V_z^2$$

C_D and C_L vary with Reynolds' number as well as shape of element or surface. surface.

Table 3.5 Logarithmic profile for a range of windspeeds. Factors to be applied to winds at the stated height to reduce them to a height of 10 m (Shearman, 1985). Conversely, multiplying the speed at 10 m by the reciprocals of these values will give the velocity at a given height

Speed			Height (m)		
(m/s)	20	40	60	80	100
4	0.948	0.9	0.875	0.858	0.845
6	0.944	0.895	0.868	0.85	0.837
8	0.941	0.889	0.862	0.843	0.829
10	0.939	0.884	0.856	0.836	0.822
12	0.936	0.88	0.85	0.83	0.815
14	0.934	0.875	0.845	0.824	0.809
16	0.931	0.871	0.839	0.818	0.803
18	0.929	0.867	0.835	0.813	0.797
20	0.927	0.863	0.83	0.808	0.792
22	0.924	0.86	0.826	0.803	0.786
24	0.922	0.856	0.821	0.798	0.782
26	0.92	0.853	0.817	0.794	0.777
28	0.918	0.849	0.813	0.79	0.772
30	0.917	0.846	0.81	0.785	0.768

3.8.2 *Extreme wind speeds*

The estimation of extreme wind speeds is necessarily based on available data, and maps giving recommended values have been produced for areas such as NW Europe (British Standards Institution, 1982). In this case, gust speeds of 3 seconds duration with a return interval of 50 years are given. Longer duration gusts are recommended for various structural members or parts of the superstructure, the windspeed being reduced accordingly. US practice is to use the so-called fastest mile speed i.e. the fastest speed at which the wind travels one mile (the 5 second gust speed is 1.18 times the fastest mile speed, according to data quoted by Gaythwaite (1981)). However adequate measured data are not available in many cases and resort has to be made to data produced from unmeasured observations at sea.

3.8.3 *Estimation from sea states*

Descriptions of the sea state are used to produce estimates of windspeed, generally by the Beaufort scale but some caution is needed in such instances, (Babbedge and Lynagh, 1984). The specific conversion in general use is the World Meteorological Organisation (WMO) Code 1100 scale but there is another preferred by some bodies, the Commission for Maritime Meteorology, CMM-1V code. These codes do not give the same values, the WMO Code 1100 scale over-estimating speeds below 10 knots but under-estimating speeds above 10 knots. The CMM-1V scale on the other hand, relative to the WMO code, increases speeds at Beaufort forces below 5 and reduces them at Beaufort forces above 5. It is believed that the CMM-1V scale is the more reliable, particularly for calculating extreme wind speeds.

Table 3.6 Comparison of the WMO 1100 and CMM-1V Wind Scales

Beaufort Force	1100 Code knots	CMM-1V code knots
0	0	0–2
1	1–3	3–5
2	4–6	6–8
3	7–10	9–12
4	11–16	13–16
5	17–21	17–21
6	22–27	22–26
7	28–33	27–31
8	34–40	32–37
9	41–47	38–43
10	48–55	44–50
11	56–63	51–57
12	>64	>58

Irrespective of the merits of the different methods of calculation, it is clearly essential to ensure that sets of data are consistent in their mode of acquisition and processing. The two scales referred to are given in Table 3.6.

Making use of such observations together with measured data, extreme values have been predicted for NW Europe, and the 50 year return period values plotted (Noble Denton, 1985).

3.9 Marine growth/fouling

All structures in the sea are likely to be subject to some type of marine fouling. The problem has been well known to the shipping industry for a very long time of course (as victims of the old punishment of keel-hauling under ships must have learned to their harm). In relation to cost, one figure quoted (Freeman, 1977) is that an increase of 100 μm in surface roughness corresponds to a 10% loss of power. It was also estimated in 1976–77 that the total annual cost of fouling to the UK shipping industry alone was approximately £50 m at that time. So far as fixed structures are concerned, however, the problem is perhaps more complex in that not only running costs are involved (in respect of inspection, maintenance and countering corrosion), but some allowance for increased loading has to be made in designing them in the first place.

3.9.1 *Types of growth/fouling*

A short description of the processes which take place, useful particularly to the non-biologist, is given by a UK Department of Energy Working Party (DofEn, 1980). Extracting briefly from this might be helpful. 'The biological nature of the fouling, its distribution and rate of growth are dependent upon the geographical location of the installation and, to a lesser degree, on local activities which perturb the natural environment.' Marine organisms tend to 'grow out from the slowly moving boundary layer into the faster-moving food-rich waters, and... to occupy the maximum area of the food or energy bearing water... any new surface may be fouled rapidly... Marine fouling organisms... are drawn from both the plant and animal kingdoms. Some encase themselves in a hard outer skeleton, such as a shell or calcareous tube, whereas others have no such protection and so marine growth is often referred to as either hard or soft fouling... The normal succession of growth starts with the formation of bacterial slime in which micro-organisms appear within two or three weeks. Barnacles and soft foulers may then attach themselves to the strucfure and reach maturity within three to six months. These "primary foulers" appear to alter the surface conditions so that "secondary foulers" such as mussels, sponges and anemones are able to settle. Mussels will normally appear within two seasons and may eventually produce

Table 3.7 Categories of fouling organism (North Sea)

Hard growth	
Mussels	frequently dominant; contribute to loading on the structures in the upper zones
Barnacles, tubeworms, limpets, etc.	firmly attached growths; mostly one organism thick; difficult to remove; hamper inspection
Calcareous (red or brown) algae	potential fouler in deeper waters, thin covering, hard and very adherent
Soft growth	
Kelps	potentially dominant in areas where mussels are absent; long frond-like growths may introduce new considerations
Algae (other than kelps)	branching; large and small
Sponges, anemones, sea-squirts, alcyonium	soft bodied, but often bulky organisms
Rock borers	sponges and molluscs capable of boring into concrete
Hydroids, bryozoa	soft, branching and compressible growths

a mussel-dominant colony, beneath which many of the earlier foulers will die. Although mussel fouling develops more slowly than barnacles, it is a more characteristic permanent growth around the British Isles.'

For the North Sea (and other regions will have their equivalents) the categories of fouling organism are listed at Table 3.7.

The presence and rate of growth of each organism depends on several variable factors, including depth, distance from shore, temperature, season, exposure, salinity, water current, water quality, light penetration, availability of food, other organisms present, presence of anti-fouling systems, surface characteristics, corrosion-protection systems, and imposition of other environmental stresses such as cleaning (Oldfield, 1980). Thus with increasing depth (and consequent pressure increase), temperature, light penetration, and food availability all decrease and so metabolism and rate of growth are reduced. This leads to reduced thickness and, say, a mussel-dominant community being replaced by a bryozoan-dominant community. Algae need light and so are restricted to the surface region. Below depths of 150 m the same organisms are found in any ocean.

Clearly growth depends on both location and climate which are not necessarily inter-related: the location, for example, will determine the influence of currents. By and large, surfaces are not fouled by coastal organisms beyond the 40 m depth contour (depending on currents); maximum growth occurs at around 30 to 35°C, although the longer growing season in

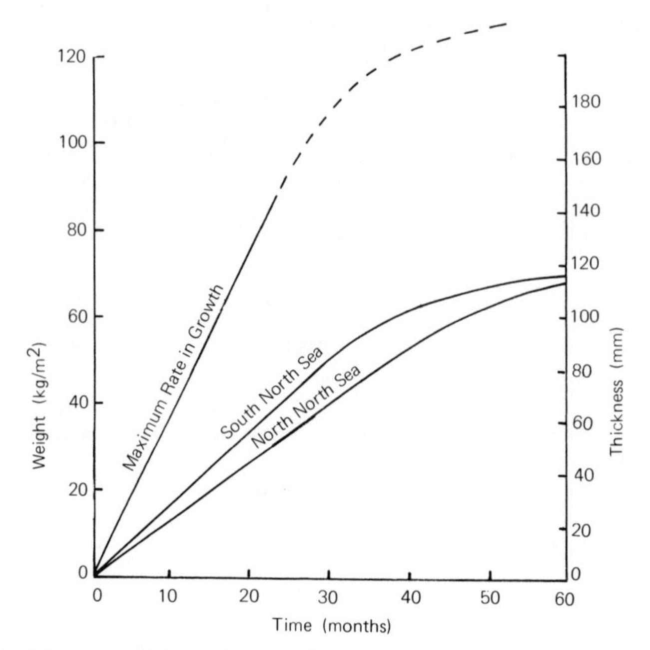

Figure 3.27 Weight and thickness of mussel-dominant fouling against time for surface waters in the British Isles

warmer waters will also have an effect; more exposed sites result in more attrition of growths; increased salinity reduces bulk and diversity of growth. In the North Sea, mussel growth is predicted (Oldfield, 1980) to follow a pattern as shown in Figure 3.27 to a depth of around 30 m. On the other hand, in some areas, kelps could grow to a thickness of 2 m and more in the long-term over the same depth range, with hydroids 100 mm thick over the full depth of the structure.

It is also clear that micro-climate is important, since neighbouring structures do not necessarily exhibit the same growth patterns (Forteath et al, 1984).

What then is the effect on load calculation?

3.9.2 *Influence on load calculation*

Hard growth in particular has two primary effects (if one ignores increase in weight):

(a) the diameter of structural member is increased by up to 250 mm, perhaps more in some instances;
(b) the surface roughness and hence drag coefficient C_D is increased (there may also be some effect on inertia coefficient).

3.9.2.1 *Effect on size* Estimates of the effect on a typical steel platform have been made (Oldfield, 1980) and it is clear from Figure 3.28 that variation in thickness has a diminishing effect with increasing diameter. Hence the conclusion is that, for typical offshore concrete platforms, there is little or no problem so far as load is concerned. However, for drag-dominated structures of smaller dimensions as in some coastal structures, it has to be recognised. Moreover, subsea structures may have to be reviewed carefully.

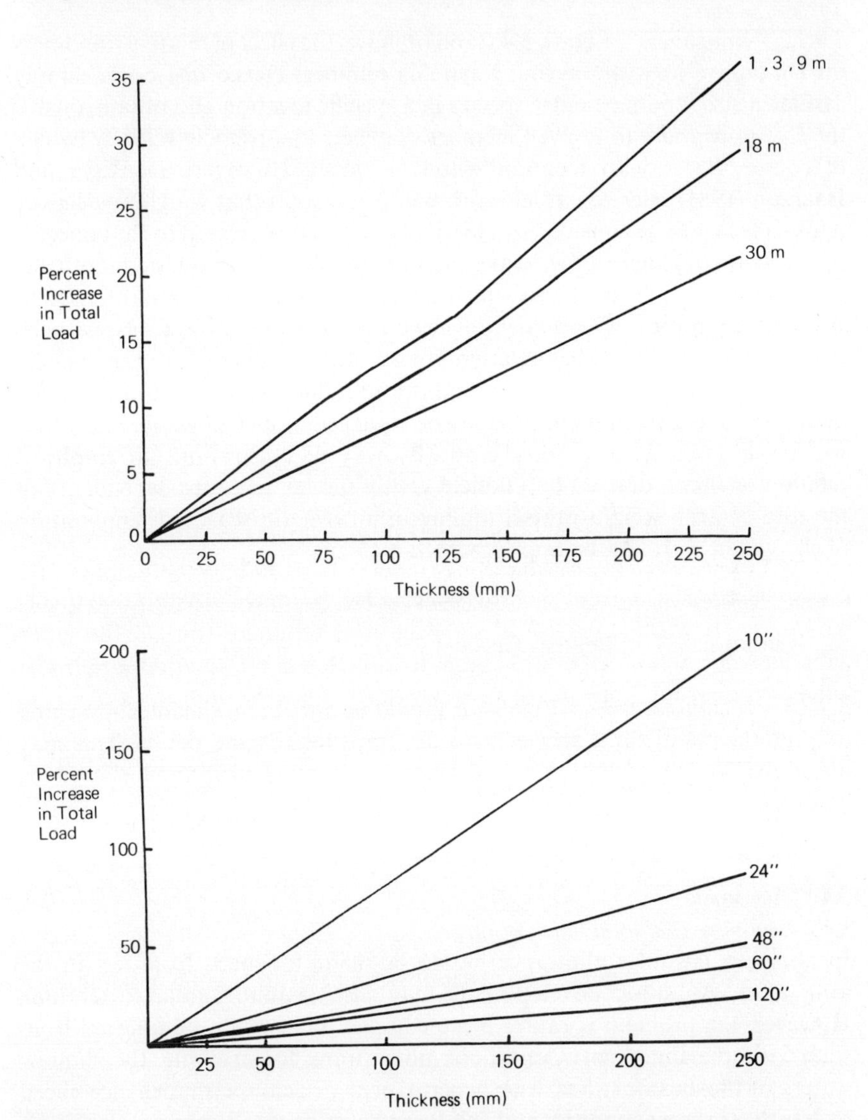

Figure 3.28 (a) Percentage increase in load with marine growth thickness for various wave heights (b) Percentage increase in load from 30 m design wave for various member diameters

It is also the case that the apertures in protection walls such as the Jarlan type may become choked. But the primary reservation must concern the effect of growth such as kelp on the inertia coefficient, the question being what effect does a dense growth have on the added mass? In the absence of evidence to the contrary one must conclude that it might well be significant and so an appropriate cleaning programme should be implemented (which regular inspection would probably demand in any case) to avoid the problem.

3.9.2.2 *Roughness* Figure 3.4a (and b) shows the effect of relative roughness on C_D (and C_M) with varying Reynolds number. Hence one could simply assume a size for a particular species in a specific location and obtain a value for C_D appropriate to a given member diameter and particle velocity (which of course varies with location within a wave). However, Sarpkaya and Isaacson (1981) refer to earlier work which indicates that roughness **density** is also a factor to be considered. This of course must be related to the concepts of isolated-roughness flow, wake-interference flow, and quasi-smooth (or skimming) flow (Chow, 1959) which introduce the inter-relation of element size and separation. Consequently it would seem advisable to assess such considerations carefully before attempting to adjust values of drag coefficient, C_D. Moreover, location will be an important influence as has been indicated above: marine growth might usefully be regarded as a crop whose variations will be as great as any other crop. (By way of illustration, an empirical relation has been derived by Oldfield giving the temperature dependence of the rate of fresh weight mussel fouling in air, M_a on seasonal temperature of the water, T. It can be approximated by

$$M_a = 2T - 1 \tag{3.68}$$

M_a is in kg/m^2, T in degrees Celsius)

3.9.2.3 *Weight increase* Finally, it should be noted that the interbranching and interlacing of some species provides traps for silt and debris. This may affect the weight of individual members and so its likelihood should be assessed.

3.10 Ice loads

In northern latitudes allowance has to be made for loads from ice (in the long term, Antarctic developments may also require similar attention). However, the problem is rather more complex than might be inferred from such a simple statement. Conditions range from, for example, the shallow waters of the Beaufort Sea with a permanent or semi-permanent ice sheet, to North Sea type conditions of the Bering Sea or the Western Atlantic off Newfoundland with the addition of a seasonal ice sheet and icebergs which can be several million tons in weight. Bergy bits the size of an office desk can be an additional hazard when travelling at water particle velocities. The much slower movement of an ice sheet several metres thick can exert considerable thrust, while low atmospheric temperatures can result in severe temperature gradients and engender other problems in a porous material such as concrete unless they are recognised and accounted for (although concrete has a number of commendable features for low-temperature applications).

There is a considerable literature on ice, too voluminous to cover in detail here. Not all is confined to marine or Arctic problems of course: ice thrust in the design of dams is a consideration in many parts of the world, for example. However, the most severe problems and greatest literature coverage concern Arctic and sub-Arctic regions (more correctly perhaps, boreal and sub-boreal regions). A few references are given in the text below but there are certain 'multiple sources' which should be consulted for detailed guidance. (These are all English-language):

(a) Proceedings of the various conferences on Port and Ocean engineering under Arctic Conditions (POAC now has a secretariat based at the Technical University, Trondheim, Norway);
(b) Proceedings of the various Ice Symposia of the International Association for Hydraulics Research;
(c) Proceedings of several conferences on Cold Regions Engineering of the American Society of Civil Engineers;
(d) Proceedings of several conferences on Offshore Mechanics and Arctic Engineering of the American Society of Mechanical Engineers.

Many papers have been published elsewhere in journals, specialty conferences, more general conferences (such as the Offshore Technology Conferences of the Society for Petroleum Engineers in Houston, Texas) and there are also the reports published by organisations such as the Cold Regions Research and Engineering Laboratory of the US Army (CRREL), the Centre for Cold Oceans Resources Engineering (C-CORE) in Canada, the Centre for Frontier Engineering Research (C-FER) also in Canada, and the former Arctic Petroleum Operators Association (APOA) (now the Canadian Petroleum Association) Calgary, Canada. Regular abstracting bulletins have also been produced for a number of years by the Arctic Institute of North America at the University of Calgary and the Scott Polar Research Institute at the University of Cambridge in the UK. The list is not intended to be, nor can it be, exhaustive, and work has been carried out in a number of countries such as Canada, USA, Norway, Finland, Germany, Japan, USSR, UK, and others. By and large, however, a considerable volume of this effort can be traced through the sources mentioned above.

Although it cannot include the more recent literature, a very comprehensive coverage of the behaviour of ice and of ice loads is provided by Michel

(1978). A slightly later state of the art report (by several authors) was produced two years later (IAHR, 1980), to be subsequently extended (IAHR, 1984, 1987). Fairly recently an authoritative general review was given in shorter compass (Croasdale, 1985) and a comprehensive and more extensive treatment still later (Sanderson, 1988). In broad terms these publications consider ice forces from a static or quasi-static viewpoint i.e. the forces are non-dynamic. This is a potentially useful distinction from dynamic forces likely to result from the impact between, say, an iceberg, bergy bits, or ice floes and a structure. The dynamic case is concerned with the momentum and energy of an ice mass when it strikes the structure: the short-term dissipation or absorption of energy within the ice mass and within the structure are important. In the **non**-dynamic case, the behaviour and mode of failure of the ice sheet are the significant factors; that is, the ice sheet is a continuum (however discontinuous it might be in practice!) under static or quasi-static load by the structure. As developments stand at present, these two cases correspond, respectively, to North Sea-type depths and conditions of the Hibernia field off Newfoundland with a winter ice sheet and icebergs approaching from the Davis Strait to the north and, in the non-dynamic case, to the shallow water depths (< 50 m) and multi-year ice of the Beaufort Sea.

Especially useful for visualisation, a photographically-illustrated introduction to some of the terminology employed is provided by Thoren (1969): the photographs range from, for example, the oily appearance of the sea with incipient ice, to an aerial view over Fletcher's Ice Island T-3 which was at that time about 7.2 × 14.5 km × 50 m thick. Such islands generally derive from shelf ice (which can be several thousand years old) while icebergs are calved from glaciers (and are thus 'of the land'). Sea ice has the typical profile of Figure 3.29. The inshore (or landfast) ice is stationary, first-year ice. Offshore the pack-ice (multi-year) rotates (clockwise in the Beaufort Sea) at a few km per day. This multi-year ice incorporates vertical channels and pockets through which brine drains or is trapped, so that the salt content

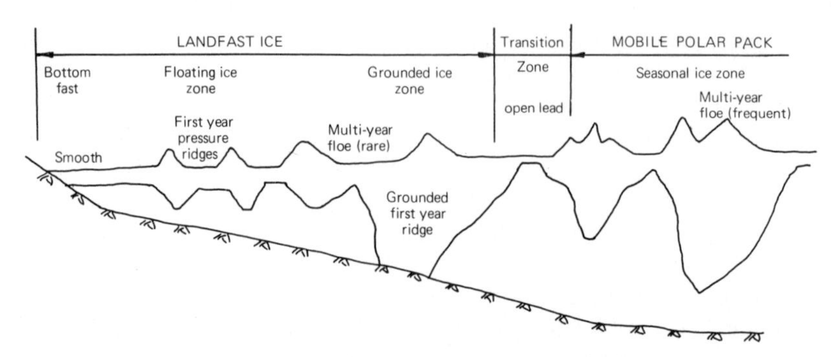

Figure 3.29 Typical nearshore ice conditions

of ice varies from perhaps 8–10‰ at the surface of new ice (formed from water with a salinity of 30‰) to < 0.1‰ in second-season ice. In other words, water melt from sea-ice more than one season old is potable. First-year ice can be up to 2 m thick while older ice is thicker.

Movement of the ice sheet, whether circulatory, wind-drawn, or thermal, causes stresses so that, on the one hand, cracks may develop while, on the other, pressure ridges (and/or keels) are formed with an overall thickness of up to 30 m or so as compared with a normal thickness of 2.5–3.0 m. Thus it would be a gross over-simplification to regard the ice sheet as a simple floating plate. Moreover, additional complexity results from the variability of ice strength (uniaxial, shear and flexural) with temperature, crystal structure, direction of loading, rate of loading, confining force, sample size, salinity, and air inclusions (IAHR, 1980).

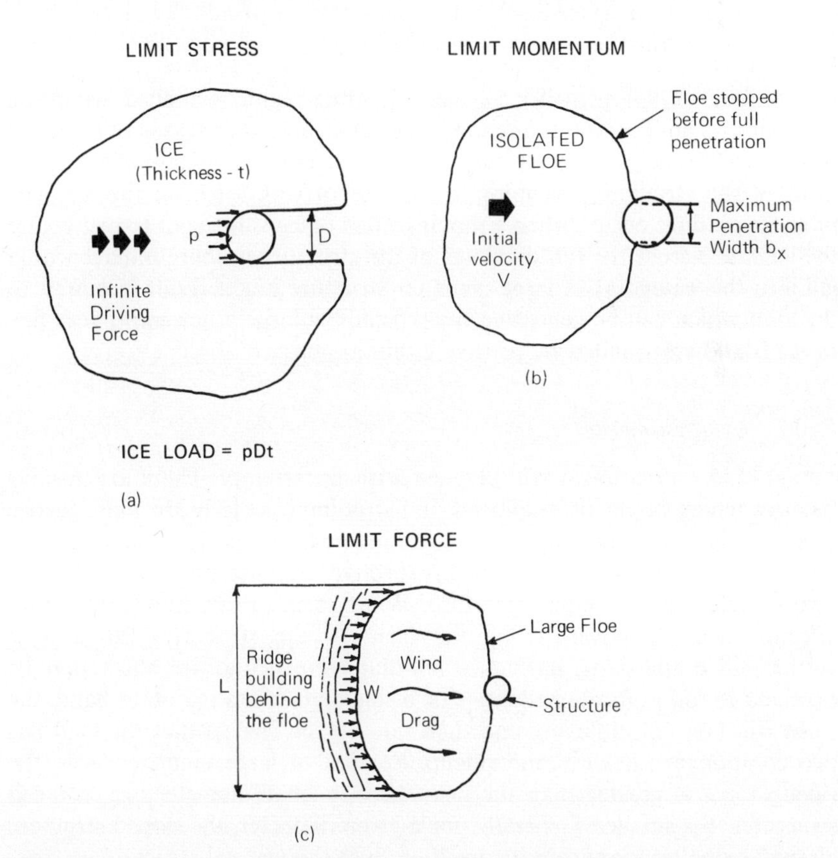

Figure 3.30 Classification of general loading regime: (a) limit stress ice load governed by local ice failure (b) limit momentum ice load governed by initial speed and mass of ice floe (c) limit force ice load governed by driving forces.

3.10.1 *Loading regimes*

Michel (1978) lists four modes of action of an ice sheet against marine and hydraulic structures. They are

(a) static pressure from thermal movement, particularly important for constrained sheets as in reservoirs and sheltered coastal areas;
(b) impact (sic) of moving ice sheets, pressure ridges and ice islands due to wind and/or currents, observed in very large rivers at break-up, in coastal waters, large estuaries and in the Arctic seas—they affect **all** types of marine structure;
(c) pressure of unconsolidated ice accumulations resulting from load transfer by friction between the ice pieces—many types of hydraulic or marine structure are affected;
(d) vertical forces resulting from adhesion between ice and structure and variation in water level—this phenomenon of adfreeze, however, is not confined to vertical surfaces alone.

Croasdale (1985), provides a general loading regime classified according to particular limiting states or conditions. He defines three cases, exemplified by Figure 3.30 (a), (b) and (c). Limit stress relates to the failure of the ice in front of the structure, assuming an infinite driving force on the ice. The momentum limit occurs when a moving mass of ice is brought to rest before the ice fails across the full diameter of the structure (iceberg impact would fall into this category). A large diameter structure might result in a limit to the load which can be generated by, typically, a large, thick multi-year floe or ice island surrounded by relatively thin annual ice.

3.10.2 *Load estimation*

Actual load estimation is still plagued with uncertainty. There are various theories which might be employed and it is impossible to do them justice here. Any formulae quoted therefore, are intended primarily to be indicative, only for initial guidance and the prospective designer must make his/her own detailed study of the literature. What seems clear, however, is that selection will be governed by the form of the structure. When it is vertical-sided and particularly of small diameter, then an ice 'sheet' may be expected to fail in local crushing. For a **sloping** face, on the other hand, the sheet will first fail in flexure and then ride up the face so that the load has two components: flexural and rideup. For flat or large diameter faces, the rideup force is greater than the flexural. For small-diameter (e.g. conical) structures it is smaller: Generally, for a given diameter, the sloped structure 'draws' a smaller load from the ice than does the vertical structure.

Simple two-dimensional theory results in the following relationships (IAHR, 1980) giving the horizontal force per unit width of structure, H_u

('width' being normal to the line of action) and vertical force per unit width, V_u:

$$\left.\begin{aligned} H_u &= \sigma_f\left(\frac{\rho_w g t^5}{E}\right)C_1 + Zt\rho_i g C_z \\ V_u &= 0.68\,\sigma_f\left(\frac{\rho_w g t^5}{E}\right)^{1/4} \end{aligned}\right\} \tag{3.69}$$

where $$C_1 = 0.68\left(\frac{\sin\alpha + \mu\cos\alpha}{\cos\alpha - \mu\sin\alpha}\right)$$

and $$C_2 = \left[\frac{(\sin\alpha + \mu\cos\alpha)^2}{(\cos\alpha - \mu\sin\alpha)} + \frac{(\sin\alpha + \mu\cos\alpha)}{\tan\alpha}\right]$$

ρ_w is the density of water, t is the ice thickness, σ_f is the flexural strength of ice, ρ_i is the density of ice, Z is the height to which the ice is pushed up the slope, x is the angle of the slope to the horizontal and μ is the coefficient of friction between ice and surface.

It is considered that this may underpredict the ice breaking component and it is recommended that the two-dimensional force be adjusted by the ratio of the length of the circumferential crack to the structure width. The length of the circumferential crack is equal to $0.25\pi^2 l$ where l is the characteristic length given by

$$l = \left(\frac{Et^3}{12\rho g(1 - v^2)}\right)^{0.25}$$

where E and v are Young's modulus and Poisson's ratio of ice respectively.

In eq. (3.69) above the first term is the flexural or breaking term and the second is the rideup term. 'Typical' values of ice properties are $\sigma = 1050\,\text{kN/m}^2$, $v = 0.33$, $E = 7 \times 10^6\,\text{kN/m}^2$, $\mu = 0.15$ for a smooth surface.

For ice crushing on a narrow vertical pier, a simplified theory (based on the action of a flat indentor at the edge of the ice sheet) results in the equation (IAHR, 1980):

$$p = \sigma_c\left(1 + 0.304\frac{t}{b}\right) \tag{3.70}$$

where p is the effective ice stress
b is the width of the pier
t is the ice thickness
and σ_c is the crushing strength of the ice

Where ice ridges are likely to exert thrust on structures then eqs (3.69) and (3.70) are unsuitable. However, the reader is urged to consult the literature in preference to further abbreviation here. It is worth noting that in shallow water (< 60 m, say) such forces, as for icebergs, are likely to be reduced due to grounding on and gouging of the sea bed.

3.10.3 *Icebergs*

Consideration of impact forces from drifting icebergs can be separated from the impact of smaller ice masses such as floes, bergy bits and growlers, although the problems of computation are analogous. Hazard from *large* icebergs is much more restricted geographically—to the seas and waters adjacent to and south of Greenland which produces about 20 to 34 thousand icebergs annually (Weeks and Mellor, 1984). Effectively, in the present context, primary concern relates to offshore oil and gas exploration areas, notably in the regions off the Baffin and Labrador coasts, the Grand Banks of Newfoundland and off the West Greenland coast. Icebergs are considered to be not much of a problem in the North Pacific, the Bering Sea or most parts of the Arctic Ocean: Greenland has eight times as much ice as all the permanent ice fields in the Canadian and Soviet Arctic and in Svalbard taken together (Weeks and Mellor, 1984). South of latitude 48°N icebergs deteriorate rapidly in warm water and air temperatures.

The size of icebergs to be expected is governed to a considerable extent by water depth. Davidson and Denner (1982) quote a maximum draft of 300 m near the source and drafts up to 200 m off Labrador but scour marks have been identified at depths up to 450 m. On the Grand Banks a mean draft of 60–75 m might be expected with a maximum of about 80 m so that the maximum possible mass should be about 10–12 million tonnes. Iceberg drift speeds are up to 0.6 m/s with mean values near 0.2 m/s. 45–55% of icebergs reaching the Grand Banks are expected to exceed one million tonnes and 30% could exceed two million tonnes. Consequently design requirements for fixed structures could be to resist a 12 million tonne mass at 0.5 knots or 2 million tonnes at 1.0 knots, perhaps. Much will also depend on whether or not provision is to be made for towing or diverting forecast icebergs.

Approximate relations for iceberg dimensions are (Hotzel and Miller, 1983)

$$H = 0.402\,L^{0.89}$$

$$D_r = 3.781\,L^{0.63}$$

where H is height, L length and D_r draft. Width is about 80% of the length.

The ratio of mass to length is size dependent in that it varies from about 130×10^3 tonnes/m at 30×10^6 tonnes to 1×10^3 tonnes/m at 30×10^3 tonnes.

The actual loads imposed by such masses on a structure, of course, will depend also on the assumptions made about shape (above and below surface water level), failure and rate of failure of the ice by crushing, and so on. The form of the structure itself will also be significant—whether or not the face is smooth or 'toothed', or energy absorbent elements are adopted, for example.

Model studies have been made of the wave-dependent motion of 'smaller' ice masses (ranging from 10^1 to 10^6 tonnes) (Lever et al, 1984). These confirmed the particle-like motion of such masses depending, however, on

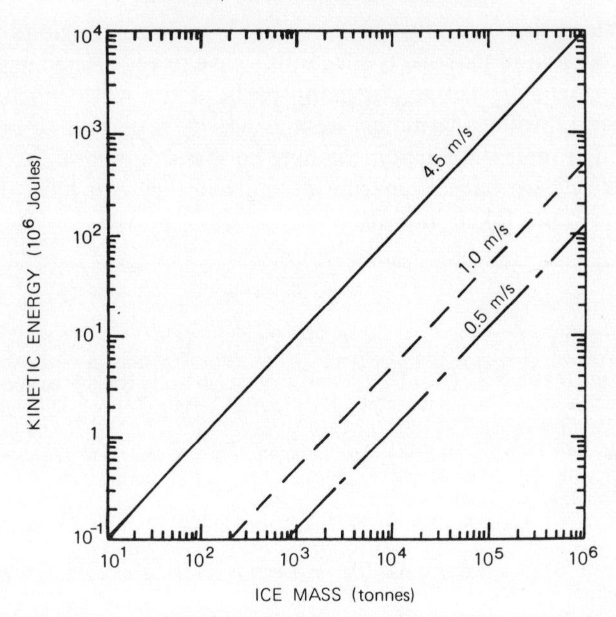

Figure 3.31 Maximum full scale kinetic energy versus mass for ice in 14 m, 12 s storm waves, as determined from wave tank tests. Straight lines representing drift velocities of 0.5 and 1.0 m/s are shown for reference

the ratio of wavelength L to characteristic ice dimension D (analogous to the diameter of the leg or tower of a structure). Particle-like motion occurred for $L/D > 13$ but not for $L/D < 10$ (tending to diffraction dominated patterns). For $10 < L/D < 13$ model shape had an influence. Full-scale estimation of the energy of ice masses for a storm wave $H = 14$ m, $L = 220$ m is shown on Figure 3.31 for particle/ice drift velocities of 4.5 m/s, 1.0 m/s and 0.5 m/s.

There seems little doubt that the problem has to be taken seriously, particularly since it will be more widespread geographically than the prevalence of large icebergs. Small icebergs and small bergy bits and growlers, apart from occurring in or adjacent to seas with seasonal ice cover, also have small drafts and so will occur over a much greater range of depths. Higher velocities also imply a more dynamic loading regime and will have global *and* local consequences. The designer clearly has to assess the situation carefully in each specific case.

3.11 Other environmental loads

To conclude this chapter a very brief mention must be made of certain other environmental loads not hitherto considered. Geographic location of structures governs the magnitude of waves, currents, winds, and ice effects. In specific places in addition there may be hurricanes or typhoons and so their possible occurrence must be considered. In Pacific regimes tsunamis may also occur and so their likely effects may have to be reflected in structural and/or hydrodynamic design. In many parts of the world (including the North Sea and northern latitudes) seismic activity may have to be allowed for. These environmental phenomena may be sparse in relation to the more dominant effects we have been considering but they can have disastrous consequences if not taken into account.

References

Abdelradi, M. and Miller, N.S. (1986) The effect of variation in hydrodynamic coefficients on the loading on jacket structures and the implications for tank testing. *The Naval Architect, RINA,* Sept: 269—290 (incl. disc.)

Babbedge, N.H. and Lynagh, N. (1984) The calculation of extreme wind speeds from visual observations with particular emphasis on the choice of Beaufort scale. *OI 2.17, Proc. Oceanology International.*

Bajpai, A.C., Mustoe, L.R. and Walker, D. (1977) *Advanced Engineering Mathematics.* John Wiley, London.

Barber, P.C. (1986) Shallow water waves. *Developments in breakwaters, Proc. ICE Conf.,* Thos. Telford. London.

Bea, R.G. (1975) Development of safe environmental criteria for offshore structures. *Proc. Oceanology International.*

Bea, R.G. and Lai, N.W. (1978) Hydrodynamic loading on offshore platforms. *OTC 3064, Proc. 10th Annual OTC.* Houston.

Bhattacharyya, R. (1978) *Dynamics of marine vehicles.* John Wiley, New York.

Bishop, J.R. (1980) *A new coefficient for total wave force. NMI R77, (OT-R-80177), Nat. Maritime Inst.*

Borgman, L.E. (1967) Spectral analysis of ocean wave forces on piling. *J. Waterways and Harbors Div., Proc. ASCE, 93, WW2.* May: 129–156.

British Standards Institution (BSI). (1982) *Code of Practice for fixed offshore structures, BS 6235.*

Chow Ven te. (1959) *Open Channel Hydraulics.* McGraw-Hill, New York and Tokyo.

Clifford, J.E. (1984) The design process. *Breakwaters-design and construction. Proc. ICE Conf.,* Thos. Telford. London.

Coastal Eng. Rsch. Center (1984). *Shore Protection Manual* (2 vols.). US Dept. of the Army, US Govt. Printing Office, Washington.

Cokelet, E.D. (1979) Breaking waves—the plunging jet and interior flow-field. *Mechanics of Wave-Induced Forces on Cylinders* (ed. T. L. Shaw), Pitman, San Francisco and London.

Croasdale, K.R. (1985) Recent developments in ice mechanics and ice loads. *Behaviour of Offshore Structures, Proc. 4th Int. Conf. (BOSS'85).* Elsevier, Amsterdam.

Darwin, C. (1953) Notes on hydrodynamics. *Proc. Cambridge Phil. Soc.* **49**: 342–354.

Davidson, L.W. and Denner, W.W. (1982) Sea ice and iceberg conditions on the Grand Banks affecting hydrocarbon production and transportation. *Proc. Oceans 82, Marine Tech. Soc. and IEEE.*

Dean, R.G. and Dalrymple, R.A. (1984) *Water Wave Mechanics for Engineers and Scientists.* Prentice-Hall, New Jersey.

Dean, R.B. and Wootton, L.R. (1987) The importance of vortex shedding in the design of offshore structures. *Proc. Oceanology International.*

Dover, A.R. and Bea, R.G. (1979) Application of reliability methods to the design of coastal structures. *Proc. Conf. on Design Construction Maintenance and Performance of Port and Coastal Structures, ASCE.*

Easson, W.J. and Greated, C.A. (1985) Forces on structures due to deep water breaking waves. *Behaviour of Offshore Structures, Proc. 4th Int. Conf. (BOSS'85).* Elsevier, Amsterdam.

Dept. of Energy. (1980) *Review of the fluid loading research programme. OTP7,* CIRIA for the Dept. of Energy. London.

Dept. of Energy. (1980) *Report of the marine fouling working party. OTP4, CIRIA for the Dept. of Energy.* London.

Dept. of Energy. (1986) Sensitivity study of offshore structures. *Offshore Research Focus*, **52**, April. Hollobone Hibbert and Associates for the Dept. of Energy.

Forteath, G.N.R., Picken, G.B. and Ralph, R. (1984) Patterns of macrofouling on steel platforms in the Central and Northern North Sea. *Corrosion and Marine Growth on Offshore Structures* (ed. J.R. Lewis and A.D. Mercer), Ellis Horwood for the Soc. of Chemical Industry, Chichester.

Freeman, J.H. (1977) *The marine fouling of fixed offshore installations. OTP1, CIRIA for the Dept. of Energy.* London.

Garrison, C.J. (1974) Hydrodynamics of large objects in the sea Part I-Hydrodynamic Analysis. *J. Hydronautics*, **8, 1**, Jan: 5–12.

Garrison, C.J. (1978) Hydrodynamic loading of large offshore structures: three-dimensional source distribution methods. *Numerical Methods in Offshore Engineering.* (ed. O.C. Zienkiewicz, R.W. Lewis, K.G. Stagg), John Wiley, Chichester.

Gaythwaite, J. (1981) *The Marine Environment and Structural Design.* Van Nostrand Reinhold, New York.

Guedes Soares, C. and Moan, T. (1983) On the uncertainties related to the extreme hydrodynamic loading of a cylindrical pile. *Reliability Theory and its Application in Structural and Soil Mechanics* (ed. P. Thoft-Christensen), NATO ASI Series E No. 70, Martinus Nijhoff, The Hague.

Hallam, M.G., Heaf, N.J. and Wootton, L.R. (1977) *Dynamics of Marine Structures Report UR8, CIRIA/UEG*, London.

Hara, H., Zienkiewicz, O.C. and Bettess, P. (1979) Application of finite elements to determination of wave effects on offshore structures. *Proc. 2nd. Int. Conf. on Behaviour of Off-shore Structures BOSS '79.* BHRA, Cranfield.

Hogben, N. (1974) Fluid loading on offshore structures, a state of art appraisal: wave loads. *Maritime Tech. Monograph No. 1. RIBA.* London.

Hogben, N. (1976) Wave loads on structures. *BOSS '76, Proc. Conf. Behaviour of Offshore Structures, Norwegian Inst. of Tech.* Trondheim.

Hogben, N. and Standing, R.G. (1974) Wave loads on large bodies. *The Dynamics of Marine Vehicles and Structures in Waves. Proc. Conf. IMechE.* London.

Hogben, N. Miller, B.L., Searle, J.W. and Ward, G. (1977) *Estimation of fluid loading on offshore structures. NMI R11, Nat. Maritime Inst. Feltham.*

Holmes, P., Tickell, R.G. and Burrows, R. (1978) *Prediction of long-term wave loading on offshore structures. Section A, OTR No. 7823, Dept. of Energy.* London.

Hotzel, I.S. and Miller, J.D. (1983) Icebergs: their physical dimensions and the presentation and application of measured data. *Annals of Glaciology* **4**: 116–123.

Hydraulics Research Station. (1979) *Extreme responses in multi-directional random waves. Report No. Ex 861 (OT-R-7902) Hyd. Rsch. Stn.* Wallingford.

Hydraulics Research Station. (1981) *Wave loads on vertical cylinders by spectral methods. Report No. Ex 982 (OT-R-8127), Hyd. Rsch. Stn.* Wallingford.

International Association for Hydraulics Research (IAHR). (1980) *Special Report 80–26, Working Group on Ice Forces on Structures, A State-of-the-Art Report* (ed. T. Carstens). US Army Cold Regions Rsch. and Eng. Lab. (CRREL).

IAHR. (1984) Ice Forces. *Proc. IAHR Ice Symp. Hamburg (7th Int. Symp. on Ice)* Vol. IV, IAHR.

IAHR. (1987) *Special Report 87–17, Working Group on Ice Forces, 3rd State-of-the-Art Report* (ed. T.J.O. Sanderson) US Army, CRREL.

John, F. On the motion of floating bodies II simple harmonic motions, *Comm. on Pure and App. Maths., Inst. Maths and Mechanics, N.Y. Univ., III, Interscience.* New York.

Keulegan, G.H. and Carpenter, L.H. (1958) Forces on cylinders and plates in an oscillating fluid. *J. Rsch. Nat. Bur. Standards, 60, 5, Rsch. paper 2857.* May: 423–440.

Lamb, H. (1962) *Hydrodynamics, 6th edn.* Cambridge Univ. Press.

Lever, J.H., Reimer, E. and Diemand, D. (1984) A model study of the wave-induced motion of small icebergs and bergy bits. *Proc. 3rd. Int. Offshore Mech. and Arctic Eng. Conf.* New Orleans, ASME.

Lighthill, J. (1979) *Waves in Fluids* (pb edn). Cambridge Univ. Press.

Longuet-Higgins, M.S. (1952) On the statistical distribution of the heights of sea waves. *J. of Marine Rsch.* **XI, 3**. 245–266.

MacCamy, R.C. and Fuchs, R.A. (1954) Wave forces on piles: a diffraction theory. *Tech. Memo. No. 69 Beach Erosion Board, Corps of Engineers.* Dec.

Marshall, A.L. (1986) Load estimation for offshore structures as affected by errors in wave period. *Underwater Technology*, **12, 1**, Spring: 9–14.

Marshall, A.L. (1988) Wave load estimation in deep water as affected by wave period errors. *Proc. 7th Int. Conf. on Offshore Mechanics & Arctic Engineering, OMAE '88, SME.*

Mei, C.C (1978) Numerical methods in water-wave diffraction and radiation. *Ann. Rev. Fluid Mech.* **10**: 393–416.

Mei, C.C (1983) *The Applied Dynamics of Ocean Surface Waves.* John Wiley, New York.

Michel, B. (1978) *Ice Mechanics, Les Presses de l'Université Laval.* Quebec.

Miller, B.L. (1980) *Wave slamming on offshore structures, NMI RR81 (OT-R-8041), Nat. Maritime Inst.*

Mitchell, A.R. and Wait, R. (1977) *The Finite Element Method in Partial Differential Equations.* John Wiley, London.

Morison, J.R., O'Brien, M.P., Johnson, J.W. and Schaaf, S.A. (1950) *The force exerted by surface waves on piles, T.P. 2846, 189, Petroleum Trans. AIME*: 149–154.

Newland, D.E. (1975) *An Introduction to Random Vibrations & Spectral Analysis.* Longman, London.

Newman, J.N. (1972) Diffraction of water waves. *Applied Mech. Reviews*, **25**: 1–7.

Noble Denton and Assoc. Ltd. (1985) *Environmental Parameters on the United Kingdom Continental Shelf, OTH 84 201.* HMSO, London.

Nolte, K.G. (1978) The effect of uncertainty in wave force coefficients for offshore structures. *Proc. 10th Annual OTC.* Houston.

Ochi, M.K. and Wang, S. (1976) Prediction of extreme wave-induced loads on ocean structures. *Proc. Conf. on Behaviour of Offshore Structures BOSS '76, Norwegian Inst. of Tech.* Trondheim.

Ochi, M.K. (1982) Stochastic analysis and probabilistic prediction of random seas. *Advances in Hydroscience Vol. 13.* Academic Press.

Oldfield, D.G. (1980) *Appraisal of marine fouling on offshore structures, OTP 6 (OT-R-8003), CIRIA for the Dept. of Energy.* London.

Pearcey, H.H. (1979) Some observations on fundamental features of wave-induced viscous flows past cylinders. *Mechanics of Wave-induced Forces on Cylinders* (ed. T.L . Shaw), Pitman, San Francisco and London.

Peregrine, D.H. (1979) Mechanics of breaking waves—a review of Euromech 102. *Mechanics of Wave-Induced Forces on Cylinders* (ed. T.L. Shaw), Pitman, San Francisco and London.

Richards, T.H. (1977) *Energy Methods in Stress Analysis.* Ellis Horwood, Chichester.

Sanderson, T.J.O. (1988) *Ice Mechanics-Risks to Offshore Structures.* Graham and Trotman, London.

Sarpkaya, T. (1978) In-line and transverse forces on cylinders in oscillatory flow at high Reynolds numbers. *OTC 2533, Proc. 8th. Annual OTC.* Houston.

Sarpkaya, T. (1981) A critical assessment of Morison's equation. *Proc. Int. Symp. on Hydrodynamics in Ocean Engineering, Norwegian Inst, of Tech.* Trondheim.

Sarpkaya, T. (1986) On fluid loading of offshore structures—after ten years of basic and applied research. *Proc. Offshore Operations Symp., 9th Annual Energy-Sources Tech. Conf.* ASME, New Orleans.

Sarpkaya, T. and Isaacson, M. (1981) *Mechanics of Wave Forces on Offshore Structures*, Van Nostrand Reinhold, New York.

Shearman, R.J. (1985) The mathematical representation of wind profiles over the sea. *Underwater Technology*, **11, 1**, Spring: 13–19.

Standing, R.G. (1981) *Wave loading on offshore structures: a review, NMI R102 (OT-R-8113). Nat Maritime Inst.*

Stoker, J.J. (1957) *Water Waves.* Interscience, New York.

Streeter, V.L. and Wylie, E.B. (1979) *Fluid Mechanics, (7th edn.).* McGraw-Hill Kogakusha, Tokyo.

Thoren, R. (1969). *Picture Atlas of the Arctic.* Elsevier, Amsterdam.

Tickell, R.G., Burrows, R. and Holmes, P. (1976) Long-term wave loading on offshore structures. *Proc. ICE, Part 2,* **61**. March: 145–162.

Vallentine, H.R. (1967) *Applied Hydrodynamics (2nd edn).* Butterworths, London.

Vinje, T. and Brevig, P. (1981) Numerical calculations of forces from breaking waves. *Proc. Int. Symp. on Hydrodynamics in Ocean Engineering, Norwegian Inst. of Tech.* Trondheim.

Weeks, W.F., and Mellor, M. (1984) Mechanical properties of ice in the Arctic seas. *Arctic Technology and Policy, Proc. 2nd. Ann. MIT Sea Grant College Prog.* Hemisphere, Washington.

Wehausen, J.V. and Laitone, E.V. (1960) Surface Waves. *Encyclopedia of Physics* (ed. S. Flugge), Vol. IX, *Fluid Dynamics III* (co-ed. C. Truesdell), Springer Verlag.

Whitham, G.B. (1974) *Linear and Non-Linear Waves.* John Wiley, New York.

Wills, J.A.B., Grant, A. and Boyack, C.F. (1986) *Offshore mean wind profile, OTH 86 226.* HMSO, London.

Zienkiewicz, O.C. with Cheung, Y.K. (1970) *The Finite Element Method in Structural and Continuum Mechanics.* McGraw-Hill, London.

Zienkiewicz, O.C., Bettess, P. and Kelly, D.W. (1978) The finite element method for determining fluid loadings on rigid structures: two and three-dimensional formulations. *Numerical Methods in Offshore Engineering* (eds. O.C. Zienkiewicz, R.W. Lewis, K.G. Stagg), John Wiley, Chichester.

4 Behaviour of plain concrete

4.1 Introduction

From considering marine environmental loading in some of its primary manifestations it is appropriate now to turn to the structures intended to withstand it. However, this presupposes proper appreciation of the materials employed in construction since their effective, efficient, and economic use governs structural form, configuration, and integrity. Therefore in this chapter we first deal with concrete as a material in its own right while the following chapter is concerned with its composite use in reinforced and prestressed concrete.

Indiscriminate reference to 'concrete' can be confusing. For example it can mean any or all of plain, reinforced, prestressed, fibre-reinforced, lightweight concrete, and so on. Frequently, however, the qualifying term is omitted where the context makes clear what is being discussed, and so that practice will be followed here, hopefully without any resulting loss of clarity.

It is no accident that there is a preponderance of references to offshore activity. The scale of investment in undersea oil and gas development has given considerable impetus to the generation of ideas and resolution of problems with their accompanying need for investigation. Many of the lessons learned thereby warrant more universal application. Such spin-off, even in the construction industry, is not a new phenomenon of course—water power developments and civil nuclear power programmes are but two examples which spring to mind of analogous spurs to progress. Admittedly not all evidence is readily accessible since much work has been industrially sponsored with consequent commercial confidentiality. By and large, however, it is to the credit of the industry that so much has been made freely available. Moreover, the international nature and scale of the business has possibly seen greater national and international coordination of effort than the construction industry has previously witnessed.

Certain limits to discussion are being imposed here. While properties such as strength, elasticity, creep and shrinkage are undeniably important they will only be considered in passing since they are amply covered in various well-known and familiar texts such as that of Neville (1981). Instead preference is given to the view that *durability* is the most important characteristic of marine concrete (assuming that the designer's requirements are complied with). In turn this implies that permeability, fatigue and thermal response are highly significant. Of course the properties mentioned above are relevant but the traditional emphasis given to them can mask the nature of the phenomena most in need of examination.

Disruption of the concrete due to reinforcement corrosion (which can be a major durability problem) is deferred to the next chapter as being more appropriate to reinforced concrete. Here we are concerned rather with the intrinsic importance of concrete, the structural material, as distinct from the protective coating for steel.

4.2 Durability

Definition of terminology makes a useful starting point for discussion. According to the Shorter Oxford English Dictionary, '*durability*' is 'the quality of being durable'. '*Durable*' is 'capable of continuing in existence; persistent; permanent.' Additionally (or alternatively) it means 'able to withstand change, decay or wear'. In practical terms, however, that does not mean that something which is durable must remain 'as new'—it should still fulfill its intended function despite *some* wear since nothing is indestructible. Furthermore 'permanence' must be capable of different meanings if economic design is to be achieved. That is, the notion of permanence must be related to the economic life of the object concerned. From this it is then but a short step to considering durability as a property which diminishes with time so that one might conceive of *residual durability*: the ability at a given time to withstand decay or wear for the *remainder* of economic life. (Hence repairs and/or restoration should perhaps be directed towards matching residual durability rather than the original as-new condition.)

Different applications also require different conceptions of durability according to nature of use and aggressiveness of the environment: the foundation in sulphate soil is likely to face different demands from nuclear reactor containment which will differ from those of the city centre office block. Generally, however, concrete should remain *sound* (although there may be loss of appearance), that is there should be no physical disruption of the material, no uncontrolled detrimental chemical reaction and mechanical properties should at all times be adequate for intended performance and structural safety.

It is hardly surprising that numerous field and laboratory studies have been made of concrete durability over the years. Several extensive surveys of performance have been reported for both land and marine structures and it is worth examining one or two of the latter in particular, before going on to more detailed matters. First, however, some primary factors should be noted.

Attention has been drawn to the significance of homogeneity (Valenta,

Table 4.1 The principal processes which threaten marine concrete (Cabrera, 1986)

(a) Physical:	(b) Chemical:
Ice formation	Hydrolysis and leaching
Water evaporation	Sulphate penetration
Crystallisation of salt in pores	CO_2 diffusion
Permeation of water and gases	Chloride corrosion
Abrasion	Silica-alkali reaction

1969). While stress tends towards concentration in regions of high quality concrete, deteriorating processes usually concentrate in the weakest, most inferior parts (presuming uniform exposure). Moisture penetration (to be dealt with later) can be decisive in deteriorating processes and adverse effects are related to permeability, porosity, moisture absorption, reduced bond between ingredients, and strength (mainly tensile) of concrete. These are concrete *qualities* or *properties* so, in other words, it is in their inadequacies that lies the source of durability problems. Deficiency permits infiltration by the 'aggressor' or at least a substantial proportion of the 'attacking forces', the principal of which are listed in Table 4.1.

Clearly all of these will result, to some degree, from exposure to a 'natural' marine environment. There are also many *un*-natural potential problems, and certainly from an environmental or conservationist viewpoint, there is considerable concern about enclosed seas and their coastal waters in particular.

'Major chronic inshore marine pollution problems can often be attributed to the local discharge of large volumes of wastes. These include materials which are partially bio-degradable, such as raw sewage, sewage sludge, food and beverage processing wastes, pulp and paper mill effluents, woollen and cotton mill wastes and sugar refinery effluents' (Keckes, 1983). In some cases therefore, maritime structures may be exposed to particular chemical environments which exacerbate exposure problems. Adjacency is not essential since currents, tides, estuarial flows and so on ensure wider distribution of potential problems.

4.2.1 *Durability: in-service experience*

4.2.1.1 *Exposure* In looking at some of the field experience and studies of working structures, two aspects should be kept in mind. Reference has been made in an earlier chapter to climatic variations and global (and seasonal) fluctuations in salinity. However, at any given location there are substantial differences in exposure over the height of a structure which have led to the classification of different exposure zones. For example, the UK Department of Energy (1977) specifies three *external* exposure zones:

(a) Submerged zone: that part of the structure below the splash zone.

(b) Splash zone: that part of the structure between the crest level of the 50-year (average) wave superimposed on the highest astronomical tide, and 10 metres below the lowest astronomical tide.
(c) Atmospheric zone: that part of the structure above the splash zone.

It is noted that structures continuously exposed to salt-laden air are often as vulnerable to attack as those in the splash zone. (The CEB-FIP Model Code for Concrete Structures (1978) categorises maritime atmospheric conditions as 'severe exposure'.)

Internal exposure zones are two in number:

(a) Submerged zone: that part of the structure in contact with oil, sea water or other liquids.
(b) Atmospheric zone: that part of the structure above the submerged zone.

Whatever the limits of the splash zone, it can be regarded as comprising several superimposed cycles of different amplitude and frequency of wetting and drying, related to the complexity of tidal variation compounded by waves. Since it seems likely that wetting of concrete is a more rapid process than drying, intermittency over periods up to 12 hours may not be too significant other than in layers close to the surface of the concrete (although climate must affect drying in particular).

In comparing the behaviour of structures then, one must ensure that the zones of exposure are also compared. (To confuse the issue it should be noted that the 'splash zone' above is not the same as the 'splash zone' often used to denote the zone *above* spring tide high water level. Low to high water is then called the 'tidal zone.' The 'submerged zone' corresponds roughly in both systems.)

Likewise it is advisable to ensure that similar concretes are being compared. This may be more difficult than at first appears because, not only must the normal implications of age, mix proportions, temperature and so on be accounted for, but so must changing properties of cement over the years. For example, over the 25 years or so from 1960, UK OPC has increased in strength by about 25% (Pomeroy, 1986). This may be reflected in other properties as well as in cement contents and water/cement ratios, of course, which only adds to the difficulty of historical unravelling.

4.2.1.2 *Ageing and symptoms of deterioration* In many ways concrete structures resemble the ageing human body. The football injury of the young man becomes the arthritic joint of later years. Over-exposure to aggressive environments shortens the life-span. And not every incident or ailment is recorded or remembered. Furthermore, in concrete (as in life) 'hidden safety margins of the system and the cleverness of the material—advantageous features of concrete structures identified by Rusch—become a disadvantage from the point of view of recognising defects at an early stage' (Jungwirth, 1984). A damaged limb re-adjusts to load (not always to the benefit of the undamaged portion) and compensates, thereby masking the injury.

'Ageing' of concrete has been defined as 'gradual deterioration of the structure and mechanical properties of concrete, even in the absence of recognisable aggressive exposure' (Idorn, 1967). Ageing makes concrete weak and mushy when moist, and brittle and friable when dry. The aggregates are loose in the matrix or are easily loosened by hand or through natural wear and tear. Ageing can be accelerated by a variety of deleterious physical and/or chemical agents and its most important symptoms have been considered to be (Idorn, 1967):

1. absence or only indistinct remains of β-dicalcium silicate and calcium hydroxide; isotropic or even impure, fibrous character of the remains of cement paste;
2. high degree of porosity due to dissolution of cement paste;
3. extensive precipitation of calcite and calcium aluminate sulphate (or similar compounds);
4. corrosion of the boundary areas of aggregate particles;
5. weathering of alkali-silica gel, when alkali-aggregate reaction has been part of the primary deterioration;
6. extensive microfracturing, drying-shrinkage, crack-formation, boundary cracking, etc.

These all relate to evidence obtained from thin sections of 'real' concretes (as distinct from laboratory specimens) and it is pointed out that no one symptom alone is an indication of ageing.

Stemming from an extensive study of structures in Denmark, the symptoms are the outcome of detailed and microscopic tests in specific cases. At the macroscopic end of the work a subjective classification of deterioration was used which enabled the identification of problem structures. Although it depends a great deal on the individual observer, conditions of observation, and so on, it is worth quoting (Table 4.2) as an example of what can be done quite simply.

Table 4.2 Classification of structure deterioration

Character	Characteristics
0	Completely undamaged structures
1	Slight, scattered surface damage
2	Severe, scattered surface damage
3	Slight, overall deterioration
4	Overall deterioration
5	Severe, overall deterioration

As to marine structures in particular the most aggressive types of marine exposure were considered likely to be:

1. severe freezing with tide in 6-hour cycles;
2. mangrove swamps in the tropics;
3. salt lagoons and lakes in arid areas and deserts.

Formation of hydrogen sulphide from decomposing vegetation can be an important factor but it could be misleading to imagine it as confined to mangrove swamps—it is to be found at many places in Denmark, for example, and at all depths greater than 200 m in the Black Sea (see below under chemical attack).

In general, Danish experience suggests the same symptoms of deterioration over many years:

map-cracking;
scaling and abrasion;
decomposition and disintegration.

However, map-cracking is ascribed to alkali-aggregate reaction and is presumably, therefore, the result of poor initial selection of materials (possibly the result of ignorance of the phenomenon at the time of construction). For concrete structures with non-reactive aggregates, scaling and raveling are the result of freezing/thawing. Advanced decomposition and disintegration are due to chemical influence of the salts in sea water and ageing, the latter due to evaporation of water from the cement paste and to carbonation and decomposition of the cement gel. Important contributory factors in this process are lean mixes, insufficient compaction and unsuitable curing conditions. These of course result in high porosity, low paste content, inhomogeneity and initial cracking due to shrinkage and carbonation, etc.

4.2.1.3 *Dutch experience* A Dutch study covered 64 normal weight reinforced concrete structures ranging in age from 3 to 63 years (Wiebenga, 1980; CUR commissie B23, 1981). Specified cement contents were traced for 53 of the structures ranging from 220 to 400 kg/m^3 with a median of 340 kg/m^3. Of these,

Table 4.3 Damage characteristics of 64 reinforced concrete structures

Amount of damage	Corrosion	Cracks	Crazing	Weathering
None	56(88)	42(66)	58(91)	32(50)
Little	2(3)	14(22)	5(8)	24(38)
Moderate	3(4.5)	7(11)	1(1)	8(12)
Much	3(4.5)	1(1)	0(0)	0(0)
Total	64(100%)	64(100%)	64(100%)	64(100%)

48 included blast furnace slag. Table 4.3 shows the number of structures displaying damage of different types. Regardless of cement content there was no visible corrosion in structures less than 30 years old and it occurred mainly where cover was less than 45 mm (it was attributable mainly to chloride penetration).

Other general conclusions were that there had been relatively little attack on the structures, weathering of the surface being most common with cracks coming second. Depth of carbonation was small, even in occasionally porous concrete, due probably to the concrete always being wet (the tidal zone was the main field of study).

Cores were drilled from five uncorroded structures and compressive strength, porosity, depth of carbonation and cement content measured with results as indicated in Table 4.4.

It would be less than wise to make too much of so few figures but the long-term strength is clearly satisfactory except in one case. Here, the *specified* cement content is much lower than, although the *measured* value is not so far removed from, the other values: perhaps there is a minimum threshold value.

4.2.1.4 *Some Norwegian experience* Four reinforced cement vessels, 35 to 60 years old, have been examined in Norway (Hoføy and Hafskjald, 1983): a dredger pontoon, a barge, a small tug/freighter and a wrecked ship. On the first two, concrete strengths ranged from 54 to 78 N/mm^2 (with some values reaching 130 N/mm^2 for internal ribs) while the third, which had been shotcreted, had lower values at 35 to 45 N/mm^2. Densities were 2250 to 2335 kg/m^3 for the first two. Visual inspection only had been carried out on the fourth vessel at the time of reporting. The brief conclusion of the study was that such (thin-walled) structures can have very good durability in a marine environment, provided that proper materials, structural form and construction methods are used.

4.2.1.5 *Port structures in Canada* This implicit reference to quality of workmanship echoes the sentiment of an earlier report about four structures in the port of Halifax, Nova Scotia (Tibbetts, 1971). 'Improved workmanship in

Table 4.4 Core sample characteristics from five uncorroded concrete structures

Core No.	Age years	Depth of carbonation mm	Comp. strength N/mm^2	Porosity % by vol.	Cement (plus trass) content kg/m^3 Measured	Specified
1	26	0–5	60.7	14.1	340	340
2	49	2–10	26.6	20.0	300	220
3	27	0–1	76.4	12.0	348	340
4	18	0–5	64.5	13.8	325	325
5	16	0–2	66.2	12.1	343	330

mixing, placing and curing of concrete to be exposed to sea-water is required. Education of the workmen actually doing the work, and in many instances, of those supervising it, would be the greatest contribution towards the successful performance of concrete structures in sea-water.' Details of the structures are accompanied by brief histories of their maintenance: repairs by shotcrete and patching do not seem to have been very successful. The tidal zones are the most critical, although frost and chemical action may affect concrete above high water levels. Furthermore, once the original surface has been damaged chemical deterioration appears to be rapid. Interestingly, the observation is made that precast concrete performs better than that cast *in situ.* The same comment has been heard recently concerning bridges, and the view was expressed that this probably has a great deal to do with the high quality control achievable in precasting works.

4.2.1.6 *France—the Rance tidal power station* Measures taken to ensure quality on the Rance tidal power station in France are outlined in a report on experience of fifteen years service (Duhoux, 1980). A blast furnace cement was used and, in general, immersed concrete had not given any trouble up to that time. Thermal contraction cracks (from heat of hydration cooling) were treated successfully during construction. Some surfaces were waterproofed and horizontal construction joints (and vertical joints of course) were sealed; leakage into the drainage system decreased from 20 litres/minute in 1972 to 7 litres/minute in 1980 (the volume of concrete involved in the power station was 250,000 m^3) so the system appears to have been successful. Such problems as did occur were in the atmospheric zone above spring tide high water level where it seems the blast furnace cement was less successful. Concrete surfaces in certain zones of the superstructure displayed a loss of fines to reveal the gravel mosaic beneath but the reinforcement was not affected. Particular care had been taken during construction over the selection of aggregates and sand which was kept regular, selection ensuring it had a low shell content and was not too fine. Cement contents were 350 or 400 kg/m^3 and the sand content fairly low. The sluices were bridged using beams with pre-cast flanges and these required remedial action. On the other hand, parts of the main structure (including treated cracks and joints) performed trouble-free under a tidal range of 13.5 m and flow velocities up to 10 m/s. Regular inspection and maintenance obviously have merit.

4.2.1.7 *Offshore performance* Installation of concrete offshore structures in the North Sea and off Brazil is providing closely monitored experience although much detail is not publicly available. One general point is worth bearing in mind, however (and that in relation to steel structures originally). Historical comparisons of behaviour suggest an increase in number of defects, but this is more likely to be due to increased inspection effort and quality control than to a time dependency of fault (Røren et al, 1986). (It is analogous to historical comparisons of crime rates which are functions of observation, changes in legislation, notifiable offences and demographic and social patterns as well as actual incidence of crimes.) Contrary to such trends, however, it has been suggested (Det norske Veritas, 1980) that the amount of in-service inspection could be reduced, maintenance may be unnecessary and repairs restricted to accidental damage. This is particular to these types of structure, of course, and is due to specific circumstances such as deliberate over-design for various reasons.

Degradation of concrete is attributed to four causes (Fjeld and Røland, 1982):

1. chemical effects;
2. freeze/thaw damage;
3. fatigue;
4. abrasion (including cavitation).

One beneficial effect to be observed is the tendency for cracks to be sealed by sedimentation of magnesium and chalk compounds. The effect seems to be quite pronounced. In contrast, and reflecting the experience in Halifax mentioned above, a 'rather common feature is spalling of gunite and epoxy coatings either in repair areas from construction, or areas where special coatings have been applied' (Røren et al, 1986). Where significant cracks and spalling occurred in one major structure, the problem appeared to be due to over-loading 'in temporary phases' and due to foundation inadequacy. In other words, there was no fault inherent in the material. The conclusion emerging from experience is that, 'although the service lives of the present structures have not been long, the overall performance of the concrete in the existing North Sea structures has been exceptional, and indications are that it will continue in that manner for many years to come.' (Hoff, 1988).

4.2.1.8 *The Middle East–experience in the Gulf* Elsewhere, experience has been very different (Normand, 1986) but circumstances also are very different. In the area of the Persian Gulf some reinforced concrete elements 'had reached a terminal condition in as little as 5 years whereas there were also items which were in good condition after 20 years.' Much of the damage can be attributed to corrosion of the reinforcement (to be considered in the next chapter) but it is also clear that the arid environment, coupled with shortage of water for mixing and curing, lower strengths (implying lower cement contents and/or higher water/cement ratios), high salt contents in the ground as well as in the sea, and poorer workmanship in places, all combine to provide a breeding ground for problems. These and other difficulties (such as availability of materials) for operations in the region have been catalogued (Walker, 1985) and it is tempting to regard the Gulf and the North Sea as epitomising opposite ends of a spectrum. Such a portrayal has many traps for the unwary and superficial judgements are to be avoided. Nevertheless the Gulf demonstrates what can happen and the North Sea what can be attained. What also seems clear from the Gulf experience is that specifications formulated for one part of the globe should not be too readily applied in another.

4.2.1.9 *A climatic approach* A 'climatic approach' to durability has been attempted, in fact, (Fookes et al, 1986) and Figure 4.1 shows the suggested climatic zones. They are described as follows:

1. Cool with freezing
   Usually with marked summer/winter temperature differences, annual average temperature < 10°C, summer temperature generally < 20°C. Fairly high relative humidity (RH) not often < 40% and usually > 50%. Sometimes windy, moderate rainfall.
2. Temperate
   Fairly marked summer/winter temperature differences, seldom freezes, annual average temperature 10 to 20°C. Fairly high RH. Sometimes windy, moderate to heavy rainfall.
3. Hot dry
   Mainly hot desert climate. No freeze/thaw action but fairly marked summer/winter temperature differences, annual average temperature > 20°C with summer maximum usually > 45°C. Usually low to moderate RH with long periods < 50%. Sometimes windy and with little rainfall.
4. Hot wet
   Mainly tropical climate. No freeze/thaw and with only moderate or small summer/winter differences, annual average temperatures usually ≯ 30°C. Moderate/high RH, usually ≯ 60%. Sometimes windy and with at least one rainy season.

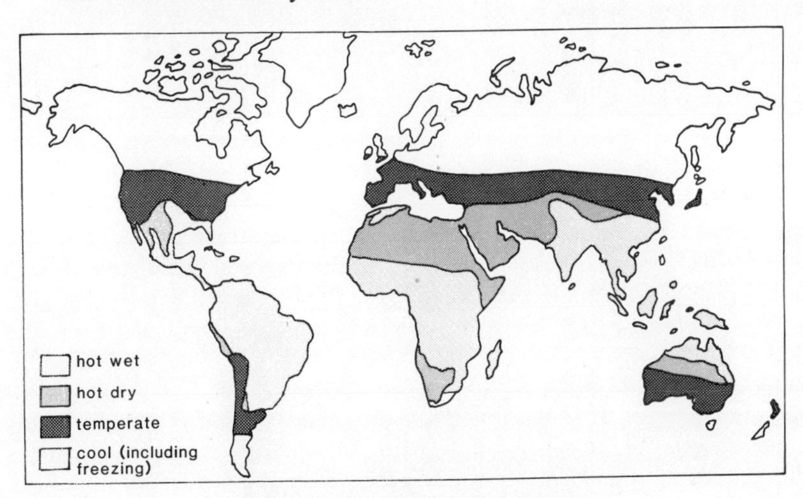

Figure 4.1 World climatic regions

Table 4.5 General aspects of reinforced concrete deterioration and their occurrence in different climates (Fookes et al, 1986)

Feature	Possible effect	Approximate age at which features become visible	Major factors	Occurrence rating in particular climate Scale: 1 lowest–5 highest				Avoidance measures
				Cool inc freezing	Temperate	Hot Dry	Hot Wet	
Defects of workmanship (eg. honey-combing, movement of formwork)	Loss of effective cover, porous concrete, loss of strength	First day	Inexperienced operatives, lack of supervision, Inadequate specification	—	—	—	—	Improve specification
Bleeding & Plastic settlement	Cracking, localised loss of bond, loss of effective cover	First day	Gap-graded fines	3	3/4	4/5	2/3	Improve specification, mix and workmanship.
			Unsuitable mix design, deep sections, excessive retardation	—	—	—	—	Improve aggregates especially grading of fines
Plastic shrinkage	Cracking or crazing loss of effective cover	First day	Porous aggregates, drying winds	1/3	3/4	4/5	2	Improve mix design, start efficient curing quickly
			Inadequate protection	—	—	—	—	
Initial thermal contraction	Cracking at restrained locations	First month but depends on section	Rapid cooling after hardening, (thermal chill) Unsuitable casting sequence, hot concrete, excessive cement content unsuitable cement	1/2	2/3	3	1/2	Reduce temperature differentials Allow for movement. Improve design. Use low heat cement.

			type, very large sections, high expansion aggregates	—	—	—	—	
Structural distress	Cracking due to movement, cracking at restrained locations, loss of effective cover	At any time, but most commonly in early months	Long-term cyclic temperature movements	1/2	2/3	3/4	1	Improve design generally. Allow for temperature movements and earthquakes where necessary.
			Faulty design, weak concrete, overloading, premature loading, impact or earthquake damage					
Drying shrinkage not controlled by reinforcement	Deep cracks, loss of effective cover	First month to first year	Temperature/moisture gradients	2/3	3/4	5	2	Improve mix, especially by reducing w/c ratio.
			Faulty design, high shrinkage aggregate, poor curing					
Excess internal chloride	Severe cracking and spalling	A few months to several years	Contaminated aggregates, salt water for mix and/or curing	1/2	1/2	4/5	1/2	Use clean aggregates, low-chloride additives and cement, and potable mixing water
			Inappropriate admixtures, unsuitable cement, very rich mixes					
Chloride Ingress	Corrosion of reinforcement leading to severe cracking and spalling	First few years	Temperature/moisture gradients	2/4	2	4	2	Improve all durability aspects, especially by providing dense, low w/c concrete and deep cover to steel.
			Unsuitable mix design, insufficient cover, porous concrete					

Table 4.5 (Contd.)

Feature	Possible effect	Approximate age at which features become visible	Major factors	Occurrence rating in particular climate Scale: 1 lowest–5 highest				Avoidance measures
				Cool inc freezing	Temperate	Hot Dry	Hot Wet	
Frost damage	Spalling of concrete, loss of effective cover, loss of section	At any age	Freezing & thawing, porous concrete saturation of surface	4/5	1/2	—	—	Provide dense, low porosity, low w/c concrete. Use air entrainment.
Physical salt weathering	Disintegration of surface, loss of effective cover to steel and loss of section	First few years	Difficult environment (e.g. subject to wind blown salty sands), unsuitable mix design, poor curing	1/2	1/2	4/5	1/3	Specify and achieve dense concrete. Use sound aggregate. Provide deep cover to steel
External sulphate attack	Disintegration of surface, loss of effective cover and loss of section	First few years	Sulphate rich environment (e.g. coastal sulphate swamps, seawater sulphates), unsuitable cement, inappropriate mix design	1/2	1/2	2	2	Specify and achieve dense concrete. Use sulphate resisting cement.
Reactive aggregate	Severe cracking	A few months to many years	Unsuitable aggregates, unsuitable cement, very rich mixes, high humidity	1	1	1	1/2	Do not use. Test first

The 'hot wet' and 'hot dry' climates correspond generally to the geographer's 'tropical rainy' and 'dry' climates while 'temperate' and 'cool' overlap with 'middle latitude', 'cold snowy forest' and 'polar' climates (Encyclopaedia of World Geography, 1974).

Case histories are presented of structures from each of the four categories (Fookes et al, 1986) and arising from these, and other work, a general assessment has been made of the effects of climate on initial defects. The data is reproduced in Table 4.5, and is a useful anticipatory guide; the approach adopted is one which warrants closer study.

4.2.1.10 *The Thames Estuary* An extensive programme of tests was carried out on a 35-year old Naval Fort located at the mouth of the Thames Estuary in the UK (Taylor Woodrow, 1980). Strength, permeability, chemical attack and

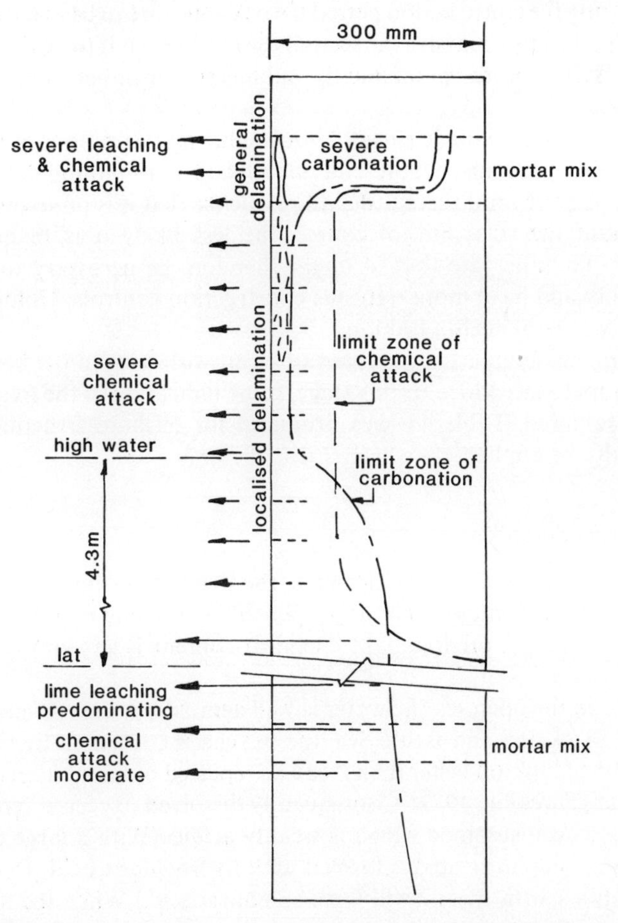

Figure 4.2 Estimated depths of chemical attack, Tongue Sands Tower

so on were all measured in the three exposure zones and various non-destructive tests carried out. The concrete itself appears, generally, to have performed satisfactorily with 'durability damage' being due primarily to reinforcement corrosion. A profile of attack was estimated and is given in Figure 4.2.

4.2.1.11 *The need for inspection* There seems little doubt from the experiences touched on above, that concrete *can* be produced which will resist the marine environment for long periods of time. There seems little doubt also, however, that it is not difficult to produce non-resistant structures, even from good quality concrete. For instance badly-formed construction joints and areas of grout loss, if not honey-combing, can be sources of rapid deterioration. On the weakest link principle, coupled with the unrelenting nature of the sea (other than for short spells), it is evident that a high standard of vigilance is required during the construction period if weak zones are to be avoided. It also emerges that, the more massive the section the less likely it is to be the source of problems. Furthermore, permanently submerged members are the least problematic.

Despite the remarks made earlier about reducing inspections, it would be wise to institute an inspection and maintenance programme for major structures unless or until there is definite evidence that it is unnecessary. The less important the structure, of course, the less likely it is to have such monitoring. To anticipate that, it might therefore be necessary to build-in higher quality and have more rigorous construction controls. Unfortunately life seldom works in such a fashion...

Inspection, maintenance and repair are dealt with later in the book but it may be helpful to introduce, at this stage, some indication of the frequency of inspection required. Table 4.6 was produced for offshore structures but it could usefully be applied elsewhere (Fjell, 1977).

4.2.2 *Chemical attack*

Sea water contains most of the elements, the majority as minute traces: the total dissolved salt content is about 3.5 percent, consisting mainly of sodium chloride. The normal predominant chemical content is shown in Table 4.7 (King, 1975).

Generally in the open sea the water is well aerated but in restricted basins, such as the Black Sea and Baltic Sea, the oxygen is consumed by biological decay of falling plankton faster than it can be replaced by circulation or inflow of fresh water (Turekian, 1972). Consequently dissolved oxygen is replaced by dissolved hydrogen sulphide which is weakly acidic. (With a large excess of chlorine water, sulphuric acid is formed with hydrochloric acid (Partington, 1951), or contact with the air will form sulphuric acid.) When the hydrogen sulphide develops through the decomposition or organic substances, it is

Table 4.6 Inspection Check List

CHECK POINTS: / LOOK FOR:	Coatings	Fixing plates flange connections	Previous repairs	Connection deckconcrete tower	Pronounced change in geometry	Points of difficult concreting	Heavily stressed details	Splash zone	Construction joints
TIME INTERVAL	1 yr	2 yr	1 yr	1 yr	2 yr	2 yr	1 yr	6 mths	1 yr
CRACKS									
Corrosion				×	×	×	×	×	×
Overload				×	×		×		
Shrinkage			×			×		×	
SPALLING									
Corrosion		×	×	×	×	×	×	×	×
Overload		×		×	×		×	×	
BAD CONCRETE	×		×		×	×		×	×
ABRASION	×	×						×	
SURFACE DEPOSITS									
Lime dissolution		×						×	×
Efflorescence						×		×	×
Rust									
DETERIORATED CONCRETE									
Freezing-thawing				×				×	
Chemical action	×		×			×		×	×
Crystallization						×		×	×

Table 4.7 Chemical concentrations in sea water

Chemical	Concentration g/kg
Chloride	19.353
Sodium	10.760
Sulphate	2.712
Magnesium	1.294
Calcium	0.413
Potassium	0.387
Bicarbonate	0.142

transformed into sulphate by aerobic, so-called sulphate bacteria (Biczok, 1964). All of these effects are or can be slowly corrosive to concrete so that long term deterioration may occur. In contact with reinforcement, iron sulphide is formed to corrode the steel. Coupled with these various reactions may also be formed calcium carbonate and calcium bisulphide which are water soluble and can thus be leached from the concrete.

There is then a biological side effect on concrete which is confined to particular types of location governed by limited circulation and animal or vegetable decay. This decay may also lead to increased concentrations of dissolved carbon dioxide which results eventually in the formation of calcium bicarbonate and gypsum in the hardened concrete, both soluble in sea water. This loss of material and weakening or mushiness of hardened cement paste is obviously detrimental, added to which is the possibility of decomposition of all the hydration products due to carbonation (Mehta, 1980).

Direct chemical attack comes from the magnesium salts in sea water. The concentrations may be small (typically 3200 ppm magnesium chloride and 2200 ppm magnesium sulphate) but they are sufficient to produce calcium chloride and gypsum (both soluble) and ettringite, associated with expansion and cracking. Other reactions may also lead to brittleness and strength loss (Mehta, 1980): substitution of magnesium for calcium in tobermorite, for example (Venuat, 1980). Concretes immersed in sea water, however, do not exhibit the expansion already mentioned, presumably because of leaching: for example, calcium hydroxide and calcium sulphate are considerably more soluble in sea water than in fresh; when combined with wave action leaching is bound to be increased.

'Magnesium sulphate has a more far-reaching action than other sulphates and decomposes the hydrated calcium silicates, in addition to reacting with the aluminates and calcium hydroxide' (Lea, 1970). The most significant reaction is with the hydrated calcium aluminates because the calcium sulphoaluminate which is formed (with magnesium hydroxide) is unstable in the presence of a magnesium sulphate solution, and gypsum, hydrated alumina and magnesium hydroxide are formed.

Evidently cements with limited tricalcium aluminate (C_3A) perform better in sea water and the limit appears to be 10% (Regourd, 1980). Other tests (Conjeaud et al, 1980) indicate, in fact, that cement with C_3A content $< 5\%$ performs better than any with C_3A content between 5% and 10%, categorised as a cement suitable for underwater use (CPA Prise Mer) (C_3A content $< 5\%$ —3.5% in the UK—defines Sulphate Resisting Portland Cement). Other limits suggested for a marine cement are $C_3A + 0.27C_3S < 23.5\%$ and $SO_3 < 2.5\%$ (Venuat, 1980). C_3A content $> 10\%$ leads to greater expansion in cements rich in C_3S (tricalcium silicate; Regourd, 1980).

Magnesium chloride reacts with hydrated lime ($Ca(OH)_2$) to produce calcium chloride but discussion on this will be deferred to the next chapter since here we are considering concrete per se. Solutions of magnesium chloride would have to be rather stronger than is normally the case in sea water to have a direct effect.

The reaction between sea water and cement is clearly complex and comprises many different reactions proceeding simultaneously at different rates. Fortunately not all are harmful. For example, a protective layer of brucite (magnesium hydroxide) and/or aragonite (calcium carbonate) is formed in the surface (Conjeaud, 1980) providing a shield against permeation of water (although in zones of intermittent immersion some of the benefits are lost through cracking). Again, the precipitation of magnesium hydroxide in the pores of the material has a blocking effect (Somerville and Taylor, 1978).

4.2.3 *Permeability*

Obviously, for chemical action to take place, water must have access to the interior of the concrete, both in sufficient quantity and at sufficient rate (this is also true for movement of chloride ions, oxygen and carbon dioxide). Any physical damage, such as cracking and progressive deterioration, will accelerate the process by providing faster routes, but it is easy to see that the permeability of the material must be a significant factor. In fact it is considered to be the most important characteristic determining the long-term durability of concrete exposed to sea water (Mehta, 1980). Since this view is echoed throughout the literature, it is clearly worthwhile examining the phenomenon more closely. Moisture dependence of the thermal response of concrete at elevated and reduced temperatures adds emphasis to this concern. Apart from that, moisture penetration of a water-excluding structure is intrinsically important.

A word of caution might be opportune, however. Graphs can be drawn, and equations derived, which correlate various physical properties and characteristics. Such empirical quantification can be invaluable in assessing which of these are the most significant or which measures the most effective. However, data arrived at in the laboratory are often not easily applied on site. That does not invalidate them of course, since data are essential to arriving at a proper understandng of behaviour. On the other hand, where the informatin **can** be used directly, it is equally important to ensure that it is being applied within the constraints imposed by the way in which it was obtained. In other words, phenomenological understanding is a basic pre-requisite to application: the user must know what is going on and must appreciate the limitations of the data at his disposal—they should not be applied indiscriminately. Consequently the aim here is for a perceptive view as much as for directly applicable details.

It is important at the outset to realise that 'permeability' and 'porosity' are not synonyms. Porosity (of which more anon) is a measure of pore volume and distribution while permeability is more a measure of pore continuity which, although governed to some extent by how many pores there are, is not necessarily directly related to their total quantity. The pore network varies in size and distribution so that transmissibility must vary, as will be shown, and individual pores may be blocked or isolated. In measuring porosity, moisture is evacuated by heat and/or vacuum and so a pore will empty if it has access to a free surface. Permeability, in contrast, generally concerns movement from one free face to another and therefore needs a path between them. Superficial consideration therefore suggests that, if there is indeed a relationship between the two properties, it is quite likely not to be straightforward. A housing estate with a single access road and many culs-de-sac branching off will have no through traffic whatsoever despite the fact that it might have an appreciable road area and regular, occasionally frequent, outflow and inflow.

The presumption is made in what follows that high quality, sound aggregates are necessary for durable marine concrete. We can therefore ignore their influence for the present and concentrate by and large on the binder, cement paste, which can be conceptualised as a matrix separating or binding aggregate particles as in Figure 4.3. Throughout the paste is a (three dimensional) network of capillaries of different diameters connecting pores of

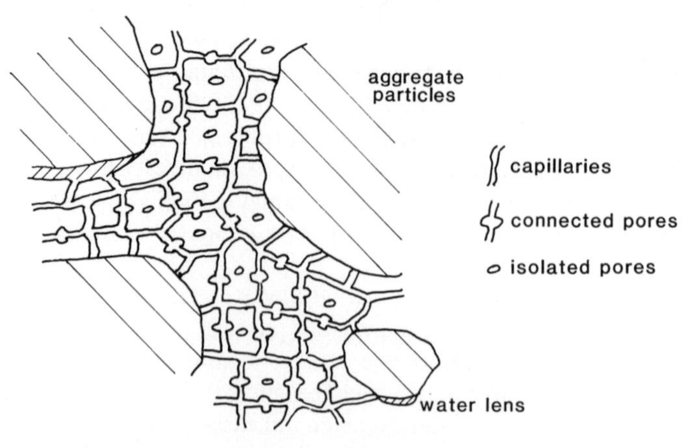

Figure 4.3 Stylised matrix

various shapes and sizes. Of course a cement paste capillary may be far removed from the smooth tube suggested. There are also isolated pores and paths around, or along, the surface of aggregate particles as well as through micro-cracks. However, the model helps in picturing the fluid mechanics of the problem.

4.2.3.1 *Pore filling* Porosity is generally defined as the ratio of the volume of voids in a material to the total volume. In a saturated cement paste the volume of voids is taken as the evaporable water content which, of course, will depend on the method of evaporation used to measure it—commonly by heating to 105°C or so. The **non**-evaporable water (primarily that in chemical combination) is a measure of the amount of cement gel present. Pore water is divided somewhat arbitrarily into capillary water and gel water with the gel water being equal to about three quarters of the non-evaporable water. The gel pores are much smaller that the capillary pores and so we are concerned mainly with the capillaries. For a more detailed account the reader should refer to Lea (1970).

While it is convenient to relate strength and other properties to age, the primary governing condition is, of course, degree of hydration which is, in turn, a function of the extent to which the pores are filled with gel. This is measured by the gel/space ratio, established by Powers some years ago (Neville, 1983) as

$$x = \frac{2.06 v_c \alpha}{v_c \alpha + \dfrac{w_0}{c}} \tag{4.1}$$

where x, the gel-space ratio, is the ratio of the volume of gel to the total space available to it, v_c is the specific volume of cement, c is the weight of cement, w_0 is the volume of mixing water and α the fraction of cement which has hydrated. v_c can be taken as 0.319 ml/g. Powers' relation for the compressive strength of concrete σ_c is then (Neville, 1983)

$$\sigma_c = 234x^3 \text{ N/mm}^2 \tag{4.2}$$

irrespective of the age or mix proportions of the concrete.

Unfortunately there does not appear to be a simple method of determining x directly and it can be expected to vary with time, water/cement ratio, effectiveness and duration of curing, and temperature. Nevertheless a causal chain can be discerned. Durability must be related to degree of hydration which may be assessed by the extent to which pores are filled with cement gel rather than water (too much water in the mix initially means that some pores can never be filled with gel). Hence there must be a connection with porosity and if this, in turn, is related to permeability (however complexedly) then a link between durability and permeability is to be expected.

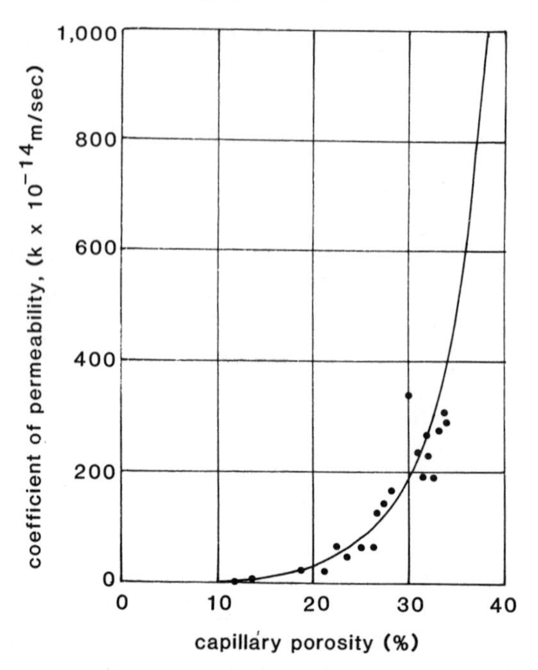

Figure 4.4 Relation between permeability and capillary porosity of cement paste

Powers' relationship between porosity and permeability is illustrated in Figure 4.4 (Haynes, 1980) which suggests that some correlation is likely for practical purposes: see also eq. (4.12) (Vuorinen 1985) and the work of Nyame and Illston (1981), for example.

4.2.3.2 *Pore size distribution* The pore size distribution of cement paste is variable, as has been suggested. Typical values are shown in Figures 4.5 and 4.6 which indicate how the distribution is affected by age and water/cement ratio. It should be noted that, for a mature paste, most of the pore volume lies

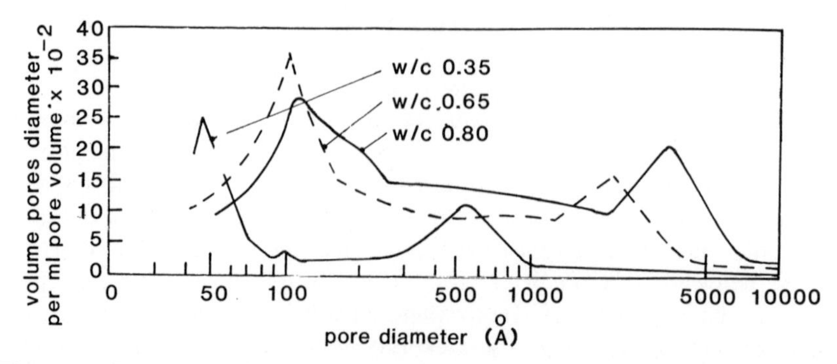

Figure 4.5 Pore size distribution in cement pastes which have been moist cured for 11 years

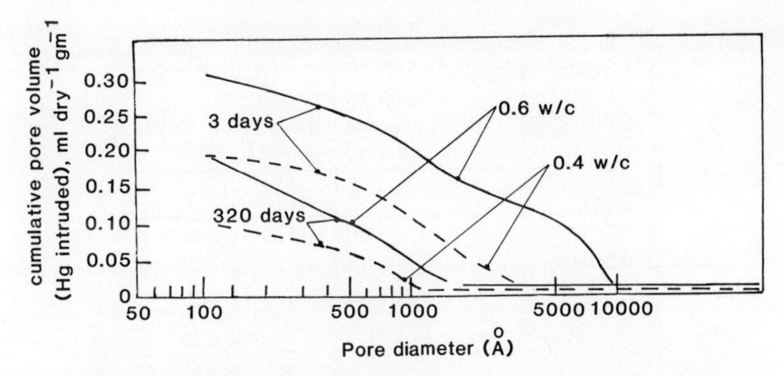

Figure 4.6 Cumulative pore size distribution curves for cement pastes

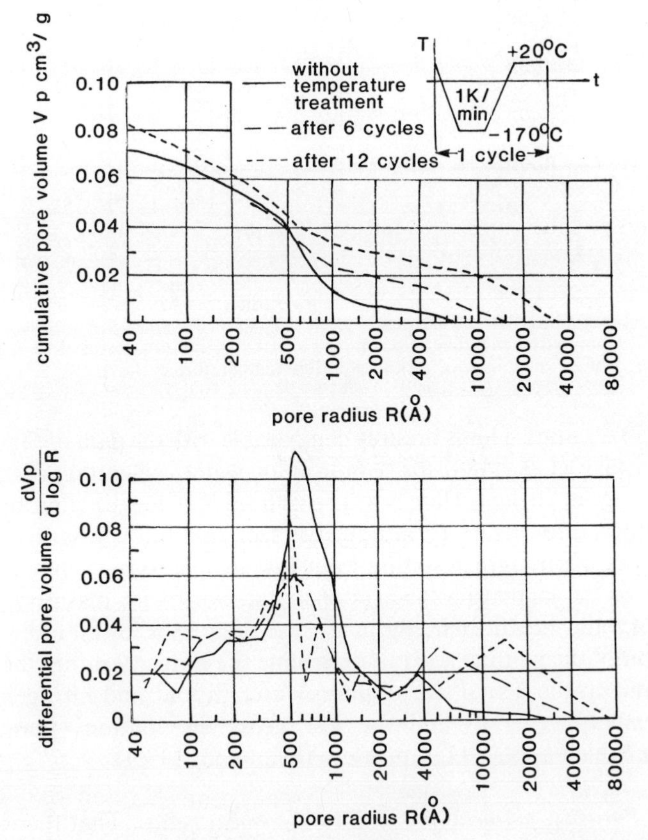

Figure 4.7 Cumulative and differential pore volume distribution versus pore radii for a portland cement mortar prior to and after low temperature cycles

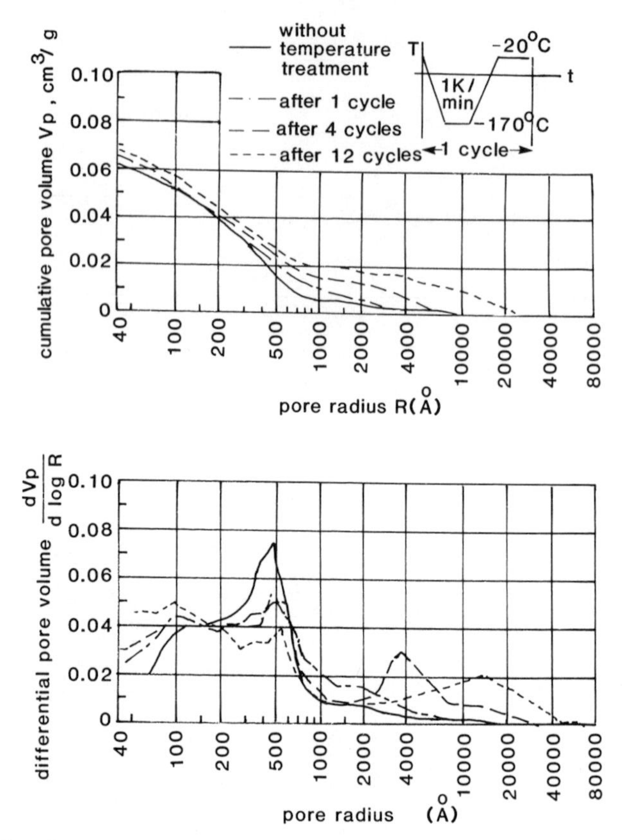

Figure 4.8 Cumulative and differential pore volume distribution versus pore radii for blast furnace slag cement mortar prior to and after low temperature cycles

below 750 Å radius. This is broadly comparable with the data of Figures 4.7 to 4.10 (Rostasy et al, 1980) for a mortar of water/cement ratio 0.5 and an apparent age of 90 days. Here it will be noticed how the distribution coarsens with temperature cycling. (It is argued, in fact, that this is what is responsible for the loss of strength resulting from low temperature cycling.) This is in contrast to the increasing fineness of the network with maturity, shown in Figures 4.5 and 4.6. Incidentally, in all these figures the lowest radius shown is a function of the method used to determine the porosity rather than being a lower limit to the actual size of pore. Water vapour and nitrogen sorption measurements generally indicate less gross distributions than mercury intrusion which applies high pressure (Diamond, 1976).

4.2.3.3 *Porosity, permeability and water/cement ratio* That there might be a relation between porosity and permeability has already been suggested (Figure 4.4). There also appears to be a relationship between permeability

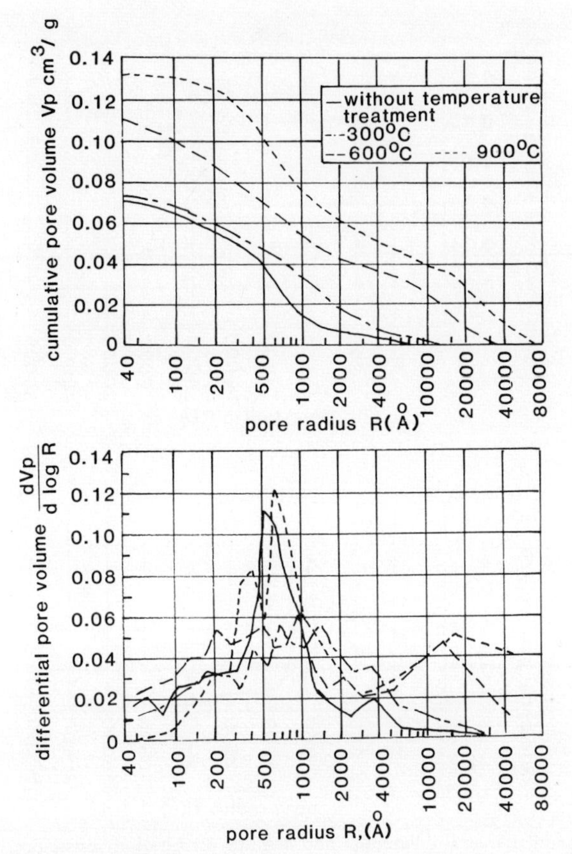

Figure 4.9 Cumulative and differential pore volume distribution versus pore radii for a portland cement mortar after high temperature treatments

and water/cement ratio: Figure 4.11 due to Powers (Browne and Baker, 1979). This indicates a maximum water/cement ratio <0.5 for permeability $<10^{-12}\,\mathrm{ms}^{-1}$. Others (Nyame and Illston, 1981), while confirming that water/cement ratio has some influence, have found an age effect as well as a relationship with the hydraulic radius of pores. (In this context the hydraulic radius is the pore volume divided by the surface area (Watson and Oyeka, 1981) which is not quite the same as the conventional value for a cylindrical tube.) Against that, amongst views summarised elsewhere (Hughes, 1985), is the consideration that permeability is due to flow in pores of radius >50 nm, although evidence is also alluded to that such a distinction is questionable.

To a degree all these effects can be accommodated in simple general terms, albeit without quantification. Increasing age means increased hydration (assuming sufficient moisture supply) which means the pores are better-filled. Reduced water/cement ratio means that there are fewer pores to be filled (when

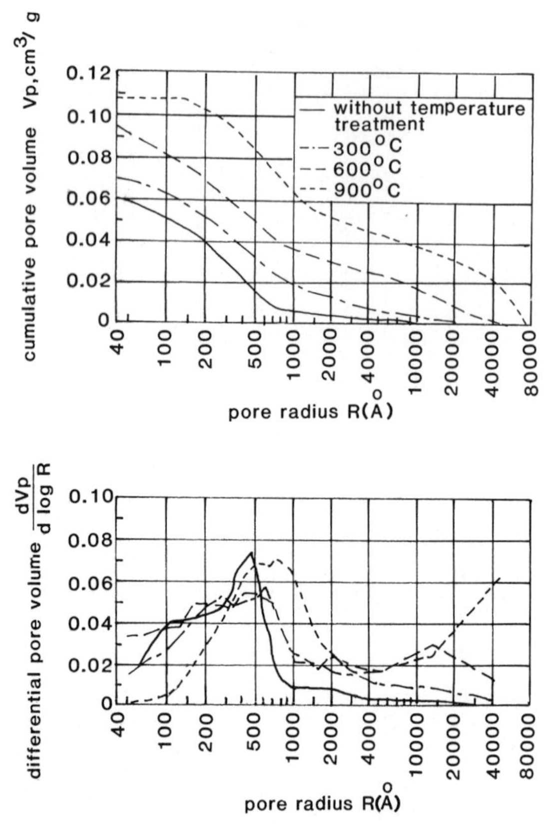

Figure 4.10 Cumulative and differential pore volume distribution versus pore radii for blast furnace slag cement mortar after high temperature treatments

one considers the low water/cement ratio actually needed for hydration) or fewer to be left empty. (It is easier to think in terms of 'number' or 'quantity' of pores than in terms of pore space.) Consquently one would expect reducing porosity with reducing water/cement ratio and increasing age. Similarly one would expect greater porosity with reduced curing. Now, in considering permeability in relation to porosity, it is helpful to think in crude statistical terms.

Suppose the pores are to be grouped according to radius ranges. Then a given size will have a certain probability that it will be continuous, and the greater that particular pore volume i.e. the more pores there are of that size, the more likely that some will be continuous. Similarly, the smaller the pore radius, the lower the probability that it will be continuous (i.e. the more likely it is to be blocked). Thus one would expect coarse pastes to be more permeable than less coarse pastes because of the increased likelihood of continuity. What difference, then, do coarse pores make to permeability?

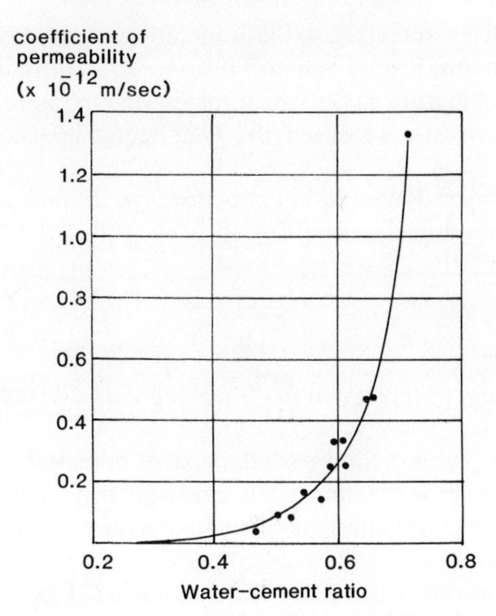

Figure 4.11 The effect of water:cement ratio on the permeability of mature cement paste

4.2.3.4 *Influence of pore size* Figure 4.12 (Nyame and Illston) gives some experimental data relating permeability and hydraulic radius for cement pastes of various water/cement ratios and maturities. Permeability appears to be sensibly constant above a hydraulic radius of 2500 Å which, assuming for the moment the conventional definition of hydraulic radius in a cylinder with axial flow, corresponds to a pore radius of 5000 Å which would have to be a *mean* radius, of course. Figures 4.4 to 4.10 suggest that this would require an impracticably high water/cement ratio, indicating incidentally the conse-

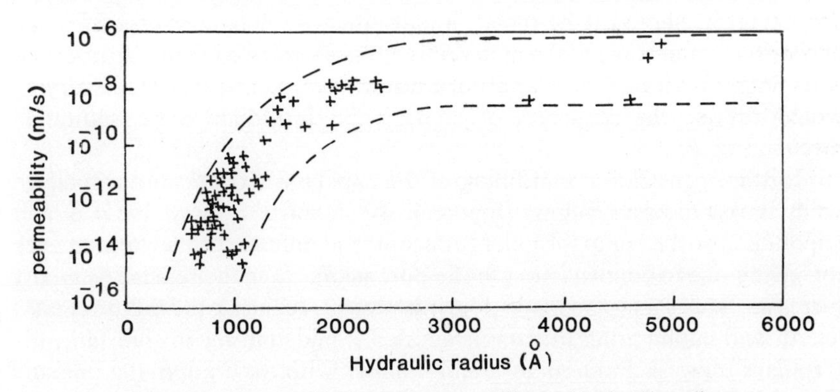

Figure 4.12 Relationship between permeability and hydraulic radius

quence of excessively wet mixes, so that generally we must be considering pore sizes smaller than this. From Figure 4.5 the pore size distribution curve shows two peaks for very mature pastes which correspond to smaller pore diameters as the water/cement ratio is reduced (the other figures also show such peaks for younger pastes).

The familiar Hagen-Poiseuille equation for flow through a horizontal tube of circular cross-section, diameter d, is, (Streeter and Wylie, 1979) after slight modification:

$$q = \frac{\pi d^4}{128\mu} \cdot i_p \qquad (4.3)$$

That is, discharge is proportional to d^4 (μ is the viscosity and i_p the hydraulic gradient).

For the mature pastes of Figure 4.5 the 'coarse' pore peak is at a diameter of approximately 10 times the 'fine' peak, consequently the discharge through such a pore will be 10 000 times that through the finer. It is quite obvious that the appropriate areas under the curve are not in such proportions. While the distributions of the other figures are different it will still be the case that large diameter pores have an influence disproportionate to their size and number. This also gets to the root of the argument quoted earlier that permeability is due to flow in pores of radius > 50 nm.

The foregoing account is helpful in understanding or appreciating the necessity for minimising water/cement ratios, maximising the effectiveness of curing, and delaying pressurised submergence as long as possible (although low pressure submergence may help curing). However, it does not help in quantifying moisture penetration on a structural scale (or indeed that of any other fluid). Fortunately there are certain well-recognised physical laws or models which may be applied to both structure and to method of assessment or measurement.

4.2.3.5 *Moisture penetration* The simplest way to begin is by presuming a dry concrete. Should it be moist initially then we enter the process at an appropriate later stage. If the material is already saturated then the process is at its simplest. An adequate supply of moisture is presumed at all stages: drying would reverse the sequence, of course, and introduce some additional mechanisms.

Moisture penetration and filling of the capillary network and associated voids is visualised as follows (Figure 4.13) (Concrete Society, 1985). Water vapour is adsorbed onto the inner surface (a) and diffuses through the network (b), giving rise to condensation in the pore necks (c) as the relative humidity increases. This short-circuits vapour movement, reducing the diffusion path length and culminating in vapour-assisted liquid transfer in thin films (d). Capillary flow is induced (e) until finally, with saturation (f), pressure controlled flow results. Thus moisture penetration may be represented by

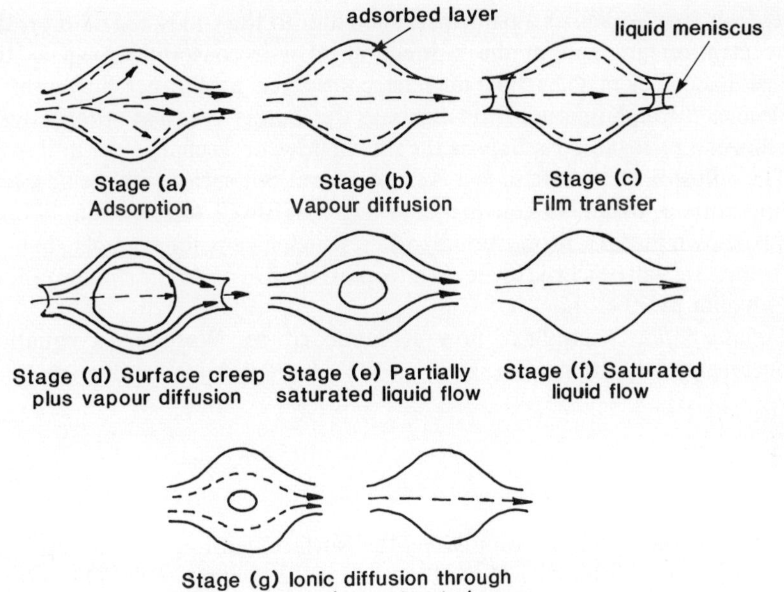

Figure 4.13 Idealised model of movement of water and ions within concrete

three principal successive processes, each of which can be modelled mathematically:

(i) diffusion;
(ii) capillary flow;
(iii) pressure flow.

While these are sequential at a given location, throughout a concrete mass they may be occurring simultaneously at different locations, particularly where a moisture wave or front is advancing. In some situations there may be accompanying drying and in others there may be insufficient moisture for the whole sequence to occur.

Vapour diffusion and capillary flow are relatively short phases whose duration is governed by the initial condition of the concrete. Pressure flow is the most significant for immersed elements whereas in the splash/atmospheric zone diffusion and capillary flow may dominate, especially in both wetting and drying. Tide-affected zones might be less critically affected than areas immediately above, since drying proceeds more slowly than wetting.

Diffusion Diffusion of vapour is modelled by Fick's law which may be expressed as

$$J = -D \cdot \frac{dn}{dx} \tag{4.4}$$

where J is the flux across a plane perpendicular to the x-axis and dn/dx is the concentration gradient in the x-direction. D is a constant known as the diffusion coefficient. The flux may be considered as the net flow rate of molecules through unit area and dn/dx as the density gradient. Alternatively the flux can be regarded simply as the rate of flow, or discharge, per unit area.

The diffusion coefficient in fact is not constant but varies with the absolute temperature T. From the kinetic theory of gases (Elwell and Pointon, 1978) it can be shown that D is proportional to $T^{3/2}$ which means, for example, that the diffusion rate will be a little higher in a tropical than in an Arctic climate (about 14% higher at 30°C than at 5°C).

Capillary flow Capillary flow is modelled by Washburn's equation (Concrete Society, 1985). The rate of flow V of a liquid in a capillary is given by

$$V = \frac{r\gamma_L}{4d\mu}\cos\theta \tag{4.5}$$

where r is the capillary radius, γ_L the surface tension, d the depth of penetration, μ the viscosity of the fluid and θ the contact angle. The surface tension γ_L varies slightly with temperature and the viscosity μ markedly so (Streeter and Wylie, 1979). The relationship between viscosity and temperature of water was formulated by Poiseuille (Gibson, 1952) and values tabulated by Streeter and Wylie can be fitted by the relation

$$\mu = \frac{1.8398 \times 10^{-3}}{1 + 0.03907T + 0.0001591T^2} \tag{4.6}$$

where μ is the viscosity in Nsm^{-2} and T the temperature in °C. The viscosity at 30°C ($0.801 \times 10^{-3}\ Nsm^{-2}$) is thus approximately half that at 5°C ($1.519 \times 10^{-3}\ Nsm^{-2}$) thereby doubling capillary flow. From other values quoted by Streeter and Wylie, capillary rise for tap water is only half that of distilled water. This implies that the presence of slight impurities can make a significant difference and so the pore water in concrete will not necessarily behave like tap water. Presumably the effect is due to changes in surface tension (rather than viscosity) and will depend on whether or not the chemicals in the water are surface active (Rao, 1972).

Pressure flow Pressure flow is probably more familiar to civil engineers since it involves the well-known d'Arcy's Law (Linsley et al, 1975):

$$v = ki \tag{4.7}$$

where v is the discharge velocity, i the hydraulic gradient and k the coefficient of permeability (or hydraulic conductivity) of the medium through which the flow occurs. This stems from the study of flow through soils (Terzaghi and Peck, 1960; Linsley et al, 1975). If the difference in head between two points distance l apart is h then the hydraulic gradient $i = h/1$ and the pressure gradient

$i_p = wh/1 = wi$ where w is the specific weight of water. For a medium not affected structurally by the fluid flowing through, the discharge velocity v is given by (Terzaghi and Peck, 1960)

$$v = \frac{K}{\mu} . i_p \tag{4.8}$$

where K is the intrinsic permeability of the medium and is a property of it only. From eqs (4.7) and (4.8)

$$k = \frac{wK}{\mu} \tag{4.9}$$

so that the coefficient of permeability depends both upon the medium and upon the fluid flowing through it. Normally in most civil engineering applications this is unimportant but, as has already been shown above (eq. (4.6)), response of a particular concrete will depend on the climate and the use to which the material is being put.

The discharge velocity of eq. (4.7) is usually taken to be given by $q = v/a$ where a is the *gross* area of the section across which the flow occurs. The actual percolation or seepage velocity v_s through the voids in the material will be given by the discharge velocity divided by the cross-sectional porosity p i.e. $v = pv_s$. If the potential leaching of certain chemicals from the concrete is to be considered then the seepage velocity v_s may be the more appropriate.

4.2.3.6 *Pressure flow and diffusion* Fick's law above derives from transport theory, as distinct from continuum mechanics which is the more familiar in civil engineering. Transport theory is a *particle* theory of matter but the well-known Euler and Navier-Stokes equations of hydro-mechanics can, in fact, be derived from it (Duderstadt and Martin, 1979) so there is a connection between the two approaches—Laplace's equation, fundamental to many engineering applications, is continuum-based but is used in heat transfer processes, normally regarded as diffusive (see, for example, Bajpai et al, 1977 and Hunter, 1983 for the so-called diffusion equation of one-dimensional heat flow, given here in equation 4.55). By analogy, *water* penetration into concrete can be regarded in the same way (Vuorinen, 1985) as it is in some soils applications. The relevant equation is

$$\beta^2 \cdot \frac{\partial^2 p}{\partial x^2} = \frac{\partial p}{\partial t} \tag{4.10}$$

where β^2 is the hydraulic diffusion coefficient ($m^2 s^{-1}$), p is the total pressure in the pores (m water) at depth x (m) (the lower bound of p is atmospheric pressure), x is the depth from the surface of the pressure action of duration t (s). β^2 is the ratio of the hydraulic conductivity (or permeability k ms^{-1} of the concrete to the water-storage capacity of the open pores in it. It is shown that

(Vuorinen, 1985)

$$\beta^2 = \frac{kp^2}{10\alpha'} \text{ m}^2\text{s}^{-1} \tag{4.11}$$

where α' is the volume of open pores in the concrete which, in the absence of measurement, may be taken to be about 80% of the air content of the freshly mixed concrete.

Numerical solution of eq. (4.10) has established a close correlation between k derived this way and values obtained from d'Arcy's equation, eq. (4.7), although they are not quite equal:

$$(\log k = 0.583 \log k_{\text{d'Arcy}} - 4.70) \tag{4.12}$$

In many practical situations involving, say, straightforward moisture movement through a wall which is wet on one side and dry on the other, d'Arcy's equation is very easy to apply and therefore very useful. However, it is not *universally* suitable and a weapon such as diffusion theory is a handy one for the armoury. More than that, though, it is valuable in its confirmation that permeability *measurement* by surface penetration methods is likely to be reliable.

4.2.3.7 *Permeability measurement* Permeability measurement has a particular value beyond its inherent informativeness. Accepting that durability is linked to permeability, then measuring the latter should provide an indication of the condition of a structure or part of it. Moreover, development of an *in situ* test has obvious benefit. Consequently it is hardly surprising that considerable effort has been devoted to the subject.

A comprehensive review has already been published accompanied by an extensive bibliography (Concrete Society, 1985) so we can confine ourselves to a short summary here. Test methods are divided into two main categories: *in situ*, or on samples removed from site (this will also cover laboratory prepared specimens). Of course *in situ* tests can be carried out in the laboratory but laboratory tests cannot normally be performed *in situ*.

***In situ* tests.** There are various methods but generally they involve applying a fluid to the surface of the concrete or into a hole drilled in the face, and noting the rate at which the fluid is absorbed (it may or may not be under pressure). In some cases this has been correlated with strength and water/cement ratio.

Laboratory tests. Straightforward absorption of water is used in various standards (the sample or specimen is dried to constant weight, immersed in water for a specified time and re-weighed). Capillary rise is used in some tests, where one surface only is placed in contact with water and the depth of penetration observed (the concrete darkens, for example). The pore size can then be estimated from $d = \sqrt{r^2 P_0 T / 4\mu}$ where d is the depth of penetration, r is the pore radius, P_0 atmospheric pressure, T duration of test and μ viscosity. Porosity P can also be measured from weight gain M, $p = 1000\,M/Ad$ where A is the area of penetration. Pressure-induced flow can be used by passing a fluid (liquid or gas) through a specimen sealed on all but two opposing faces and measuring the rate of flow between them. Depth of penetration tests are also used for low permeability concretes in special cells under pressures up to several atmospheres. Other methods are being studied or developed such as radiation attenuation, resistivity, high pressure, transient pressure, and osmotic pressure.

Inter-test correlation seems as yet to be questionable, one of the problems being change in permeability with duration of test. Surface absorption tests have obvious application to *in situ* measurements and, in general, absorption tests are much quicker than through-flow tests which can take several days. However, there is clear evidence that the latter show a substantial reduction in permeability as the test proceeds tending towards a constant value after several days. This goes well beyond what one might expect from the initial wetting process and is attributed to hydration of unhydrated cement due to more effective water penetration as a result of pressure. There may also be a blocking effect consequent upon the movement of loose particles. It is therefore difficult to estimate how this will affect results since it must be a function of the maturity of the paste and the pressure employed, despite the anticipated agreement of eq. (4.12). Broadly what can be concluded is that areas subjected or likely to be subjected to pressure flow will have permeabilities lower than those indicated by surface measurement. In other words some factor of safety may be inherent in *in situ* tests—this might apply, for example, to structural surveys where immersed areas are inaccessible and measurements are made elsewhere.

4.2.3.8 *Sea-water effects* There is an added bonus for concretes immersed in sea-water. The formation of a surface layer of brucite and aragonite has already been mentioned (Conjeaud, 1980) and surface absorption and resistivity measurements indicate that this reduces permeability quite substantially (Buenfeld and Newman, 1986 and 1984). Deep ocean tests also show that high hydrostatic pressure in cold sea water (5°C) much reduces, and possibly even eliminates, permeability (Haynes, 1980). However, material in the splash and atmospheric zones will not have these benefits.

4.2.3.9 *Chloride, oxygen and carbon dioxide penetrations* Water is not the sole permeating agent of concern. As will be discussed in the next chapter, carbonation of cement paste reduces protection against the corrosion which results from the penetration of chloride ions. Furthermore, for the corrosion to be sustained, there must also be a sufficient supply of oxygen and/or water. Consequently, concrete must have a low transmissibility of chloride ions as well as reducing the permeation of water and oxygen below the level which

Table 4.8 Typical values for concrete permeability, absorption and diffusion

Test method	Units	Concrete permeability/absorption/diffusion		
		Low	Average	High
Intrinsic permeability	m^2	$<10^{-19}$	$10^{-19}-10^{-17}$	$>10^{-17}$
Coeff. of permeability to water	m/s	$<10^{-12}$	$10^{-12}-10^{-10}$	$>10^{-10}$
Coeff. of permeability to gas	m/s	$<5\times10^{-14}$	$5\times10^{-14}-5\times10^{-12}$	$>5\times10^{-12}$
ISAT 10 min	$mL/m^2/s$	<0.25	0.25–0.50	>0.50
30 min		<0.17	0.17–0.35	>0.35
1 h		<0.10	0.10–0.20	>0.20
2 h		<0.07	0.07–0.15	>0.15
Figg water absorption 50 mm	5	>200	100–200	<100
Modified Figg (air) (Ove Arup)	5	>300	100–300	<100
Water absorption 30 min	%	<3	3–5	>5
DIN 1048 depth of penetration (4 days)	mm	<30	30–60	>60
Oxygen diffusion coeff. (28 days)	m^2/s	$<5\times10^{-8}$	$5\times10^{-8}-5\times10^{-7}$	$>5\times10^{-7}$
Apparent chloride diffusion coeff.	m^2/s	$<1\times10^{-12}$	$1-5\times10^{-12}$	$>5\times10^{-12}$

Notes: (i) ISAT—Initial Surface Absorption Test
(ii) Figg—named after the originator, these tests involve drilling a hole into the surface and then injecting water or air

would support corrosion. Clearly any limits on permeability will be governed by climate and use: it has already been pointed out that it is dependent on temperature, as are rates of reaction (most double with a 10° rise in temperature). It can therefore be expected that a particular concrete will be much more at risk in tropical waters although one might equally expect more rapid protection should it be immersed. Much depends too on whether the water temperature fluctuates as much as the atmosphere but certainly splash and atmospheric zone concrete is more likely to have problems. Chloride ion and oxygen transmission are diffusive processes.

4.2.3.10 *Typical values* Important as it is, there does not appear to be a great deal of site data available on permeability. The Concrete Society report (1985) therefore gives typical values as a guide and so the relevant data are reproduced here in Table 4.8. For more detailed discussion the reader is referred to the report and to the other papers accompanying it.

According to the report, the German standard DIN 1045 specifies maximum allowable depths of water penetration as shown at Table 4.9 (this might then provide some form of scale for Table 4.8).

An idea of what even 'low' permeability represents can be obtained from simple calculation. A value of $10^{-12}\,ms^{-1}$ for concrete 1 m thick corresponds to a

Table 4.9 DIN 1045 maximum allowable water penetration depths

Impermeable to water	50 mm
Frost resistant	50 mm
Resistant to weak chemical attack	50 mm
Resistant to strong chemical attack (plus a coating on the concrete)	30 mm
Resistant to sea water	50 mm

throughflow of about 0.9 ml/m²/day for a pressure head of 10 m. Let us assume that this evaporates into an unventilated shaft 12 m diameter which will have a contained volume of $[\pi(12)^2/4]/\pi(12) = 3\,m^3$ per square metre of surface area. At 20°C saturated air holds $17\,g/m^3$ of water vapour so if the air were bone-dry to start with it would become saturated after about 50 days after which the walls would 'sweat'. At a more normal 70% relative humidity it would take just over a fortnight.

4.2.3.11 *Influence of bleeding/water gain* It is to be expected that bleeding or water gain might increase the permeability of concrete due to the formation of water passages and lenses beneath aggregate particles and reinforcement. However, there is evidence to indicate that this is also affected both by the direction of applied stress and by its magnitude (Mills, 1984). It was concluded from laboratory tests that maximum permeability occurred when the direction of flow was the same as the direction of bleeding or the direction of sedimentation of the solids. However, biaxial prestress normal to the direction of flow reduced permeability, the reduction increasing with increase in stress. The example of practical implication drawn from this was that a cylindrical vessel constructed by slip forming would benefit from vertical and horizontal prestress. This is clearly relevant to typical offshore storage platforms.

4.2.4 *Pore pressures*

Discussion on moisture in concrete cannot omit reference to pore pressures. Theoretical consideration (Morley, 1979) indicate that radial tension is probable in thick-walled cylinders and may give rise to substantial through-wall tensile stresses due to pore pressures where saturated concrete is subjected to rapid loading. This can be typified to some extent by installation of gravity platforms in deep water, and so the situation must be checked. If the tensile strength is exceeded then laminar (so-called tangential or in-plane) cracking can occur. This is not visually obvious, of course, although it can be detected by tapping with a hand-held hammer—biaxially prestressed concrete is particularly likely to have such cracking due to various causes and appropriate precautions must therefore be taken (Gerwick and Berner, 1984).

The theoretical argument referred to above follows that familiar from consolidation theory in soil mechanics (or geotechnics) in which the total normal stress σ at any point on a section through a saturated soil is the sum of a neutral stress u (the hydrostatic pressure) and an effective stress σ' (the pressure transmitted through the points of contact between the soil grains) (Terzaghi and Peck, 1960). Experimental evidence appears to confirm the applicability of this to concrete (Butler, 1982) although it is stated that Skempton's equation is more accurate:

$$\sigma' = \sigma - [1 - (1 - f)C]u \tag{4.13}$$

where f is a coefficient often described as the uplift area coefficient (as applied to the design of dams) or the superficial, effective, or boundary porosity. Although the evidence varies f has a value approaching unity as failure is approached i.e. with a presumed increase in crack formation. C is a material constant which for concrete is 0.18. The same author (Butler, 1982) also questions the applicability of d'Arcy's law to concrete, observing a reduction in permeability with duration and 'self-sealing' as discussed above.

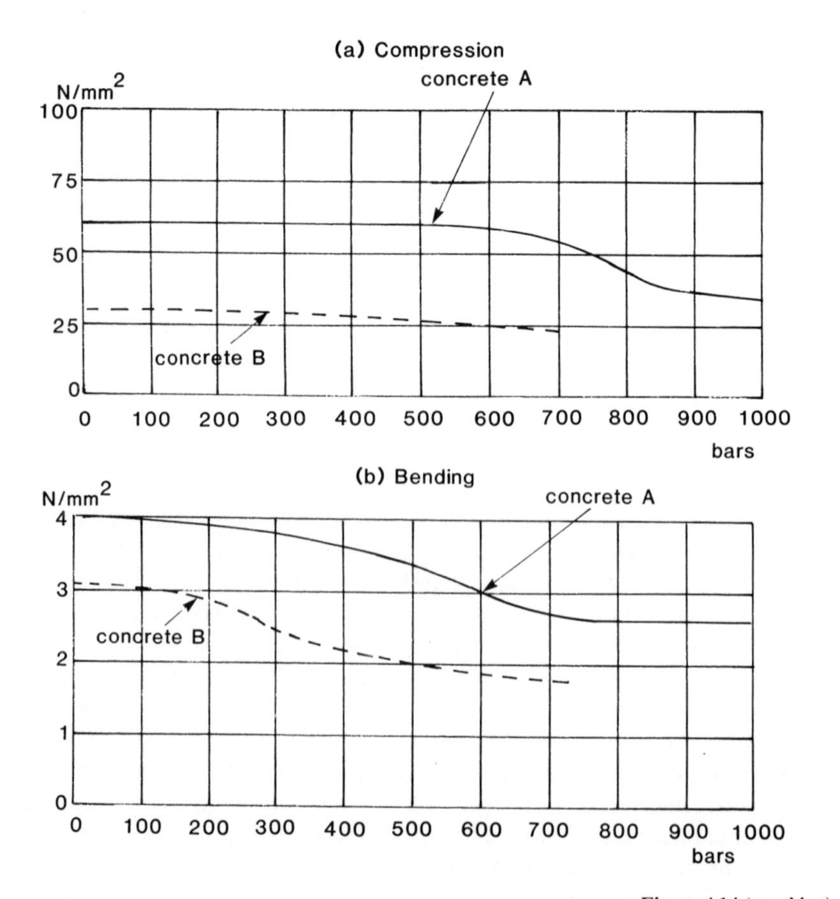

Figure 4.14 (cont'd...)

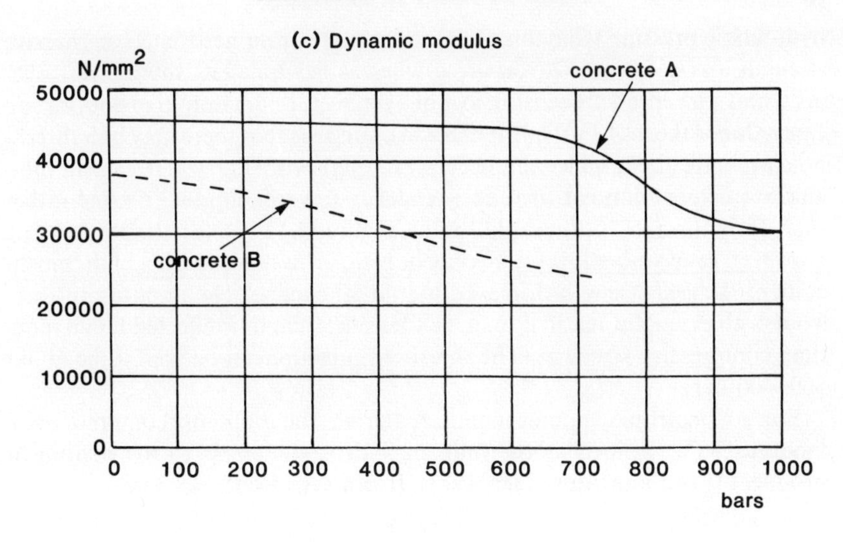

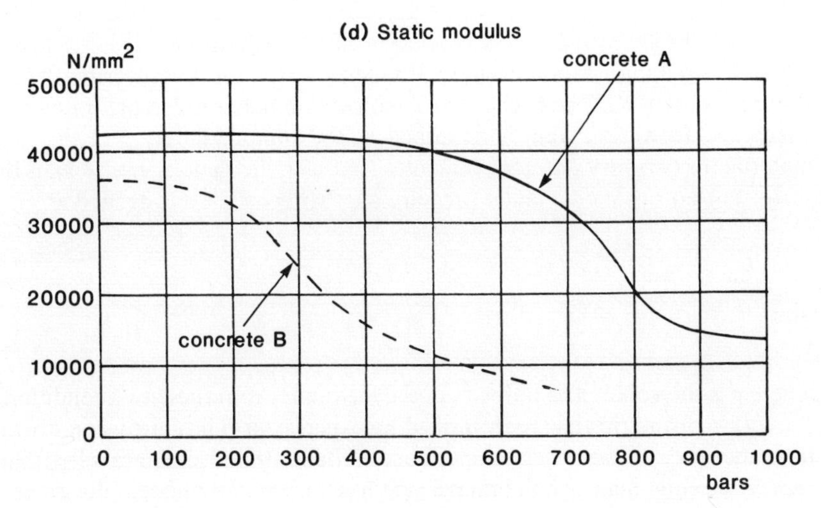

Figure 4.14 Properties of concrete subjected to hydrostatic pressure: concrete A 60 N/mm^2 concrete B 30 N/mm^2

It is well-known that, all other things being equal, the strength of saturated concrete is less than that of air-dried concrete (Neville, 1983). This has been attributed to dilation of cement gel by adsorbed water which results in a reduction in the forces of cohesion of solid particles. However, pore pressure theory may lead to a more quantifiable explanation in an engineering sense, whatever the physical phenomena involved, especially for structural members under hydrostatic pressure. There is certainly evidence that there is a reduction in concrete strength and elastic properties after being subjected to high hydrostatic pressure (Clayton, 1986). After 600 bar immersion a compressive strength loss of about 10% occurred whereas flexural tests showed just over 50%, and a gas pressure tension loss of over 80%. Reduction in pressurisation rate reduced the loss. Earlier French work suggests that there may be a threshhold pressure above which this occurs, although with high water/cement ratio and low cement content there is a tendency towards gradual decline rather than formation of a threshold, as in Figures 4.14 (a)–(d) (Ilantzis, 1980). However, even with a low water/cement ratio, in concrete with a high cement content *flexural* strength displayed a gradual decline. The implication then, from both examples referred to, is that tensile strength is affected much more than compressive strength. (The threshold pressure can be seen to be about 500–600 bar).

For an isotropic, homogeneous material, the bulk or compressibility modulus K_b is defined as the ratio of hydrostatic stress to the dilation it produces (Ford and Alexander, 1977). It can then be shown that

$$K_b = \frac{E}{3(1 - 2v)} \tag{4.14}$$

where E is Young's Modulus and v Poisson's Ratio; the compressibility k_b is the reciprocal of K_b. The French work indicates that k_b is different for different concretes. However, what is described as the compressibility δ of the *solid* material for two very different concretes (but with the same aggregates), is the same. For porous media under pressure another constant α is defined as

$$\alpha = 1 - \frac{\delta}{k_b}$$

and

$$p < \alpha < 1 \tag{4.15}$$

where p is the volumetric porosity of the medium. From these two conditions $\delta < k_b(1 - p)$ and this has been verified by experiment. The conclusion drawn from the work is that the compressibility of the saturated concrete is less than that of the solid material and therefore it has a higher elastic modulus i.e. it is stiffer.

Other work (Bouineau, 1980) has also shown that prestress reduces permeability, in other words, compressive loading reduces water penetration, presumably due to the closing up of cracks.

In general, therefore, one might conclude that pore pressures can have a disruptive effect if allowed to rise too quickly. However, where the material is prestressed, permeability and therefore presumably pore pressure are reduced. The concrete will be stiffer on the other hand and may have significantly lower tensile strength if the pressure is high enough. This does not matter where the material is prestressed but in that case through-wall tension could cause problems unless guarded against.

4.3 Fatigue

'Fatigue may be defined as a process of cycle by cycle accumulation of damage in a material undergoing fluctuating stresses and strains. A sigificant feature is that the load is not large enough to cause immediate failure. Instead, failure occurs after a certain number of load fluctuations have been experienced i.e. after the accumulated damage has reached a critical level' (Berge, 1985). 'A fatigue loading is a repeated pattern of dynamic loads (which) is likely to be repeated one hundred or more times during the lifetime of the structure and to result in failure at material stresses considerably less than those for failure under static loading' (Hawkins, 1976). Fatigue loading can thus be distinguished somewhat arbitrarily from other dynamic loads such as those due to earthquakes or single impacts from collision or explosion or whatever.

Other arbitrary distinctions are into low cycle fatigue (<5000 cycles), high cycle fatigue and fatigue limit regions as exemplified in the familiar so-called S–N or Wöhler curve of Figure 4.15. If the stress S is plotted against the number of cycles to failure N at that stress (generally on a log–log scale) then the shape shown will be typical of steel. The fatigue (or endurance) limit means that the material will have an infinite life at stresses cycling below that value. It is generally considered that concrete does not have such a limit but when its fatigue strength is mentioned without qualification it normally refers to the strength at 10 million cycles (around 55% of the static). Low cycle fatigue implies that the applied stress is a high proportion of the static strength of the material.

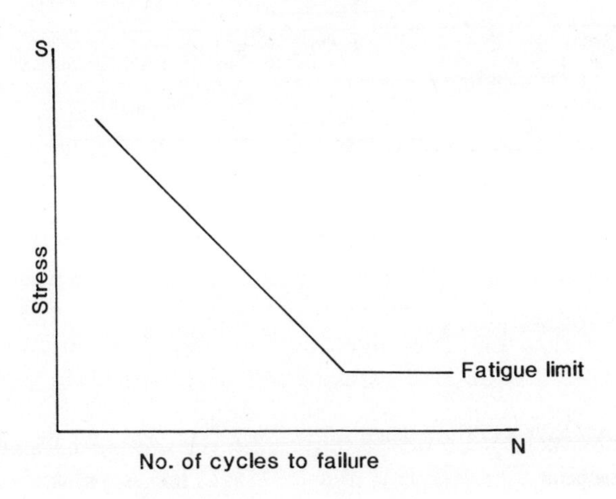

Figure 4.15 S–N (Wöhler) curve

4.3.1 *Cyclic loading*

Cyclic loading can take many forms which may be regular or random, fast or slow, high or low cycle; it may be uni-directional or reversing (in real-life it might be multi-directional of course); it can be uni-or multiaxial, flexural or torsional or have some combination of them. Such variants and their consequent effects are clearly recognised for many materials although the availability of data is by no means equal for all: it is governed by practical difficulties in testing, the nature and application of the material and the extent of its use, for example. Unfortunately, recognition of consequences does not necessarily lead to ease in interpretation.

At the outset it may be useful to set down some general definitions and considerations prior to considering concrete *per se* and then reinforcement and reinforced and prestressed concrete in the next chapter. Figure 4.16 indicates some types of cyclic loading. In all cases the horizontal axis (abscissa) is time; the vertical axis (ordinate) is frequently taken as stress (often expressed as a fraction of static strength) but could be strain (for strain-controlled testing) or simply load. The loading might be compressive, tensile, flexural or torsional; nor should combinations be excluded. Figure 4.15 typifies the *S–N* curve but it is important to recognise that any one point will have a scatter of results about it. If this scatter is, say, normally distributed, then it is possible to produce a family of (more or less parallel) curves, each with a particular probability of failure (Marin 1962).

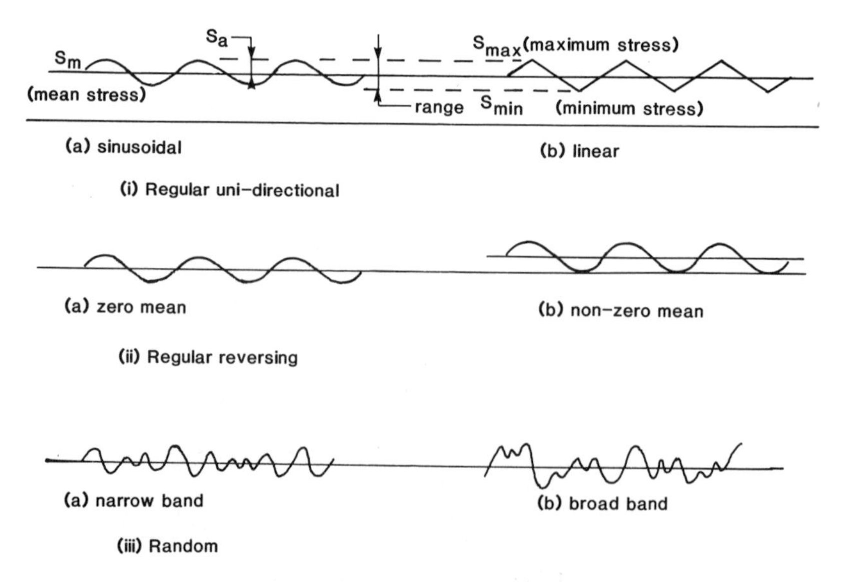

Figure 4.16 Types of cyclic loading

4.3.2 *Load spectrum*

Apart from such inherent variability, marine structures can be expected to be subject to a spectrum of loads. This can be reflected in fatigue testing, leading to a degree of complexity in interpretation. If for example, the spectrum is simulated by a succession of blocks of pulsating loads of different levels, then material response may be affected by the sequence: a high pulsating load succeeded by a low one is more damaging than the converse, where the low pulsating load can have a beneficial strain-hardening effect (Marin 1962). On the other hand completely random loading is more difficult (and expensive) to simulate as well as to assess: counting cycles can be awkward for instance—various methods are outlined elsewhere (Berge 1985). An occurrences spectrum or an exceedances spectrum may be prepared (i.e. from the number of times a particular event such as a peak or range occurs or is exceeded within a given time interval). Level crossing (such as up or down crossing of the mean load) or peak or range counting may be adopted, so may more sophisticated so-called rainflow counting methods. The latter is said to be preferable for low cycle fatigue analysis while for narrow or moderately wide band load responses the counting method is not important. It is only for very wide band histories that it matters—for irregularity factor $I < 0.5$. I(the ratio of zero crossings to peaks) is given by

$$I = (1 - \varepsilon^2)^{0.5} \tag{4.16}$$

Where ε is the spectral bandwith parameter.

$$\varepsilon = \left(1 - \frac{m_2^2}{m_0 m_4}\right)^{1/2} \tag{4.17}$$

and m_n is the nth moment of the spectrum.

It is stated (Berge, 1985) that the load spectrum can often be approximated by the Weibull distribution which can be expressed as

$$S_r = S_{r,0}\left[1 - \frac{\log n}{\log n_0}\right]^{\kappa} \tag{4.18}$$

where S_r is the stress range which is exceeded once out of n_0 stress cycles, n_0 is the total number of stress cycles, n is the number of stress cycles equal to or exceeding S_r and κ is a shape parameter, which is the reciprocal of the more frequently used Weibull shape parameter. Guidance is provided elsewhere (Lotsberg and Almar-Naess, 1985) on selection of a suitable value for $1/\kappa$ which depends on wave climate, water depths, type of structure, dynamic amplification and location within the structure. (Further discussion on the Weibull distribution in testing is also to be found elsewhere (Kapur and Lamberson, 1977).) Accepting that this distribution is applicable—and in certain conditions it approximates to the normal distribution (Kapur & Lamberson, 1977)—one might look to it as the reference distribution for assessing fatigue test results. (It should be noted that a procedure is available to calculate an equivalent *constant* stress range to give the same damage (Latsberg and Almar-Naess, 1985).)

4.3.3 *Stress range and level*

Both stress range and mean stress affect fatigue life and various expressions have been produced to show their inter-relationship. Three of the best known are (Boresi et al, 1978).

Soderberg relation
$$\frac{S_a}{S_f}+\frac{S_m}{S_y}=1 \quad (4.19)$$

Gerber relation
$$\frac{S_a}{S_f}+\left(\frac{S_m}{S_u}\right)^2=1 \quad (4.20)$$

Goodman relation
$$\frac{S_a}{S_f}+\frac{S_m}{S_u}=1 \quad (4.21)$$

Where S_f is the fatigue strength for N cycles for zero mean strength, S_y is the yield stress, S_u is the ultimate strength, S_m is the mean stress and S_a is the stress amplitude. Figure 4.17 expresses these graphically with the dashed line indicating yield failure for the Gerber and Goodman relations. For a particular value of S_m the ordinate(s) give(s) S_a. The Gerber relation is said to give fairly good estimates for ductile materials while the Goodman relation is good for brittle materials (such as concrete) but conservative for ductile materials. The Soderberg relation is also conservative for most metals. The

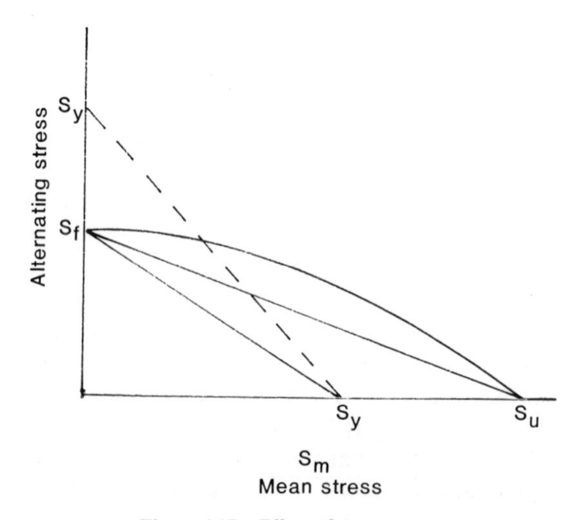

Figure 4.17 Effect of mean stress

validity of such expressions has been questioned (Berg, 1985) since the stress–strain relationships for monotonic and cyclic loading can be very different.

4.3.4 *Cumulative damage*

Where a load *spectrum* is applied to a specimen or structure the notion of cumulative damage is introduced. While this practice also is questioned in most texts, its practicality is generally welcomed, the most common application being known variously as Miner's or Palmgren's or Miner-Palmgren's rule or law. At a given load level the number of cycles to failure may be N_r and so the amount of damage (or 'use') D_r per cycle will be

$$D_r = 1/N_r \tag{4.22}$$

The total amount of damage D will then be 1 since $D = N_r D_r$. Consequently, if the actual number of load cycles at that particular load level is n_r then the amount of damage will be $n_r . (1/N_r)$ and if load is applied at various levels until failure occurs,

$$\sum n_r/N_r = 1 \tag{4.23}$$

For marine structures, fatigue is a high cycle phenomenon with most damage being caused by many cycles of small stress ranges (Vughts and Kinra, 1976). It is the response of the structure to short period, low height waves which is more important in this context than the effect of severe storms with return periods over one year. For low to moderate seas, of course, inertial forces will predominate and so accurate assessment of the inertia coefficient is very important. While this strictly may be more relevant to design than to material behaviour, it is worth bearing in mind in assessing the relevance of test results and their variability. Indeed it has been emphasised (Marshall and Luyties, 1982) that 'the overall uncertainty in fatigue life prediction is so large as to render the actual numbers almost meaningless'. The objective of designing against fatigue failure is to make sure the appropriate parts of structures are strong enough rather than to predict how long they will last.

4.3.5 *Fatigue life and fatigue testing*

There are certain implications in the emphasis on 'small' high frequency waves (a frequency of 0.15 represents over 4.7 million events in a year and would be equivalent to a deep water wave height of around 5 or 6 metres). For example, structures in very shallow water may be designed against a reduced height design wave of long return period. They thus become more sensitive to the smaller waves which are not reduced so much. This is the so-called 'shoaling paradox' (Marshall and Luyties, 1982). Then again, to obtain a practical duration to fatigue testing, a lower limit may be imposed on the stress level (along with probable clipping of the upper level to eliminate rare event

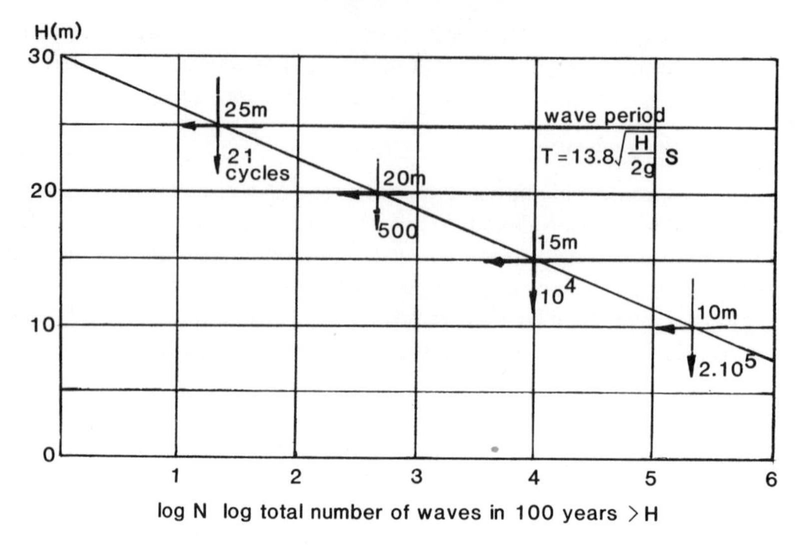

Figure 4.18 Distribution of wave heights

stresses). It has been shown (Lenschow and Seguin, 1982) that this can in effect exclude an unduly high proportion of stress time. A typical North Sea wave exceedance diagram is shown in Figure 4.18. Assuming an average period of, say, 15 seconds. the total number of waves in 10 years would be 21×10^6. If the load spectrum excluded all waves < 7.5 m high, only 10^6 would be 'needed' thereby eliminating 20×10^6—equivalent to 9.5 years 'life'. In other words, only 6 months of load life out of 10 years would be tested. Reducing the lower limit by half to 3.75 m would still only retain 10×10^6 waves, eliminating just

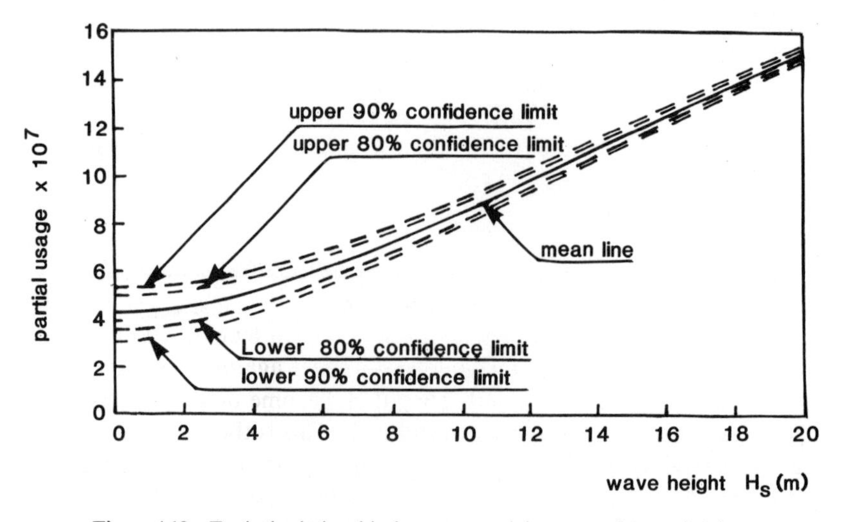

Figure 4.19 Typical relationship between partial usage and wave height

over half the load life. On the other hand, full load life simulation demands accelerated fatigue testing if inordinate test durations are to be avoided: there are sound reasons to question the validity of such an approach to *marine* concrete, as will be indicated later.

It also appears that the lower wave heights account for a disproportionately higher 'partial usage' (i.e. damage), than larger waves, as Figure 4.19 shows (Doucet et al, 1986). Although based on a particular formulation (due to Det norske Veritas) applied to an existing North Sea platform, it does suggest, for example, that a 2 m wave has as much as half the effect of a 10 m wave. Admittedly there is greater scatter at lower heights, but it is arguable whether or not that matters: it has been concluded that load scatter due to the random nature of the seaway is usually negligible for service times in the range of 20 years (Söding, 1974).

4.3.6 *Geographical variation*

We are concerned for the moment primarily with material behaviour but it cannot be isolated entirely from design processes and requirements. These after all establish the reference framework against which experimental data should be set if it is to have real value or if its immediate relevance is to be established. Indiscriminate application of test results can be misleading at the very least. Likewise, design techniques must have some connection with what is actually likely to happen. Without embarking on a lengthy discourse on the

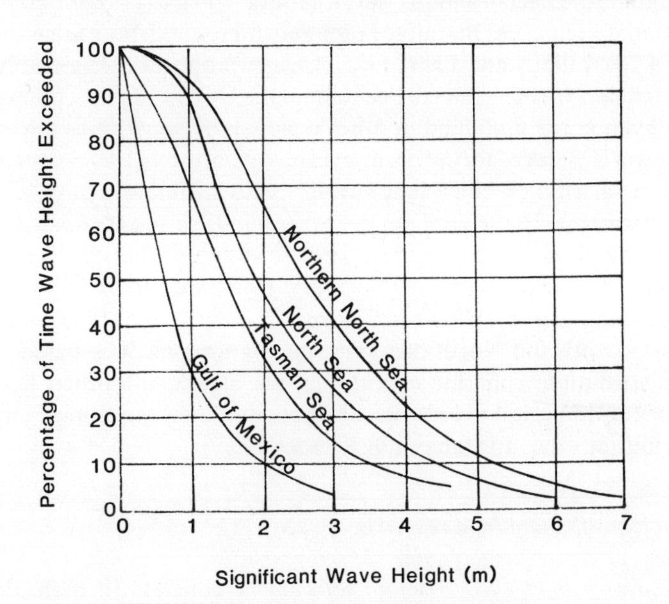

Figure 4.20 Fatigue premium for four offshore areas

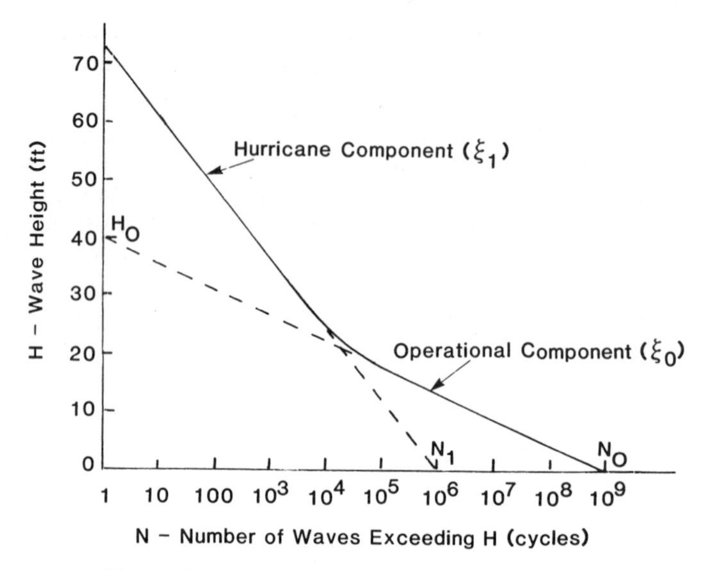

Figure 4.21 Two-component wave height distribution

interrelation between research and practice, however, let us simply state that the assumption and preconditions underlying both should be clearly understood before specific quantities or values are applied too freely or transferred from one to the other. For example, different aspects of behaviour may have varying significance according to the situation.

An attempt to represent the fatigue premium for four offshore areas is shown in Figure 4.20 (Kallaby and Price, 1976). Clearly fatigue is considerably more significant in the Northern North Sea than in the Gulf of Mexico, for example. However, even in the Gulf (and in other regions) the situation is not entirely straightforward. Severe storms occur every year in the North Sea: there is a degree of regularity or consistency which may facilitate a non-statistical approach to fatigue. In the Gulf, on the other hand, 'severe storms' are in fact hurricanes which are less regular, and so fatigue analysis there must accommodate hurricane statistics (Nolte and Hansford, 1976). A typical wave height exceedance curve is shown in Figure 4.21 (Geyer Stahl 1986) which may be contrasted with the North Sea curve of Figure 4.18. It is based on *two* Weibull distributions, one for operational waves and the other, hurricane waves. The return period is 100 years. Hence simple assumptions even about load distributions (eq. 4.18) may not be adequate.

4.3.7 *Factors affecting load response*

4.3.7.1 *Form of load cycle* Before looking at concrete in more detail a further general point may be worth considering. So far no comment has been made on the *form* of load cycle in fatigue loading. The test choice is often (probably *most* often) sinusoidal, between specific maximum and minimum loads or stresses (or strains or deformations). While for various structures this has the advantage of resembling idealised waves, according to linear theory at any rate, it carries with it the drawback from a material viewpoint of variable *rate* of loading—also sinusoidal but with a quarter cycle phase difference from the load. It is well known that rate of loading is an important influence on material response, a point discussed later, in which case it is clearly preferable to adopt triangular cycles with linear ramps between maximum and minimum, as has indeed been done in at least one case (Sparks and Menzies, 1973). However, there may also be some logic in considering the question in relation to energy which is perhaps a more significant parameter than load *per se*. Load can provide a measure of energy supplied of course, and its range and level then can be significant factors. Not all energy is necessarily absorbed as strain energy or even in material disruption through crack formation. High frequency cycling can generate appreciable quantities of heat—increasing temperature to 38°C in one case at 150 Hz (Raithby, 1979). It is illustrative, nonetheless, to assume that energy input manifests itself primarily as strain energy.

Suppose a cylindrical specimen to be loaded within its elastic range uniaxially and sinusoidally without stress reversal. The specimen has length L and cross-sectional area A; the mean load is P_m, amplitude P_a and $P_m \geqslant P_a$ so that $r \geqslant 1$ where $r = P_m/P_a$. The elastic modulus is E and stress σ is produced by load $P = P_m + P_a \sin\theta$ where θ represents a given point in time during the cycle (the period of which is then equivalent to $\theta = 2\pi$).

In fatigue testing (Berge 1985) the ratio of minimum to maximum stress is commonly expressed as $R = (r-1)/(r+1)$. Consequently, the range often adopted for R, $0 \leqslant R \leqslant 0.5$ is equivalent to $1 \leqslant r \leqslant 3$.

It is well known (Higdon et al., 1978) that the strain energy U of a material under load is given by

$$U = AL\left(\frac{\sigma^2}{2E}\right) \tag{4.24}$$

$$= \frac{L}{2AE}\cdot P^2$$

with

$$P = P_m + P_a \sin\theta$$
$$= P_a(r + \sin\theta)$$

Hence

$$U = \frac{L}{2AE}\cdot P_a^2(r + \sin\theta)^2 \tag{4.25}$$

and

$$\frac{\mathrm{d}U}{\mathrm{d}\theta} = \frac{L}{2AE}\cdot P_a^2(2r\cos\theta + \sin 2\theta) \tag{4.26}$$

For a given specimen and load amplitude then, if $(r + \sin\theta)^2$ $(2r\cos\theta + \sin 2\theta)$ are plotted against θ, we will have an indication of how (strain) energy and energy rate vary with time during a sinusoidal load cycle. This has been done on Figure 4.22 for three different values of r corresponding to the range given above. It is immediately clear that the curve for $r = 1$ (i.e. $P_{min} = 0$) is very different from the others and that the strain energy is effectively zero for a short spell i.e. there is a brief 'rest period' during the cycle. This is reflected in the plateau on the energy rate curve. As P increases the variation becomes more

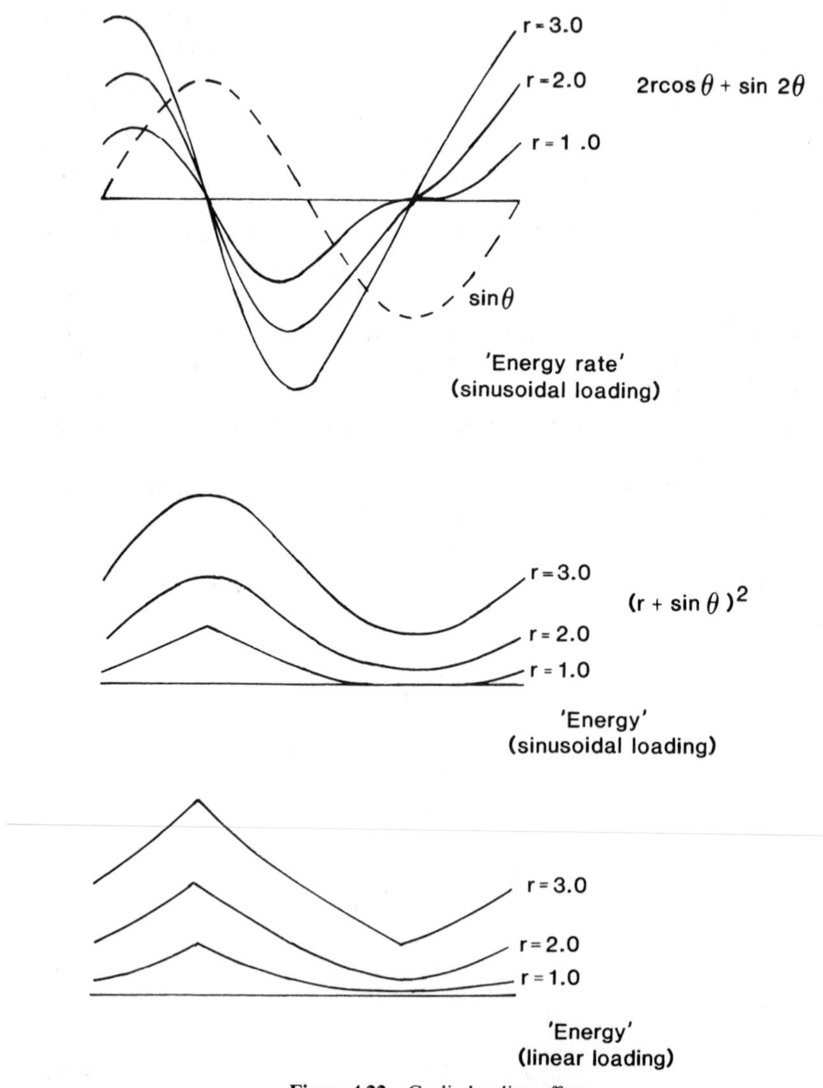

Figure 4.22 Cyclic loading effects

sinusoidal although the rate of change curve shows some skewness. For comparison, similar curves are shown for a 'triangular' cycle which also has a 'rest period' for $r = 1$, but displays a tendency towards linear energy variation as r increases—the rate curves are linear and so are not drawn. The load variation may be represented by the equation.

$$P = \begin{cases} P_m + \dfrac{2P_a}{\pi}\cdot\theta & 0 \leqslant \theta \leqslant \dfrac{\pi}{2} \\ P_m + P_a\left[1 - \dfrac{2}{\pi}\left(\theta - \dfrac{\pi}{2}\right)\right] & \dfrac{\pi}{2} \leqslant \theta \leqslant \dfrac{3\pi}{2} \\ P_m - P_a\left[1 - \dfrac{2}{\pi}\left(\theta - \dfrac{3\pi}{2}\right)\right] & \dfrac{3\pi}{2} \leqslant \theta \leqslant 2\pi \end{cases} \tag{4.27}$$

This implies that the **form** of load cycle adopted may affect results on a scale governed by the range and level of loading. By quite how much is indeterminate at this stage but it does provide a further indicator of the need for caution in comparing results from different sources, especially when contemplating their application to structures. Compatability is not always easy to ensure. While regular (or apparently regular) loading has some advantage in comparing the effect of varying specific parameters or factors, randomised loading (as occurs in nature) might go some way towards obviating problems such as have just been suggested, despite the scepticism which has been expressed about some of its aspects (Swanson, 1976), especially in relation to pseudo-randomisation.

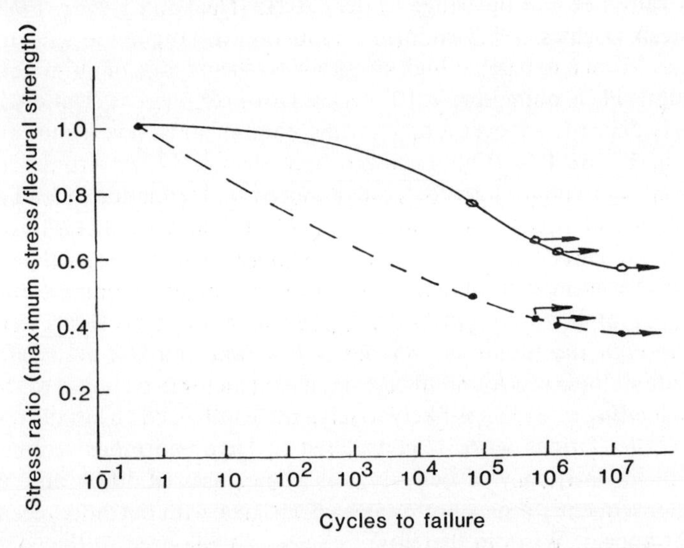

Figure 4.23 Typical endurance curves for flexural loading at 20 Hz. Each point is the log mean of 5 or more tests. →, includes some runouts; —, conventional endurance curve; –·–·–·, true endurance curve

4.3.7.2 *Rate of loading* Material response to load *rate* must also be considered. Rate-sensitivity of many materials, including concrete, at high rates of stress or strain is well-established (Mainstone, 1975). Its relevance to impact and collision is obvious and we shall return to the subject later in that connection, but for the moment it might be expected to have some bearing on fatigue—high rates of load cycling signify high rates of loading and unloading. One example cited (Raithby, 1979) is that of ultimate load being reached in 0.025s at the equivalent of 20 Hz test frequency, yielding a strength some 50% greater than under standard static stress conditions where failure would take about 2 minutes. Consequently one might well question the conventional practice of presenting fatigue results as a proportion of normal static strength. The two values are not usually rate compatible: a static loading rate equivalent to the fatigue loading rate would produce a higher apparent strength so that the 'correct' fatigue strength ratios would be lower. In other words, the conventional presentation gives an over-estimate. Figure 4.23 (Raithby, 1979) shows a 'corrected' fatigue curve. However, the convention is convenient since standard strengths are easier to obtain than high load rate strengths, so the practice is unlikely to change.

4.3.7.3 *Frequency* Remarks in the preceding paragraph would lead one to expect concrete fatigue strength to be frequency sensitive. However, the evidence is somewhat anomalous. For example, it is considered that variations between about 1 and 15 Hz have little effect on fatigue strength, provided the maximum stress is less than 75% of the static strength (Hawkins, 1976). Earlier quoted values extend the range to over 70 Hz (Neal and Kesler, 1968) while recent tests (Siemes, 1982) confirm a reduction in fatigue life with reducing frequency (from 6 to 0.06) at high stress levels (70 and 80% of ultimate). In this case fatigue life is quite low: $< 10^5$ cycles. However, there is clear evidence of frequency dependence over a much wider band of stress levels and ranges, as shown in Figure 4.24 (Van Leeuwen and Siemes, 1979). Around ten-fold reduction in fatigue strength for a one-hundred-fold reduction in load rate has also been found (Sparkes and Menzies, 1973). While it is difficult to compare this with frequency-based work, load rate might be a more logical parameter to employ, utilising a triangular wave form and thereby gaining some of the advantages implied by Figure 4.22. Again maximum stress levels were quite high although the minimum was fairly low (one third of static strength). Aggregate stiffness was found to be a significant factor in response to variation in static loading rates so it is likely to have some influence on fatigue response.

The earlier fatigue work is considered to have enormous scatter and a degree of inconsistency of behaviour in some cases (Raithby and Whiffin, 1968). Consequently it may not be at such variance with the more recent work as might appear. Wisdom therefore indicates paying heed to the conclusion that high cycle fatigue testing leads to overestimates of fatigue life. However, much less data are available for low cycle rates resembling typical wave frequencies. Such frequencies are also sufficiently slow to allow other processes

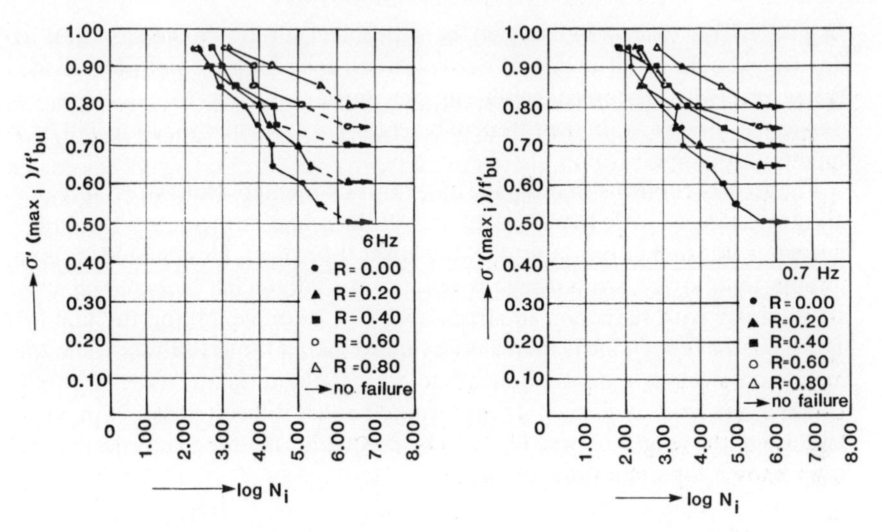

Figure 4.24 Wöhler curves for the same concrete at different frequencies

to proceed such as crack blocking (which may be beneficial), water pumping or wedging in cracks, and corrosion (detrimental). High cycle tests may provide indicators of *relative* performance but their direct application to marine structures warrants a fair degree of caution. Unfortunately low cycle work is very time consuming and introduces side effects—10^6 cycles at a frequency of 0.15 would last 77 days, in which time concrete properties could change appreciably unless specimens were quite mature at the outset; creep will also be readily apparent on such a time scale.

4.3.7.4 *Stress level* A note on stress levels may not be out of place, since it is clear that, the lower the stress level (i.e. ratio of applied stress to static strength) the greater the endurance of the material—as well as the test duration, of course. Generalisation can bristle with pitfalls, to mix metaphors, but a simple example might illustrate the point—individual structures have to be checked for their particular circumstances.

Assuming that the stress we are concerned with is related to the total horizontal force on a vertical cyclinder, then obviously it will vary with wave height. Ignoring dead and working loads and confining ourselves to wave loads, suppose a 20 m design wave, steepness 1 in 14, acting on a 2 m diameter cylinder in deep water so that we can use linear wave theory and Morison's equation (eq. (3.9)). The ratio of inertia to drag coefficient is assumed constant at 3. Then it can be shown that the maximum force from a 2m wave of the same steepness is 8.5% of the maximum design wave force and the force from a 4 m wave is 17% (the relationship between maximum force and wave height is *not* linear and the phase angle depends on the relative size of drag and inertia forces). In other words, and in this very specific case, typical maximum fatigue stresses are a small proportion of design stresses. From the data of Figure 4.24

fatigue in this case is unlikely to be a problem although the situation as described could result in stress reversals if the cylinder were not prestressed. Stress reversals (discussed below) in fact appear unlikely to cause greater damage, although again the extent of reversal in a particular location will be a function of design.

Figure 4.24 confirms that fatigue life reduces with increase in stress level for all stress ranges with the slope of the Wöhler line reducing as the range decreases (the higher the value of R the lower the range). This change in slope implies that, as the stress range is reduced, so fatigue life is increased very significantly with reduction in stress level. For example, taking the line for $R = 0.80$, the fatigue life is increased by more than a thousand-fold when the maximum stress is reduced from 95 to 80% of the ultimate static strength. Lower stress levels with such a range would be very difficult to test in practice because of the length of time likely to be needed for failure. Such effects have been known for some time, of course.

4.3.7.5 *Moisture condition* Due to the manner in which results are usually presented (with load as a fraction of static strength) there is little variation in the fatigue limit with composition, age, curing, and aggregate type for normal concretes (Hawkins, 1976). However, moisture condition does or can make an appreciable difference and dry concrete has a higher fatigue strength or limit than wet (Hawkins, 1976) (Van Leeuwen and Siemes, 1979). Acoustic emission tests suggest that dry concrete fails by fracturing throughout the material, while wet or submerged concrete failure is initiated near the surface due to the wedge action of water in the micro cracks (Nishiyama et al, 1987). In compression the fatigue limit could be at least ten-fold greater for dry concrete.

4.3.8 *Goodman diagram*

The effects of stress range and stress level can be combined in the so-called (modified) Goodman diagram as in Figure 4.25 (Neal and Kesler, 1968). Each diagram must be constructed for a specific fatigue life, in this case 10×10^6 cycles in flexure. The fatigue strength was 55% of ultimate giving point A. It was 65% when the minimum was 20% giving B. When the loading was completely reversed at a minimum (compressive) strength of 55% of the ultimate (compressive), the tensile fatigue stress was 55% of ultimate, giving C. Such points established from experimental data thereby fix the line. Figure 4.26 (Cornelissen, 1984) gives data for several lines for all tension, all compression and tension-compression (i.e. stress reversal) loading. 'Concentric tension' refers to direct uniaxial loading while 'flexural tension' as might be expected, results from flexural tests. In general flexural testing gives higher strengths than direct tensile testing while the splitting test tensile strength tends to be even higher. Similarly flexural testing gives longer fatigue lives than concentric testing, attributable to the presence of stress gradients

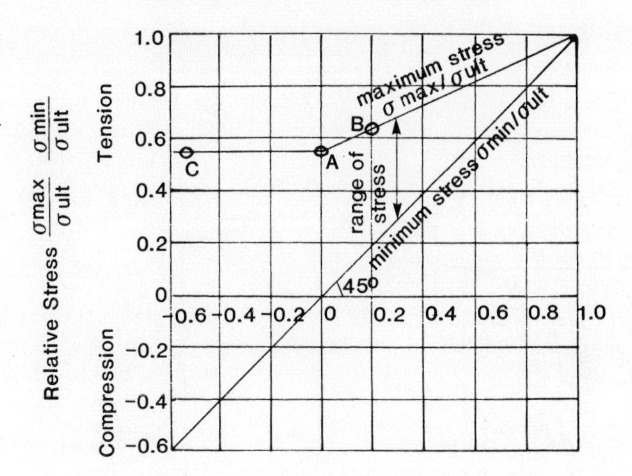

Figure 4.25 Plotting test results on a modified Goodman diagram

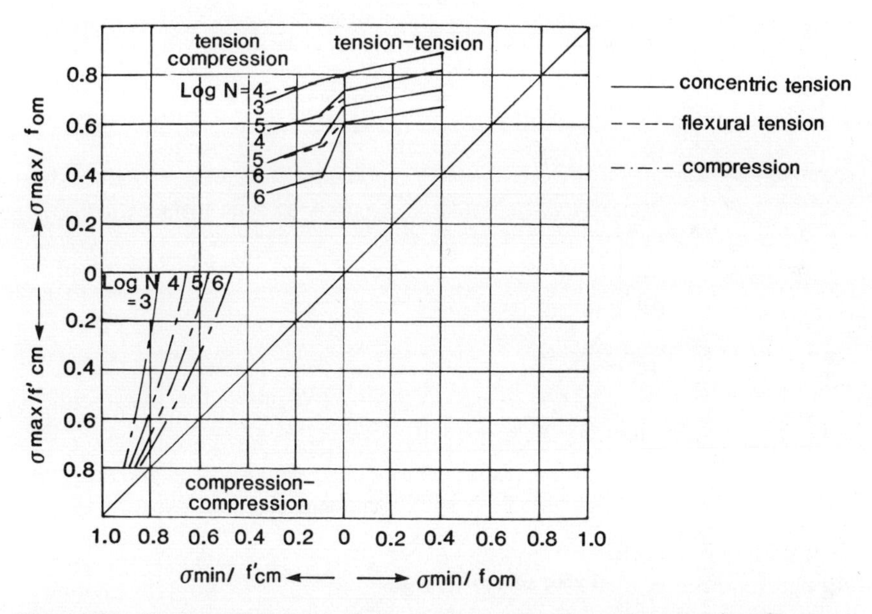

Figure 4.26 Goodman diagram for tension, tension-compression and compression of concrete

which are known to increase fatigue life. An initial stress gradient produces a strain gradient resulting in stress relaxation in the fibres with maximum strain (as in creep). This in turn produces a redistribution of stress so that fibres further in from the outermost participate increasingly in carrying the load (Cornelissen, 1984). As load cycles increase the maximum stress in bending shifts towards the neutral axis.

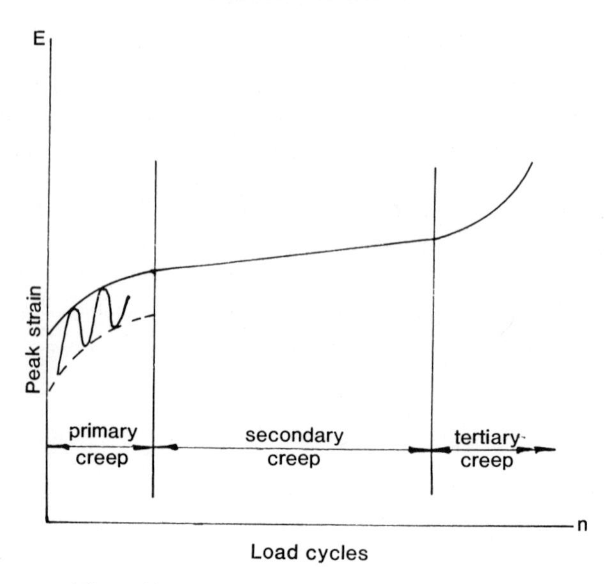

Figure 4.27 Creep of concrete under cyclic loading

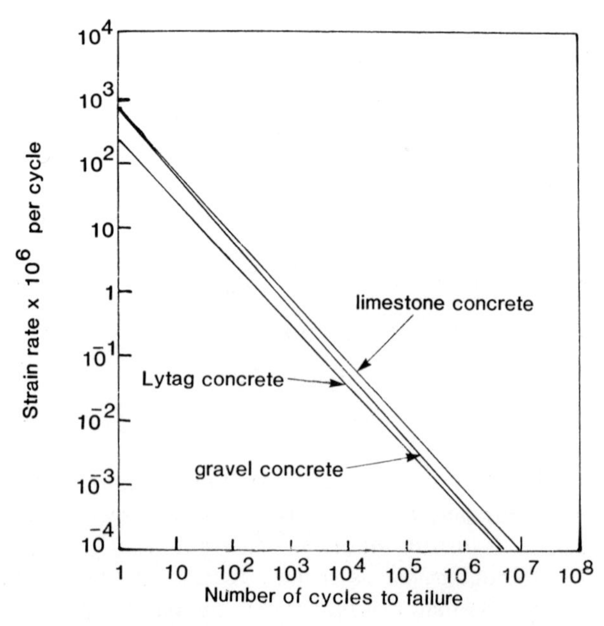

Figure 4.28 Relationship between secondary creep rate and number of cycles

4.3.9 *Creep*

Creep under constant load is a familiar phenomenon in concrete but it also manifests itself under compressive and tensile load cycling. The typical form is shown in Figure 4.27 and is conceived as occurring in three stages: primary, secondary and tertiary creep. The secondary creep *rate* in compression (i.e. the slope of that part of the curve) has been found to be independent of the rate of loading, with some distinction for different types of aggregate, and gives a straight line when plotted against fatigue life as shown in Figure 4.28 (Sparks and Menzies, 1973). Similar behaviour is apparent in tension, Figure 4.29, and there is said to be no difference with stress reversal. It appears that secondary creep rate is a more accurate predictor of fatigue life (i.e. number of cycles to failure) than the Wöhler or *S–N* diagram (Cornelissen, 1984): it may be more directly related to accumulating damage. Furthermore it seems likely that, for equal secondary creep rate, the failure *time* is independent of frequency (the number of cycles to failure will then be different for different cycle rates).

4.3.10 *Rest periods*

Rest periods have a beneficial effect on fatigue strength although how much depends on how long—over five minutes produces no further improvement

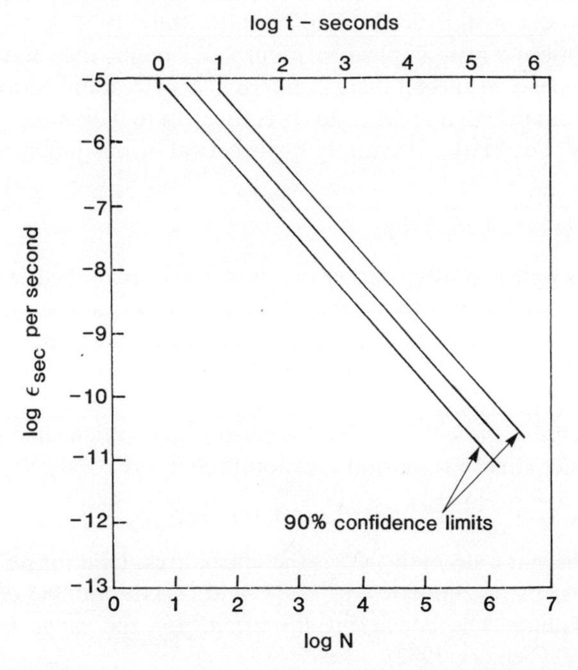

Figure 4.29 Relation between secondary cyclic creep velocity and number of cycles to failure

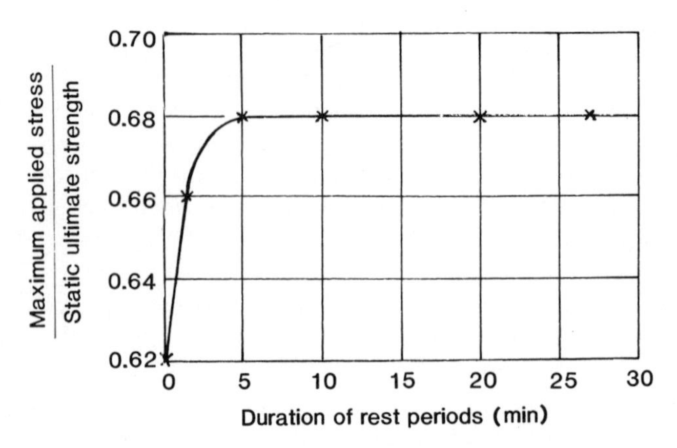

Figure 4.30 Effect of rest periods on fatigue strength at ten million cycles

according to Figure 4.30 (Neal and Kesler, 1968; from flexural tests). The maximum improvement in fatigue strength is around 10% which is not of major significance. Frequency of rest appears to be more important than duration (Raithby, 1979). It is not clear whether or not there is a *minimum* rest period. There is some evidence to suggest that rests of less than two seconds reduce fatigue life, in fact, but this was based on low cycle fatigue and the differences were not statistically significant (Raithby, 1979) so the question is open. Within-cycle rests implied in Figure 4.22 might then have a varying frequency-dependent effect if there is indeed a 'demarcation' period. For $r = 1$ the rest is about 0.25 of a cycle: at 20 Hz equivalent to 0.0125 seconds, at 2 Hz, 1.25 seconds. The matter is entirely hypothetical at this juncture, however.

4.3.11 *Empirical relationships*

Reservations were expressed earlier in this chapter about the use of empirical relationships. Undeniably they can be useful, however, and several have been produced or derived in connection with fatigue, some of which are reproduced here for reference.

4.3.11.1 *Tension and compression* A generalised straight line relationship for the Wöhler curve for tension and compression is

$$\sigma_c^{max}/f_c = 1 - \beta(1 - R)\log_{10}N \tag{4.28}$$

where f_c is the static strength, σ_c^{max} is the upper stress limit for pulsating load, σ_c^{min} is the lower stress limit, $R = \sigma_c^{min}/\sigma_c^{max}$ and N is the number of load pulses to fatigue failure. The coefficient $\beta = 0.0685$ for the range $0.3 \leqslant (\sigma_c^{max} + \sigma_c^{min})/2f_c \leqslant 1.0$ (Tepfers, 1982).

The load level expressed as a percentage and corrected for rate effects has been related to N:

$$S = 95.7 - 4.66 \log N \text{ (gravel concrete)} \tag{4.29}$$

and

$$S = 102.6 - 9.60 \log N \text{ (Lytag concrete)}$$

where

$$S = \sigma_c^{\max}/f_c = 0.957[1 - 0.0685(1 - 0.29) \log N] \tag{4.30}$$

Since this has been derived from tests where the load **rate** was maintained constant, it gives some measure of the possible correction to be applied to eq. (4.28). However, the apparent value of $R = 0.29$ is rather misleading and should also be corrected. In the tests, load was cycled between maximum loads of 70% to 90% of the strength obtained at a logarithmic mean rate of 5 MN/m^2 s while the minimum load was one third of the static strength at the standard rate of loading of 0.25 MN/m^2s. Static strength F as a percentage of cube strength was related to load rate P_t (MN/m^2s) by

$$F = 67.89 + 3.96 \log P_t \text{ (gravel concrete)} \tag{4.31}$$

Hence the standard rate static strength is 92.7% of the logarithmic mean rate and the ranges quoted correspond to R values of 0.43 to 0.34. The implication then is that β in eq. (4.28) is load rate dependent and eq. (4.30) could be re-written as

$$\begin{aligned} \sigma_c^{\max}/f_c &= 0.957 - 0.0758(1 - R) \log_{10} N \\ &= 0.957[1 - 0.0792(1 - R)] \log_{10} N \end{aligned} \tag{4.32}$$

taking $R = (0.43 + 0.34)/2.$

However, this is intended rather to illustrate the complexities of the problem than to make a specific recommendation.

4.3.11.2 *Time dependence* The normal two-dimensional strength/fatigue life/load level relationship ($\sigma - N - R$) has been expanded into three dimensions by the inclusion of time via the period of cycling T (Hsu, 1981). The resulting σ-N-T-R relationship is represented by two equations, for high cycle and for low cycle fatigue:

$$\text{high cycle } \frac{\sigma_c^{\max}}{f_c} = 1 - 0.0662(1 - 0.556R) \log N - 0.0294 \log T \tag{4.33}$$

$$\text{low cycle } \frac{\sigma_c^{\max}}{f_c} = 1.20 - 0.20R - 0.133(1 - 0.779R) \log N - 0.053(1 - 0.445R) \log T \tag{4.34}$$

The boundary between high and low cycle fatigue depends upon frequency of loading and may be defined by

$$\log N = 3 - 0.353 \log T \tag{4.35}$$

although preference is given to a relationship of similar form to eqs (4.33) and (4.34):

$$\frac{\sigma_c^{\max}}{f_c} = 1 + (0.0229 + 0.0490R)\log N \qquad (4.36)$$

Validity is claimed for compressive and flexural loading (but **not** load reversal) over a frequency range of 0 to 150 Hz and N up to 20×10^6 for normal-weight concretes with cylinder strength $< 55\ \text{N/mm}^2$ and for 'high-strength' light-weight concrete of around $41\ \text{N/mm}^2$.

4.3.11.3 *Secondary creep or strain rate* Less complication and more consistency should result from the secondary creep or strain rate approach with its apparently reduced scatter. Defining the secondary strain rate as $\partial\varepsilon/\partial n$ per cycle, the following relationships were obtained for concrete in compression (Sparkes, 1982):

$$N = 2.19 \times 10^{-3}\left(\frac{\partial\varepsilon}{\partial n}\right)^{-0.94} \quad \text{for gravel concrete}$$

and (4.37)

$$N = 1.64 \times 10^{-4}\left(\frac{\partial\varepsilon}{\partial n}\right)^{-1.06} \quad \text{for Lytag concrete}$$

While clearly dependent on the type of concrete, the equations are independent of rate of loading. Combining them with eqs (4.29) results in the following relationships between true load level and secondary strain rate:

$$S = 108.1 + 4.38\log(\partial\varepsilon/\partial n) \text{ for gravel concrete}$$

(4.38)

and

$$S = 138.9 + 10.18\log(\partial\varepsilon/\partial n) \text{ for Lytag concrete}$$

A relationship has been produced for tension and tension-compression loading, this time taking the secondary creep rate with respect to time i.e. as a strain velocity, $\dot{\varepsilon}_{\text{sec}}$ (Cornelissen and Reinhardt, 1984):

$$\log N = 3.25 - 0.89\log\dot{\varepsilon}_{\text{sec}} \qquad (4.39)$$

for a frequency of 6 Hz. A more general relationship in terms of t, the time to failure (seconds) is

$$\log t = -4.02 - 0.89\log\dot{\varepsilon}_{\text{sec}} \qquad (4.40)$$

These equations are independent of stress and apply to wet and dry concrete.

Equation (4.40) is remarkably close to a similar equation derived for sustained tensile creep tests and which is not affected by temperature, cement type or concrete quality within the range of the experiments:

$$\log t = -4.05 - 0.91\log\dot{\varepsilon}_{\text{sec}} \qquad (4.41)$$

It is also noticeable that the slope of the line log N versus log $\dot{\varepsilon}_{sec}$ of eq. (4.39) for tensile fatigue is close to the slope of the equivalent line of eq. (4.37) for compressive fatigue (in this case $\partial\varepsilon/\partial t \equiv \partial\varepsilon/\partial n$) and to that for monotonic tensile creep, at least at high load levels.

4.3.12 *Random loading*

So far the discussion has been concerned primarily with regular cyclic loading which is invaluable in studying the phenomenon of fatigue. However, in practice many, if not most, fatigue-susceptible structures are subjected to irregular or random loading. This is particularly true of marine structures and so the applicability (or otherwise) of a rule such as Miner's (eqs (4.22) and (4.23)) is of considerable practical importance in making fatigue assessments. Unfortunately, although not altogether unexpectedly, the problem then becomes rather more complex.

4.3.12.1 *Miner's number* It is convenient and, perhaps, more apposite to restate Miner's rule as a summation of damage due to a succession of individual cycles:

$$M_s = \sum_{i=1}^{c} \frac{1}{N_i} \tag{4.42}$$

where M_s is the Miner's sum after c cycles and N_i is an individual cycle corresponding to $i = 1,2,3\ldots$c. At failure the Miner's sum is known as the Miner's number which we can denote simply as M so that, if c_F cycles result in failure,

$$M = \sum_{i=1}^{c_F} \frac{1}{N_i} \tag{4.42a}$$

In the form of these two equations, Miner's rule can be more easily visualised as being applicable to any pattern of loading, from uniform cyclic to completely random.

4.3.12.2 *Variability of Miner's number* Constant-amplitude tests on concrete have confirmed that N_i is a stochastic variable in tests to failure, and has a log-normal distribution (i.e. the distribution of log N_i is normal). It has been shown, in consequence, that the Miner's number M also has a log-normal distribution (van Leeuwen and Siemes, 1979). The Miner's sum from a constant-amplitude test is then given by

$$M = n\,10^{-m(\log N_i)} \tag{4.43}$$

where n is the number of constant amplitude cycles and $m(\log N_i)$ is the mean value of log N_i; $10^{m(\log N_i)}$ is then the value of N_i corresponding to the mean log. From this it is shown that the median value of the Miner number med(M) is

$$\text{med(M)} = 10^{1.15 s^2(\log N_i)} \tag{4.44}$$

where $s(\log N_i)$ is the standard deviation of $\log N_i$, and $s(\log M) = s(\log N_i)$ by argument.

Equations (4.43) and (4.44) are generally valid for log-normal distributions since they are theoretically based. Empirical relationships have been formulated as part of the same work (Siemes, 1982). Since they are based on specific circumstances and linear regression of experimental results, the general form only will be quoted here:

$$m(\log N_i) = \frac{a}{\sqrt{1 - R_i}} \cdot \frac{\log S'_{\max i}}{m(f'_{bu})} + b \tag{4.45}$$

$$s(\log N_i) = \frac{C \cdot \dfrac{s(f'_{bu})}{m(f'_{bu})}}{\sqrt{1 - R_i} \cdot \dfrac{S'_{\max i}}{m(f'_{bu})}} \tag{4.46}$$

$m(\ )$ and $s(\ )$ signify means and standard deviations respectively; $S'_{\max i}$ is the maximum compressive stress and $S'_{\min i}$ the minimum; $R_i = S'_{\min i}/S'_{\max i}$; f'_{bu} is the static cylinder compressive strength. [a, b and c are constants which, for the particular conditions, are -13.158, 1.478 and 5.714. Extrapolation of the results outside the range of investigation could lead to erroneous conclusions (CUR–VB, 1984).]

The primary reason for quoting eqs (4.45) and (4.46) here is that they clearly indicate that the variability of N_i, and hence of the Miner number, is closely linked to the variability of the material (as well as to the stress-ratio). In fact it is concluded that material variability is the primary cause of Miner's number variability.

It is argued that, since $S(\log N_i)$ can have different values for different stress levels, eq. (4.42a) cannot strictly be applied to random loading when $M = 1$ (which is the value one would normally expect). A form equivalent to eq. (4.43) is therefore developed:

$$M_F = \sum_{i=1}^{c_F} \frac{1}{\text{med}(N_i)} = \sum_{i=1}^{c_F} 10^{-m(\log N_i)} \tag{4.47}$$

Now, it is already clear that the distribution of the Miner's number is governed by that of N_i. If then the assumption is incorporated that Miner's rule is not strictly applicable to concrete in compression, an expression may be proposed:

$$M_F = M_N \cdot M_M \tag{4.48}$$

where M_N is a factor indicating the influence of N_i, and M_M is a so-called model number, giving the **actual** Miner sum at failure. Each term in the equation is initially presumed to be a stochastic variable but in the event, from the Dutch experimental evidence, M_M proves to be either deterministic or has very little scatter. Hence variation in M_F is primarily due to variation in material quality (represented by M_N) and so Miner's rule *can* be applied to concrete under compressive random loading.

The anticipation of fatigue failure depends on anticipating M_F which in turn depends on M_N, given that a value can be assignd to M_M: ideally it should be 1 but in reality it is likely to be different. A relationship has been established to calculate M_N based on the assumption that the load be composed of n different stress combinations $S_1, S_2, \ldots S_k \ldots S_n$, each occurring in the proportions $p_1, p_2, \ldots p_k \ldots p_n$. Since $\log N_i$ has a normal distribution it may be represented by the sum of its mean and its standard deviation multiplied by an auxiliary quantity u having a standard normal distribution (whose mean is then zero and its standard deviation 1). It can then be shown that

$$M_N = \frac{\sum_{k=1}^{n} p_k \cdot 10^{-m(\log N_k)}}{\sum_{k=1}^{n} p_k \cdot 10^{-m(\log N_k) - u \cdot s(\log N_k)}} \tag{4.49}$$

By substituting different u values, different values of M_N are obtained giving its distribution according to the distribution of u. (Siemes, 1982) (CUR–VB, 1984).

As to the application of Miner's rule to tensile and tensile-compressive loading, this has been verified for regular and program loading (i.e. regular stress blocks at different levels), especially if partial damage is deduced from secondary strain rate (Cornelissen, 1984). $M = 1$ appears to be a safe criterion in general. However, it is not clear that an equation such as eq. (4.39) is valid over a wide range of frequencies.

4.3.12.3 *Empirical relationship* Modification of the Miner number by a factor, as in eq. (4.48), has also been suggested elsewhere (Holmen, 1982) where an empirical relationship has been established. The argument is developed differently and the modifying factor w is related to the ratio S_{min}/S_c where S_c is the characteristic stress level taken as being equal to the mean stress plus the rms value of the fluctuations above the mean. Figure 4.31 gives the empirical relationship between $\log w$ and $\log(\bar{S}_{min}/S_c)$ where $\bar{S}_{min}$ is the mean minimum stress. The statistical distribution of stress is thus taken into account in a rather different way from that just discussed. A warning is given that further experimental justification of Figure 4.31 is needed to establish its applicability but it is also particularly useful in indicating how increased stress ratio reduces the Miner number i.e. reduces the fatigue life, as was discussed earlier for constant-amplitude loading. It is also the case that variable-amplitude tests at a relatively high stress level are more damaging than at a relatively low stress

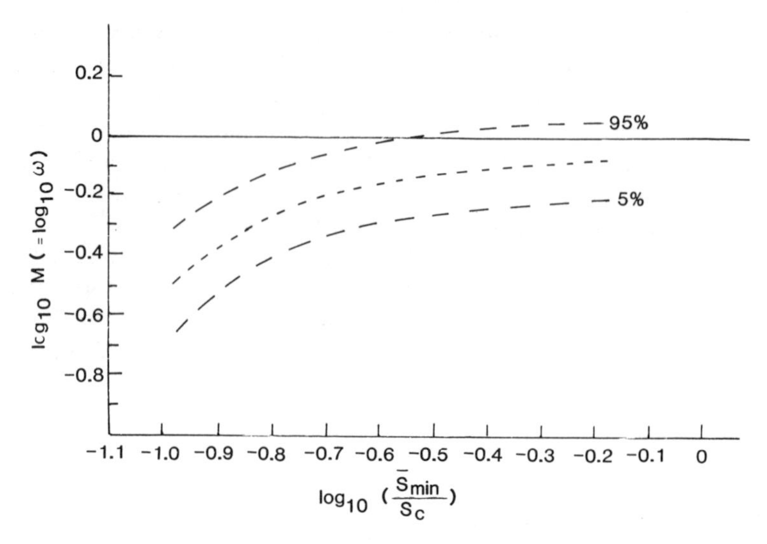

Figure 4.31 Empirical relationship between loading parameters and the factor w

level, as perhaps one might expect from constant-amplitude testing (van Leeuwen and Siemes, 1979).

4.3.12.4 *Load sequence* Sequencing of stress amplitudes has some influence on Miner's number. It is worth noting at this juncture that various forms of random loading can be adopted in tests and they may not all give the same results. Good sense then dictates that it is important when comparing data also to compare load patterns. Apart from completely random loading it is possible to have, for example, random variation above a fixed minimum, below a fixed maximum or about a fixed mean or, indeed, about a succession of fixed means. The Dutch work referred to indicates that, if the stress amplitude is gradually increased from zero to a maximum then back down to a minimum, it is more damaging with stationary minimum and less damaging with stationary maximum stresses than is the corresponding random sequence. Sequencing appears to produce Miner's numbers closer to the ideal.

4.3.13 *Biaxial loading*

Some work has been done on biaxial fatigue loading of concrete and it has been concluded, for example, that biaxial fatigue strength is greater than uniaxial (Traina and Jeragh, 1982). This was for very low cycle fatigue tests, however. Fatigue strength appears to be governed both by the absolute stress level and the relative stress levels: a monotonic or confining stress in one direction increases the fatigue strength in the orthogonal direction, the increase being greater the greater the confining stress. When the confining stress also was cyclic, there was still an increase but that appeared to be less when it equalled the 'loading' stress than when it was only half. Other work (Buyukozturk and Zisman, 1982) indicates also how increase in confining stress increases stiffness under cyclic loading. Confinement inhibits micro-cracking and it is suggested that degradation is not attributable solely to what occurs at the mortar-aggregate interface. Stress at failure appears to be increased. Again the number of cycles was very low.

The diminishing benefit of moving towards equalisation of stresses referred to above is confirmed by tests covering low to high cycle fatigue (Nelson et al 1987). Figure 4.32 gives typical *S–N* curves for biaxial stress ratios of 0.5 and 1.0. S_m is the maximum stress as a percentage of the static uniaxial compressive

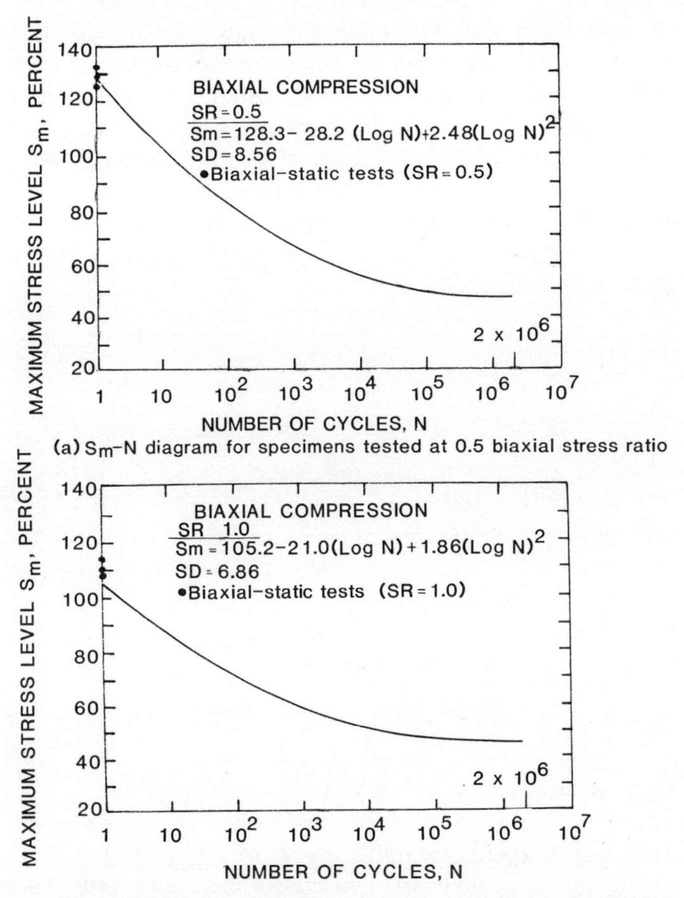

Figure 4.32 (a) S_m–N diagram for specimens tested at 0.5 biaxial stress ratio (b) S_m–N diagram for specimens tested at 1.0 biaxial stress ratio

strength and the stress ratio S_{min}/S_{max} was kept constant at 0.10 so the stress range was quite high relatively. A failure envelope was derived from such data (Figure 4.33) from which it is clear that stress level is a significant factor governing performance: above $S_m \approx 80\%$ a small lateral load has a very striking influence increasing (low cycle) fatigue life significantly. At lower stress levels, however, (high cycle) fatigue life is little affected (at $S_m \approx 50\%$ for example). The concrete tested was high strength (63 N/mm^2 at 56 days), low water/cement ratio (0.30) so one must be wary of drawing general conclusions. The authors suggest that high strength concrete has a lower fatigue limit than normal strength concrete and its endurance limit decreases with increasing biaxial stress ratio: at 2 million cycles it is 52, 51, 49 and 47% for ratios of 0.0, 0.2, 0.5 and 1.0 respectively. This should be compared with the normally accepted value of 55% at 10 million cycles mentioned earlier.

It does seem likely that low cycle and high cycle fatigue are affected differently by biaxial loading and the relative size of the lateral (orthogonal) stress affects low cycle (or high stress level) response more than it does high cycle (or low stress level). To what extent frequency is influential is indeterminate: presumably relative frequencies and whether or not they are synchronous, harmonics, or whatever, will have an influence. However, the most likely practical situation may simply (!) be uniaxial fatigue with lateral confinement at a given time and that is potentially easier to assess at the present state of knowledge.

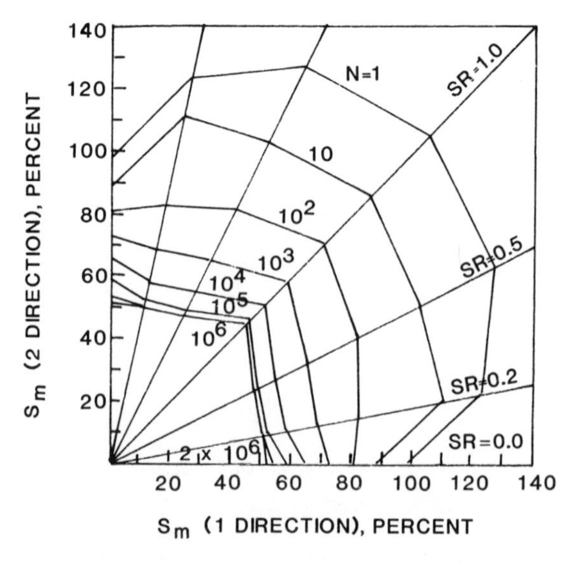

Figure 4.33 Failure envelope for high strength concrete subjected to biaxial cyclic compression

4.3.14 *Fatigue and stress/strain*

Stress-strain is an important aspect of material behaviour and constancy may often be assumed by the designer. However, it has been well-established for a long time now that, under fatigue-type loading, the shape of the stress-strain curve changes from being convex, through linearity to concavity as the load cycles increase, exemplified in Figure 4.34 (Raithby and Whiffin, 1968). The

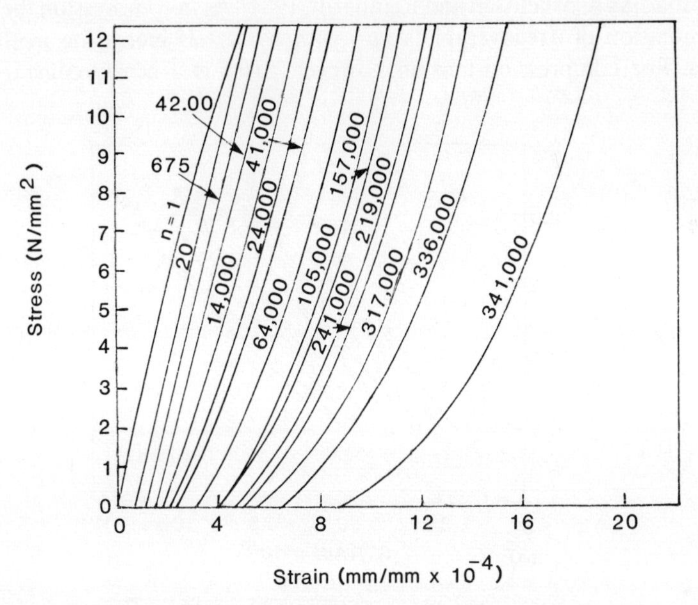

Figure 4.34 Effect of number of load cycles (n) on stress/strain curves for cylinders in compression

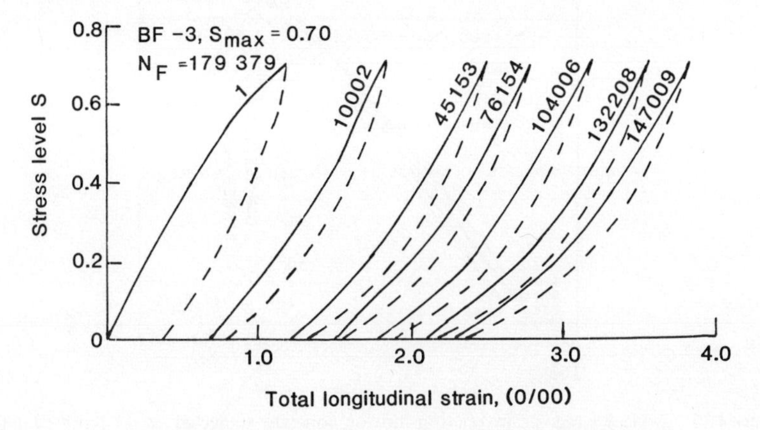

Figure 4.35 Effect of repeated load on concrete strain

pattern is rather more complicated in fact and Figure 4.35 (Holmen, 1982) shows the effect of *un*loading as well as loading: a series of loops is formed and it can be seen that the area within them changes as loading progresses. Each loop, of course, provides a measure of the energy dissipated within a particular load cycle. It is noticeable that the strain change per cycle increases considerably as failure approaches, yielding the pattern typified in Figure 4.27. Stress-strain relationships are almost linear for most of the fatigue life in tensile cyclic loading (Cornelissen and Reinhardt, 1984). As in compression there is an accumulation of irrecoverable strain with hysteresis increasing in the final stages. For compression-tension however, there is a nearly constant non-

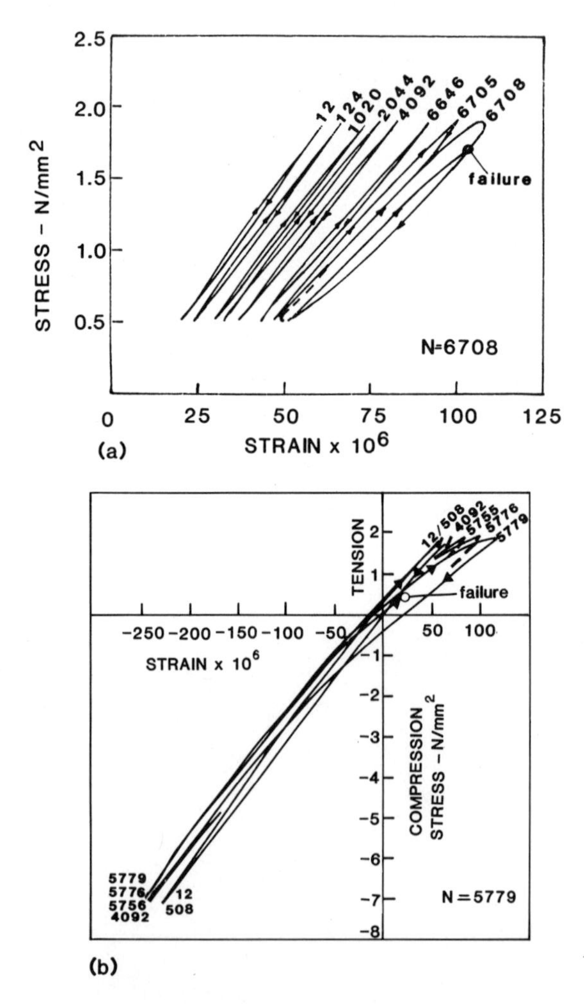

Figure 4.36 Typical stress-strain relationship for concrete subjected to (a) repeated tensile stresses and (b) to stress reversals

recoverable strain with an increasing tensile strain as in Figure 4.36 (Cornelissen and Reinhardt, 1984).

Plotting the change in secant modulus demonstrated by Figure 4.35 produces curves such as those of Figure 4.37 (Holmen, 1982). Here, the duration of load is represented by the number of cycles endured to a given point in time as a proportion of the cycles in due course required to failure. The effect of stress level is quite perceptible. If total strain also is plotted

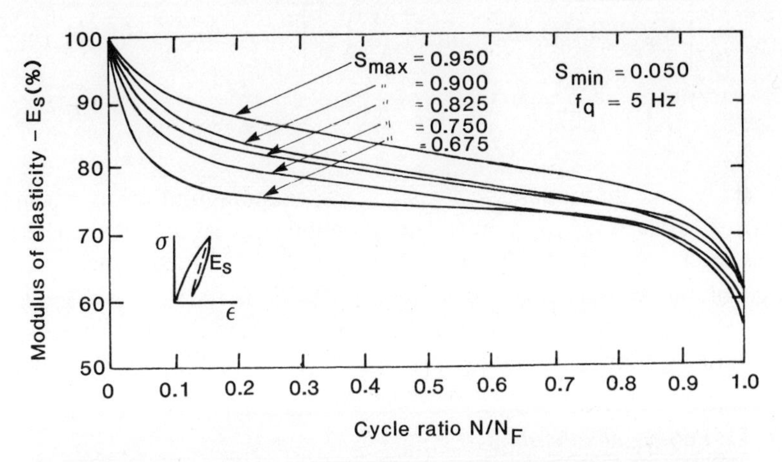

Figure 4.37 Percentage reduction of the secant modulus with the cycle ratio

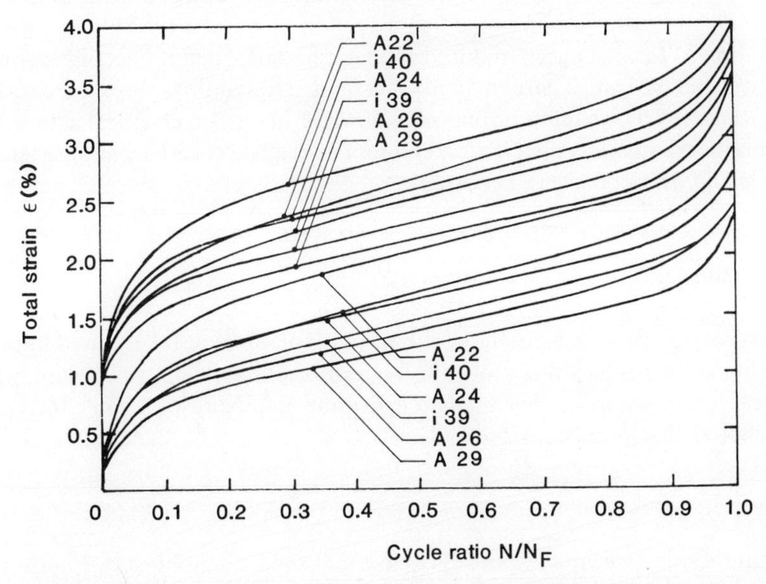

Figure 4.38 Measured variation of total longitudinal strain with the cycle ratio

against cycle ratio for different fatigue lives then the strain rate can be seen to be virtually constant for a given stress level, as in Figure 4.38. In fact equations have been derived showing that the maximum longitudinal strain variation comprises two components, ε_e which is a function of consumed endurance life and ε_t which is a function of the loading time:

$$\varepsilon_{max} = \varepsilon_e + \varepsilon_t$$

$$\varepsilon_t = 0.413\, S_c^{1.184} \ln(t+1)$$

$$\varepsilon_e = \frac{1}{tg\alpha}\left[S_{max} + 3.18(1.183 - S_{max})\left(\frac{N}{N_F}\right)^{0.5}\right] \quad \text{for} \quad 0 < \frac{N}{N_F} \leqslant 0.10$$

$$= \frac{1.110}{tg\alpha}\left[1 + 0.677\left(\frac{N}{N_F}\right)\right] \quad \text{for} \quad 0.10 < \frac{N}{N_F} \leqslant 0.80 \qquad (4.52)$$

where $tg\alpha$ = secant modulus $= \bar{S}_{max}/\varepsilon_0$; ε_0 = maximum total strain in a single (first) load cycle to $\bar{S}_{max}$; $\bar{S}_{max}$ = average maximum stress level; S_c = characteristic stress level = S_m + RMS; S_m = mean stress level; RMS = root mean square value of the variable loading ($=\sqrt{(1/T_0)\int_0^T x^2(t)\,dt}$ for constant loading); $x(t)$ = stress as a function of time t. These relationships are valid for constant- and variable-amplitude loading.

4.3.15 *Fatigue and durability*

Finally, on fatigue, a little more conjecture may be worthwhile.

Increasing disruption or micro-cracking through fatigue implies increasing permeability with a consequential reduction in durability. This diminished quality, in turn, suggests reduced fatigue resistance again affecting permeability and so on. There may thus be an accelerating cycle of damage, exacerbated by wedging or pumping of water in cracks, chemical action, etc. Good sense suggests then that structures be designed so as to minimise fatigue effects in areas of severe exposure.

4.4 Impact

Parts of marine structures may be at risk also of direct or transmitted impulse or impact loads as a result of collision, explosion, accident, wave slam, wave breaking, or whatever. Such dynamic loading and the response of concrete to it should therefore be considered.

4.4.1 *Load rate*

That there is a varying response is well-established, exemplified by Figure 4.39 (Mainstone, 1975) which shows data from several sources. Clearly compressive

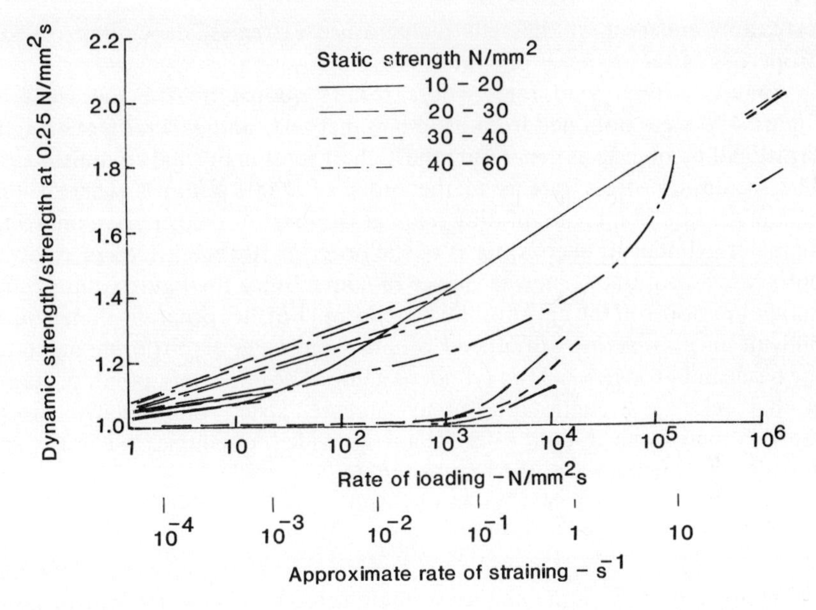

Figure 4.39 Variation of ultimate strength of concrete in direct compression with average rate of loading or straining

strength varies quite significantly with rate of loading: an equation for one set of data has already been quoted (eq. (4.29)). Another, based on stochastic theory for fracture of concrete and incorporating temperature is

$$\frac{\bar{\sigma}}{\bar{\sigma}_0} = \left(\frac{\dot{\sigma}}{\dot{\sigma}_0}\right)^{1/(\beta+1)} \tag{4.53}$$

where $\bar{\sigma}$ and $\bar{\sigma}_0$ are the strengths at high and low rates of loading ($\dot{\sigma}$ and $\dot{\sigma}_0$ respectively) (Mihashi and Wittman, 1981). β is a material constant which is inversely proportional to the absolute temperature. For concrete in compression at normal temperatures it is of the order of 25, somewhat less in flexure. On this reckoning, doubling the loading rate increases the strength by about 3% while a ten-fold rate increase produces an additional 10%, one hundred-fold gives 20% and one thousand-fold yields over 30% 'extra'. Small variations in temperature do not make much difference: a 15° shift from ambient results in less than 1% variation, even for a one hundred-fold increase in load rate. However, the differences are appreciable for more extreme temperatures, such as those of hot oil or liquefied methane (natural gas). For the 'nominal' one hundred-fold increase in load rate and a normal ambient of 20°C, the hot oil could produce a strength increase of over 20% while the methane would have less than 10% (assuming the material's characteristics do not change). A one thousand-fold increase in rate would increase the difference between the two to 25%, so response to impact might be expected to be noticeably different at different temperature extremes, depending on the properties of the material of course.

What is a high load rate? Those results quoted on the top right of Figure 4.39 were obtained from drop-hammer tests and generally one might regard falling objects as generating the highest rates in normal circumstances. The maximum stress rate is of the order of $2000\,kN/mm^2s$ whereas the maximum rate of increase of drag *force* at the crest of a 20 m wave on a 2 m diameter cylinder in deep water is of the order of 10 kN/s. (*Average* rates in both cases would be somewhat lower of course since the figures quoted are maximum slopes of the appropriate wave form.) Further generalisation is very difficult since *stress* rate will depend, of course, on water depth (for moments at the base) and properties of the cylinder section. The important point, perhaps, is that relative strength appears to increase 'faster' than relative stress rates so there is an in-built safety factor: provided, of course, that loads are adequately anticipated in the first instance.

4.4.2 *Moisture influences*

There appears to be some impact strength sensitivity to water/cement ratio in that higher water/cement ratio concretes have higher relative impact strengths (Hughes and Watson, 1978). Moisture content also seems to be significant: dry cement pastes and mortars have a much smaller gain in strength with increasing loading rate than do moist materials (Kaplan, 1980). This is attributed to changes in pore water pressure. It is also noted that impact ultimate strain decreases with compressive strength and increasing concrete age and with higher rates of strain (Hughes and Watson, 1978). That is, the material becomes more brittle as rate of loading is increased.

4.4.3 *Tensile impact*

An equation similar in form to (4.39) and (4.53) has been established for impact **tensile** loading (Zielinski, 1984):

$$ln\left(\frac{\sigma}{\sigma_0}\right) = A + B\ ln\left(\frac{\dot{\sigma}}{\dot{\sigma}_0}\right) \tag{4.54}$$

$A = 0.082$ and $B = 0.042$

It is conjectured that under 'static' or very slowly increasing tensile loading, failure occurs through crack growth along weak interfaces between aggregate particles and the cement matrix, on the weakest link principle. At higher rates of loading, however, crack growth occurs at a much greater number of microcracks and is so rapid that aggregate particles fracture. Other studies on other materials refer to static fracture mechanics approaches which concern crack tip stress intensity causing crack length growth until the intensity reduces below the value needed for failure. Dynamic behaviour is contrasted with this in that time becomes a key factor, with the stress wave travelling along the crack so rapidly that stress intensity alone is insufficient to describe behaviour (Kalthoff et al, 1984).

4.4.4 *Material breakdown*

Clearly crack growth is a major factor in the breakdown of concrete under dynamic load and a limiting density of about 4 cm/cm^2 has been suggested, above which concrete can no longer remain intact (Watson and Sanderson, 1984). The progression of damage is detectable by ultrasonic pulse velocity measurements of course, and measurements were taken in this case. However, generalised relationships between measurement and damage remain to be established satisfactorily.

Ultimate strength, naturally, is only one aspect of the behaviour of concrete and it is the effect of a load somewhat short of failure which may be the most difficult to assess—complete destruction manifests itself quite clearly, partial damage is not always so easy to recognise. There is some increase in elastic modulus under dynamic load (Mainstone, 1975), varying linearly up to about 40% increase at a stress rate of 10^5 N/mm^{2}s from zero at 0.25 N/mm^{2}s. So for a given load the strain energy should be reduced with increasing load rate, assuming elastic behaviour. However, at least at high projectile velocities, energy transfer depends also on the shape of the impactor: a blunt projectile produces larger damage velocities than one with a conical nose (Hornemann et al, 1984). At low velocities the situation might be different since much depends on the rate at which, and how far, cracks propagate. Setting $N = 1$ in equation 4.39 suggests that tensile strain rates less than $2230 \times 10^{-6}\,s^{-1}$ will not cause failure but that may be to carry speculation too far.

Impact damage on a macro- or even meso-scale is perhaps easier to assess than on the micro-scale. Unfortunately it may be particularly insidious at the micro-scale since local durability might be affected and that can take time to become apparent. A weak spot can be created and so an inspection programme (if one exists) should pay particular attention to areas where registerable impact is known to have occurred.

4.5 Thermal behaviour

Marine concrete may be exposed to a wide range of natural environments, from tropical to Arctic. Particular uses or applications, such as oil or refrigerated gas storage, can extend that range substantially, while the need to cater for occasional events such as spillage or accidental rupture can impose even wider localised limits. Oil from the reservoir can be at a temperature of 100°C, although storage may be at 45° or so to reduce thermal stresses. Liquefied natural gas (methane) has a temperature of about −165°C if unpressurised. Thus, if we adopt a temperature range from $-170°C$ to $+50°C$ we can include most normal applications as well as embracing the natural environmental extremes. Fire hazard will increase the upper limit enormously, of course, but that may be best left to be treated as a special case elsewhere.

4.5.1 *Thermal properties*

Within the specified limits properties can vary considerably. However, it is convenient (and normal) to examine elevated and sub-zero temperatures separately since the changing state of moisture at low temperatures can mark major departures from the norm. It is also helpful to distinguish between thermal properties specifically and thermal effects or influences on mechanical and other properties. First, though, a brief reminder of the thermal properties of concern is appropriate. Three are linked: thermal conductivity k, specific heat c, and thermal diffusivity $h^2 = k/\rho c$ where ρ is the density. The diffusivity may be considered as the ratio of thermal conductivity to thermal capacity and is the constant in the thermal diffusion equation (which is equivalent to the moisture diffusion equation 4.10). If θ is the temperature at time t at a point x in a bar, then

$$\frac{\partial \theta}{\partial t} = h^2 . \frac{\partial^2 \theta}{\partial x^2} \tag{4.55}$$

a simple expression of the basic heat flow equation. The fourth primary thermal property is the coefficient of thermal expansion (or contraction) and which, in structural terms, is the thermal strain due to unit change in temperature. It is *restrained* thermal, movement of course, which gives rise to thermal stresses, exemplified in bridge decks subject to solar radiation. The thermal conductivity is the ratio of heat flux (heat flow per unit area) to the temperature gradient, but for a wall, say, heat exchange with its environment is governed by the *surface* conductivity. Hence the overall coefficient of heat transfer (or U-value) through a material is a function of the conductivity of the material *and* the surface conductivities. As for a moisture bearing material such as concrete, there may be also heat exchange with the environment through evaporation or condensation but it could be quite minor: for the moisture throughflow of $0.9\,\text{ml/m}^2\text{/day}$ postulated in considering permeability, the loss of heat through evaporation would cool just under 1 mm thickness of concrete by one degree in one day. Other conditions would produce other results, of course.

4.5.2 *Influence of moisture content*

Moisture content *per se* does influence most thermal properties. For example calculation by the simple method of mixtures shows that moisture content, as a percentage by weight of dry concrete increases specific heat by slightly more than three times the moisture content up to 6%. That is, a moisture content of 4% increases the specific heat by 12% for instance. Should the moisture be frozen then 1% change in 'moisture' content increases the specific heat by only 1%, so 4% increase in moisture gives 4% increase in specific heat.

4.5.2.1 *Moisture and thermal conductivity* Thermal conductivity of most materials varies considerably with moisture content (Jakob, 1949). Concrete is no exception and has a virtually linear variation at normal temperatures—conductivity is increased by the order of 28% at 10% moisture content by volume and by 53% at 20% moisture content (these are mean values for different aggregate type concretes) (Campbell-Allen and Thorne, 1963). Converting moisture content by weight to moisture content by volume for concrete is equivalent to multiplying the values by a factor of about 2.5 so a moisture content of 10% by volume increases specific heat by about 12% and 20% moisture content increases it by about 24%. The corresponding density changes will be about 4% and 8% so thermal diffusivity will change by factors of around $1.28/(1.04)(1.12) = 1.10$ and $1.53/(1.08)(1.24) = 1.14$ for the 10 and 20% moisture contents respectively. Hence the diffusivity of moist concrete is virtually constant.

4.5.2.2 *Moisture and thermal expansion* Thermal expansion of cement paste is affected significantly by moisture, at its lowest when dry or saturated and at its highest at intermediate humidities (60–70%). Under test, specimens subjected to a temperature change and then held at a constant temperature, continued to change in length for some time afterwards before stabilising (Helmuth, 1961). This is attributed to interchange of moisture by diffusion between capillary and gel spaces. Self-desiccated specimens, however, exhibited higher expansions without delayed movement. Two contrary movements take place: warming causes thermal expansion and the concurrent moisture diffusion from the gel produces a (smaller) contraction; this contraction is absent from self-desiccated pastes resulting in their higher expansions. Presumably the two processes occur at different rates, giving the time lag already noted. Three coefficients of thermal expansion have therefore been proposed for cement pastes (Verbeck and Helmuth, 1969):

- α_i — the 'instantaneous' coefficient used to measure a process too rapid to permit an appreciable redistribution of moisture;
- α_w — measures a process slow enough to permit continuous redistribution of moisture but without change in total moisture content;
- α_{sat} — for a process in which equilibrium is maintained in a paste which is saturated and has access to water.

α_w is the coefficient which will apply in most normal situations while α_{sat} will be relevant to immersed concrete and may be of the order of one half of α_w. In

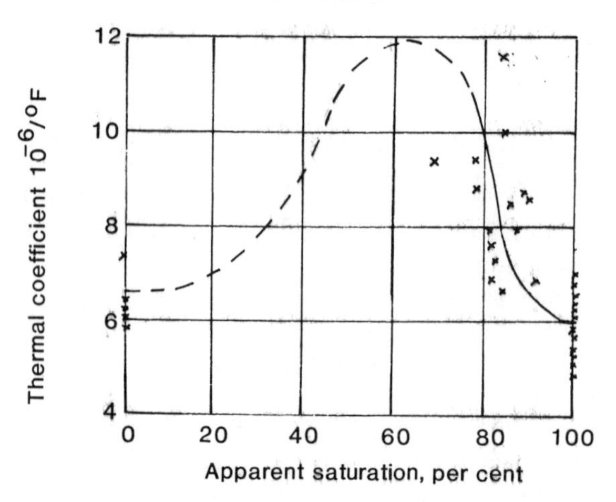

Figure 4.40 Thermal coefficient neat cement varied with moisture content

concrete, of course, the effect is likely to be masked by the greater bulk of sand and aggregate. Generalisations are difficult but it is possible to illustrate potential trends and so to obtain some idea as to whether or not moisture gradients are likely to exacerbate temperature gradients.

Figure 4.40 (Mitchell, 1953) shows the variation in thermal coefficient of cement paste with moisture content (there is also some variation with fineness) ranging from 21.6×10^{-6} to 10.8×10^{-6} per deg C. Equation (4.56) gives a relationship between the coefficients for concrete, paste and aggregate (α_c, α_p and α_a respectively) and the volume fraction g of the aggregate in the concrete (Dougill, 1968); n is an experimental exponent which may be found by measuring the coefficients and plotting $\log(\alpha_p - \alpha_a)/(\alpha_c - \alpha_a)$ against $\log 1/(1 - g)$: n is the slope of the straight line so obtained.

$$(\alpha_c - \alpha_a) = (\alpha_p - \alpha_a)(1 - g)^n \tag{4.56}$$

Now, assuming first that $\alpha_p = 21.6 \times 10^{-6}$ and then 10.8×10^{-6} we can obtain two sets of values for α_c for concrete, adopting different values of n. If then the two corresponding values of α_c are combined as a ratio and plotted against the values of n, a set of curves can be obtained as in Figure 4.41.

It is clear from this that with low aggregate thermal coefficients and low values of n, moisture content has a greater relative effect for richer mixes (i.e. those with lower g values). Note the emphasis on **relative** effects. Absolute values can give a different impression: for $n = 1$, for example, the ratio of 109 for the bottom line is obtained from two fairly high values $[(12.96 \times 10^{-6}/11.88 \times 10^{-6}) \times 100]$ while the corresponding point on the top line comes from lower values $[154 = (9.28 \times 10^{-6}/6.04 \times 10^{-6}) \times 100]$. Moreover, the range of n indicated will in practice be limited for a particular type of aggregate. Nevertheless the inference is clear. If a structure has a moisture gradient from, say, saturated **or** dry in the outer layers to 'partially

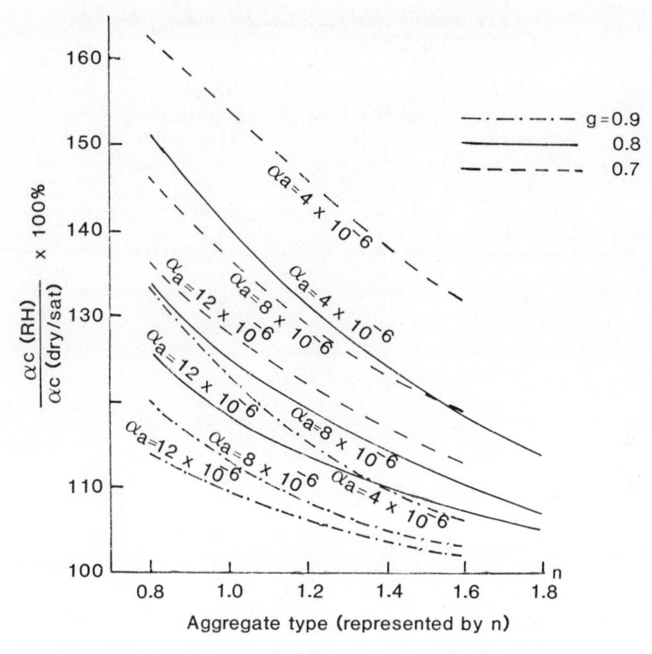

Figure 4.41 Variation of the thermal coefficient ratio, intermediate humidity:dry or saturated

dry' in the inner layers and that gradient is superimposed on a thermal gradient, then the thermal expansion (or contraction) coefficient should not be assumed to be constant. In other words, thermal stresses may be greater than might otherwise be anticipated.

4.5.3 *Influence of aggregate type*

Despite the significance of moisture content, the most important single factor affecting the thermal properties of concrete is the type of aggregate employed. Some typical values are given below for various rocks.

Table 4.10 Thermal expansion of rocks (20–100°C)(Skinner 1966)

Rock Types	Average coefficient of linear expansion (per °C)
Granites and rhyolites	$8 \pm 3 \times 10^{-6}$
Andesites and diorites	7 ± 2
Basalts, gabbros and diobases	5.4 ± 1
Sandstones	10 ± 2
Quartzites	11
Limestones	8 ± 4
Marbles	7 ± 2
Slates	9 ± 1

Table 4.11 Specific heat and thermal conductivity of rocks (1SCCOLD 1958)

Aggregate	Specific Heat kJ/kg °C	Thermal Conductivity kJm/m²h °C
Quartz	0.733	18.63
Granite	0.716	10.47
Dolomite	0.804	15.49
Limestone	0.846	11.64
Basalt	0.766	6.15
Marble	0.875	8.83
Feldspar	0.812	8.37

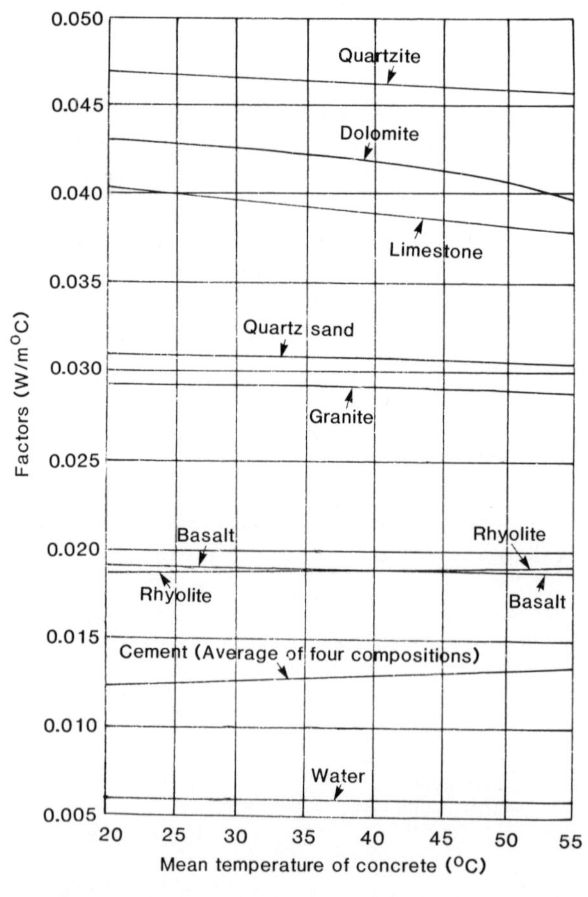

Figure 4.42 Factors for predicting conductivity of concrete

The variation is considerable but it has been reckoned that siliceous aggregates have high coefficients, limestones have low coefficients and igneous rocks are intermediate (Bonnell and Harper, 1951). Silica content may be significant—the higher it is, the higher the expansion coefficient (Browne, 1968)—but silicates do not all have the same coefficient and some are in fact anisotropic. Petrography and mineralogy are of fundamental importance so that composition and grain size, for example, need to be known in comparing aggregates which have the same generic label. Anisotropy of some rocks (and of their component minerals) also means that their cubical expansion is not necessarily directly related to their linear expansion.

A simple empirical method has been developed for the prediction of thermal conductivity and specific heat of saturated concrete for various rock types. More complex theoretical methods are referred to in at least one review (Marshall, 1972) but the older method has the merits of simplicity and a base of extensive measurement. Figures 4.42 and 4.43 (Bureau of Reclamation, 1940) plot conductivity and specific heat factors for aggregates, cement, and water

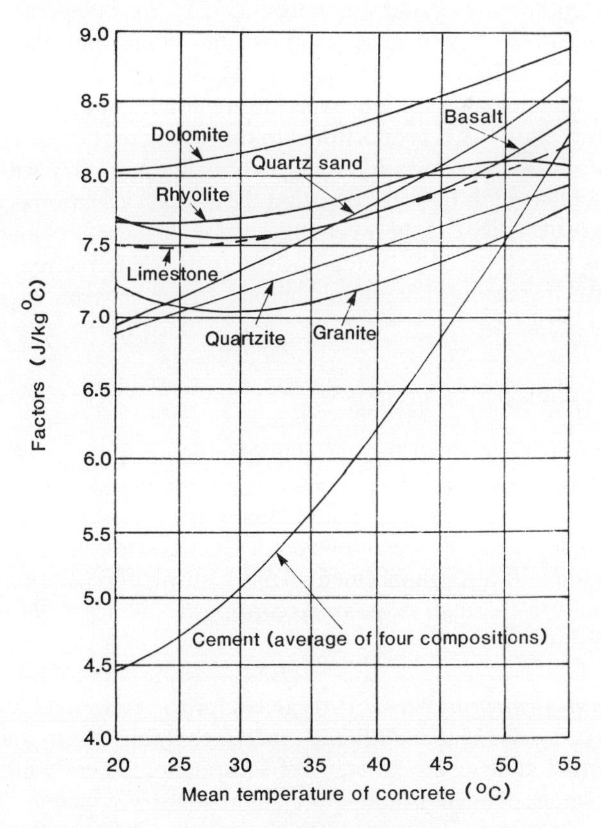

Figure 4.43 Factors for predicting specific heat of concrete

against concrete mean temperature. To calculate thermal conductivity for a particular proposed concrete, the sum is found of the factors for each component multiplied by their percentage proportions by weight of the finished concrete. That is, if the percentage proportion of cement is p_c, of water is p_w, of sand p_s and of aggregate is p_a and the respective factors are f_c, f_w, f_s, and f_a, then

$$p_c + p_w + p_s + p_a = 100 \tag{4.57}$$

and thermal conductivity $= \sum(pf)_{c,w,s,a}$

4.5.4 *Elevated temperature and mechanical properties*

Increased temperature affects the mechanical properties of concrete, and the trends are incorporated in a review already referred to (Blundell et al, 1976). For most concretes there is virtually linear reduction in elasticity and strength up to about 80°C, the size of reduction depending on aggregate type. It is suggested, for example, that the E value drops by 25–32.5% for concrete produced from limestones and dolomites, 15–20% for dolerites, basalts and andesites, and 10–17.5% for siliceous gravel and sandstones (the modulus of elasticity at 5°C is taken as the base value). Strength loss is comparable, possibly even greater, as one might expect from the common presumption that the modulus of elasticity is proportional to the square root of the compressive strength (although the 'constant' of proportionality may vary with temperature). In general it seems to be the case that the greatest variation occurs where there is the greatest disparity between the thermal coefficients of aggregate and paste. This is perhaps to be expected since more disruption or micro-cracking is likely, with increasing differential thermal strain between aggregate and matrix.

4.5.5 *Low temperature behaviour*

The behaviour of concrete at sub-zero temperatures is quite complex and can result in apparent anomalies. In broad terms, if the material is dry, then there is some minor increase in strength and modulus of elasticity, and thermal response is uniform and normal. However, if the concrete is moist or saturated, there can be a significant enhancement of mechanical properties and a varying (and non-uniform thermal) response according to the degree of weathering prior to freezing.

4.5.5.1 *Mechanical properties* Typical compressive strength variation is shown in Figure 4.44 (Marshall, 1982) from which it can be seen that there is a very substantial gain in the strength of saturated concrete which is much smaller for 'moist' concrete and quite slight for dry concrete. There is a matching increase in modulus of elasticity while tensile strength has a much

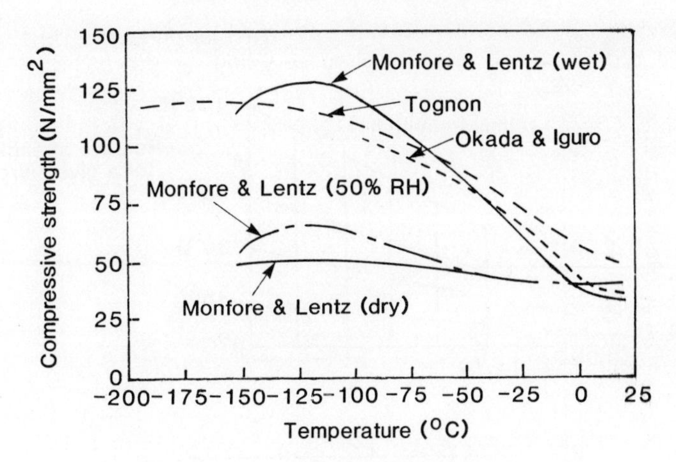

Figure 4.44 Variation of compressive strength with temperature

sharper increase below zero. Compressive strength and modulus of elasticity appear to be linked by a relationship of the form

$$E_R = 0.8\sqrt{\sigma_R} \tag{4.58}$$

where the suffix R denotes the modulus and strength at temperatures in the range -10 to -100°C relative to the values at normal temperature (20°C) (Marshall, 1983). Several empirical relationships between strength and temperature have been proposed but do not agree sufficiently to be generally applicable. However, it is evident that water/cement ratio and moisture content are significant factors.

4.5.5.2 *Water/cement ratio* Based on some Japanese data for saturated concrete (Okada and Iguro, 1978) Figure 4.45 was prepared (Marshall and Drake, 1987). Although idealised and limited in application, it does show how increasing water/cement ratio becomes more important with reduced temperature. The figure can be summarised by the relationship (Marshall, 1985)

$$\sigma_R = 1 + (w/c - 0.15)[0.54(|T| - 1.5)^{0.5} - 0.45] \tag{4.59}$$

where w/c is the water/cement ratio and T is the sub-zero temperature (deg. C). When $T \gg 1.5$ the 1.5 can be neglected.

4.5.5.3 *Freezing mechanism* The most obvious reason for the change in behaviour of concrete at sub-zero temperatures is the formation of ice in the pores: when the concrete is dry, there is negligible change in properties; when it is moist, there is some change; when it is saturated, there is a major change; water freezes below 0°C, hence, strength gain (for example) is due to ice formation. However, the phenomenon is rather more subtle than that. Why

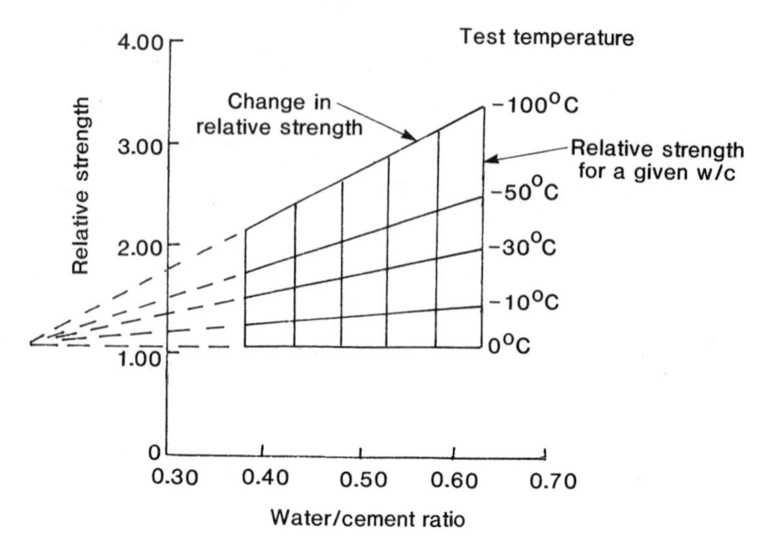

Figure 4.45 Variation in relative strength with water/cement ratio and temperature

should the concrete be strengthened, for instance? There is no certain answer to that as yet but there are various possibilities. For instance it has been suggested that the formation of ice has an internal prestressing effect on the concrete, although one might reasonably expect such action to have a disruptive effect, at least near external surfaces forming free boundaries. Benefit may come simply from filling pores and interstices so that the crack raising potential is diminished: the fracture toughness of concrete may be increased. Ice, particularly in capillaries, may act like fibre reinforcement: although the tensile strength of ice is not high it is still greater than that of air or water. Possibly, of course, there is a combination of actions which have varying effects at different stages of freezing. At any rate, whatever the uncertainty surrounding cause, the global effect is clear.

Freezing does not take place instantaneously. Due to super-cooling and the dendritic nature of ice crystal growth in capillaries, the freezing temperature reduces with size of capillary, as indicated in Figure 4.46 (Helmuth, 1960). Gel pore freezing is therefore unlikely without substantial temperature drop and it has been reported to be incomplete at − 79°C (Powers, 1953) although there is evidence to suggest that it has halted at around − 90°C (Radjy and Richards, 1969). Moreover, if access to ice crystal growth in a 'larger' diameter capillary is through a small one, freezing in it presumably cannot take place until the small one has reached its appropriate temperature, so there is then a 'throttling' effect (the question of capillary continuity has already been discussed in this chapter). What all this does suggest is that strength gain, for example, is due primarily to the freezing of *capillary* water. Consequently, the more capillary volume there is, the greater the change to be expected. Thus the

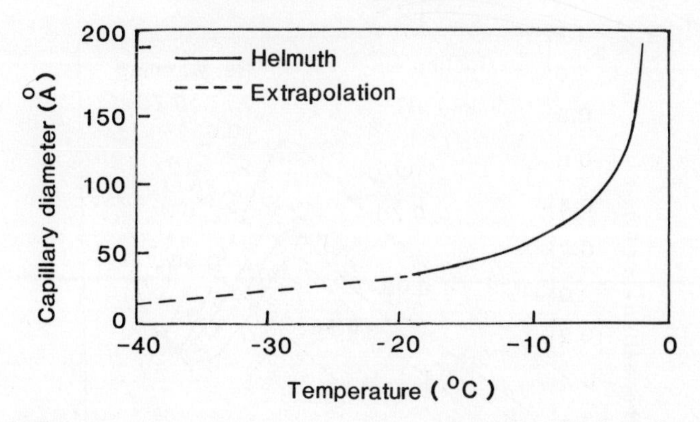

Figure 4.46 Size of capillaries required for propagation of ice formation

higher the water/cement ratio, the more capillaries there are, and the greater the possible moisture content with its greater potential for freezing 'benefits'. Water/cement ratio and moisture content are therefore not intrinsically important so far as low temperature behaviour is concerned. Their significance is in their roles as indicators of capillary volume (and distribution) and filling of that volume by water. Moisture content of course also provides a measure of *empty* capillary space. Thus it is easy to imagine that, as ice expands along a capillary, it drives moisture before it. Should the pore be filled, the water (and surrounding matrix) is put under pressure further depressing its freezing point (at 2 atmospheres the freezing temperature is about -20°C). If there is empty space, the water is not pressurised, so the property changes are less—this is where the prestressing argument above arises: presumably crack growth is hindered in a precompressed material. Migration of moisture from gel to capillaries also complicates the picture.

4.5.5.4 *Moisture effects* On the basis of the evidence referred to above and the arguments presented, and considering increased hardening with age as well as the thermal effects discussed below, it has been suggested (Marshall, 1982):

1. at 45–50% RH or less, low temperature has minimal effect—perhaps 25% increase in strength (the capillaries are empty);
2. within the range 50–80%, low temperature behaviour is governed predominantly by moisture content and is independent of age-hardening since hydration has ceased and 'porosity' cannot therefore change;
3. from 80–100% RH, low temperature effects depend on moisture content and age (since hydration rates vary according to RH and so 'porosities' will vary).

In addition, the lower the water/cement ratio, the lower the relative effect will be. Notice, however, that in all this, the emphasis is on *relative* effects. It is

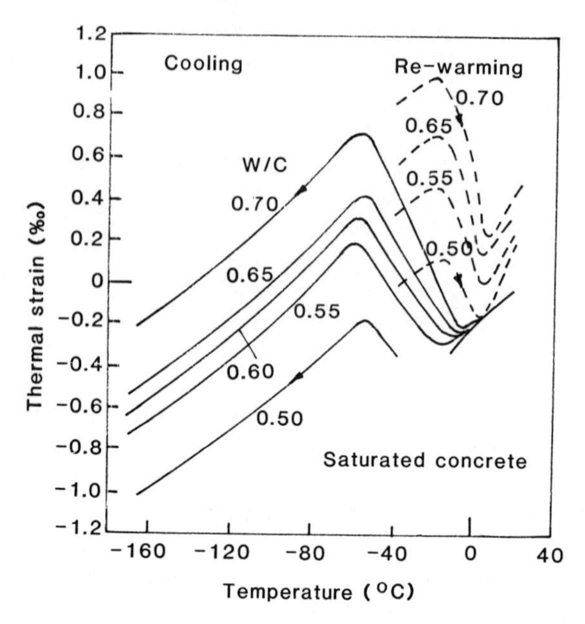

Figure 4.47 Thermal strain versus different moisture contents of concrete

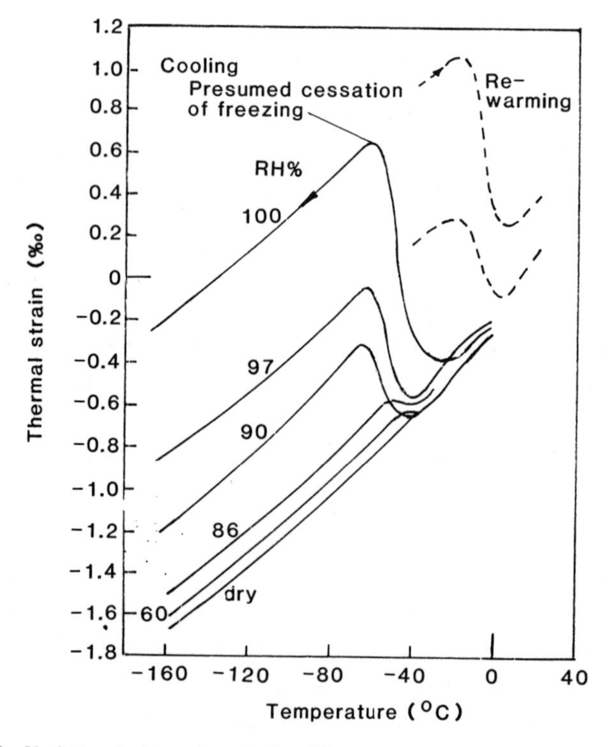

Figure 4.48 Variation in thermal strain for different water/cement ratios and temperature

important to bear in mind that lower water/cement ratios normally result in higher strength concretes so that, say, 250% of 45 N/mm^2 is greater than 350% of 25 N/mm^2 in absolute terms although the relative effect is smaller.

4.5.5.5 *Freezing expansion* Some of the more curious and interesting effects are shown in Figures 4.47 and 4.48 (Rostasy and Wiedemann, 1980) from which it can be seen that, as the temperature falls, an expansion occurs between about −10°C and −70°C, depending on the amount of moisture in the concrete. Once the expansion peak has been reached, contraction continues more or less linearly. Plotting the expansion phase of Figure 4.48 as in Figure 4.49 makes the influence of water/cement ratio clearer: as it increases so does the total expansion and the 'coefficient of freezing expansion'. In other words, pore ice has a greater effect (Marshall, 1982).

4.5.5.6 *Thermal cycling* There is some hysteresis and residual strain on re-warming. Assuming that the beginning of warming contraction shown in Figure 4.48 represents the onset of thawing, it is clear that the thawing contraction is smaller and more rapid in relation to temperature, than the

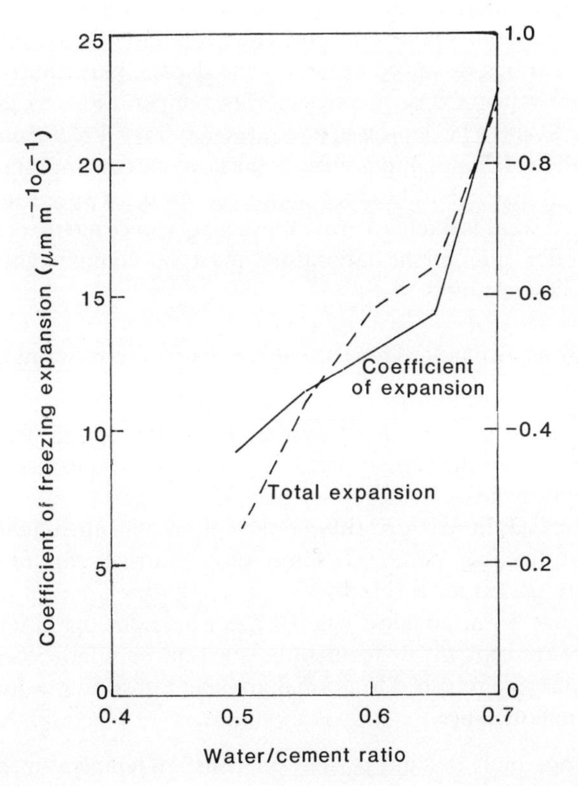

Figure 4.49 Variation in coefficient and total expansion on freezing

original freezing expansion. This is not unexpected since melting will take place in some capillaries at a higher temperature than freezing: the throttling effect of smaller 'access' capillaries only functions during cooling. Any depression of freezing and melting points due to pressure will still apply of course. Resulting thermal 'set' due to the temperature cycle is presumably due to micro-cracking and so there should be some change on refreezing. The evidence is mixed but, as in more normal freeze-thaw testing, it appears that air entrainment will have beneficial effects. Much may depend also on the rate of temperature change since thermal shock loads are unlikely to induce higher resistance in the same way as impact or fatigue loading. What is not clear as yet is whether or not there could be any harmful effect from cycling about a depressed freezing point. However, the damage potential might be minimal since any unfrozen channels will be 'surrounded' by already frozen material which is therefore stronger. Freezing of concrete is perhaps better visualised as a process of infiltration rather than as externally imposed loading.

Temperature cycling reduces the slope of the stress/strain curve at normal temperatures (FIP, 1982) but the amount will be governed by rate of cycling, water/cement ratio, moisture content and air content.

To further muddy the issue, it is perhaps as well to remember that the terms 'moisture' and 'water' do not necessarily imply pure water. As has been indicated already, the water in the pores and capillaries of hardened concrete will contain various dissolved salts in some degree, particularly should the source be sea water. Consequently freezing temperatures, to take but one example, are likely to be depressed and rates may vary. Precise quantification is therefore difficult, if not impossible, so the best one can hope for is informed estimates. It is also advisable to balance the need for speed in laboratory testing against what is likely to arise in practice. Hence a temperature rate of change of 2 deg./min in the laboratory must be compared with diurnal, seasonal or lifecycle rates.

4.5.5.7 *Thermal régimes* Three broad low temperature thermal régimes can be defined:

(a) diurnal, with atmospheric cycling about the normal freezing point, giving rise to the familiar need for protection against frost damage in small elements;
(b) seasonal, as in Arctic or sub-Arctic regions with an annual fluctuation about freezing point; at some stage diurnal variations may be superimposed for a period:
(c) life-cycle, as in liquefied gas storage where, for operational reasons, tanks are kept at low temperatures except on a few occasions where they may be returned to normal ambient temperatures for inspection and maintenance.

This does not imply that any of these is a constant temperature régime since they are all subject to regular or irregular fluctuations and a certain amount of flutter. But in relation to freezing mechanisms and subsequent thawing, they can be so grouped. It seems, therefore, that in general we are considering low-cycle thermal fatigue since, even in the diurnal régime, over a lifetime of perhaps 30 or even 50 years, only a few thousand cycles may be involved. Against that must be set the possibility of significant deterioration in only a 'handful' of cycles. Consequently low permeability would appear to be a desirable characteristic of concrete intended for low temperature use. Cycling may tend to increase it if Figures 4.7 to 4.10 are at all typical.

4.5.5.8 *General summary* Finally, brief reference should be made to some other properties, although data is generally rather sparse. Figure 4.50 (FIP, 1982) shows average values of thermal conductivity from which it can be seen that there is some increase with reduction in temperature of saturated concrete. Specific heat does not change significantly except for moisture contents of perhaps 5% by weight (or more), when the change from water to ice causes a reduction (pure ice has half the value of pure water) (Marshall, 1982). Fatigue strength is enhanced at low temperatures (Kronen and Anderson, 1983) as is impact strength and dynamic response (Marshall and Drake, 1987). Permeability is reduced substantially: of the order of 50% from 20 to −165°C and even more for lightweight aggregate concrete (Bamforth, 1984). It is also worth noting that there is likely to be a greater scatter of property values at low

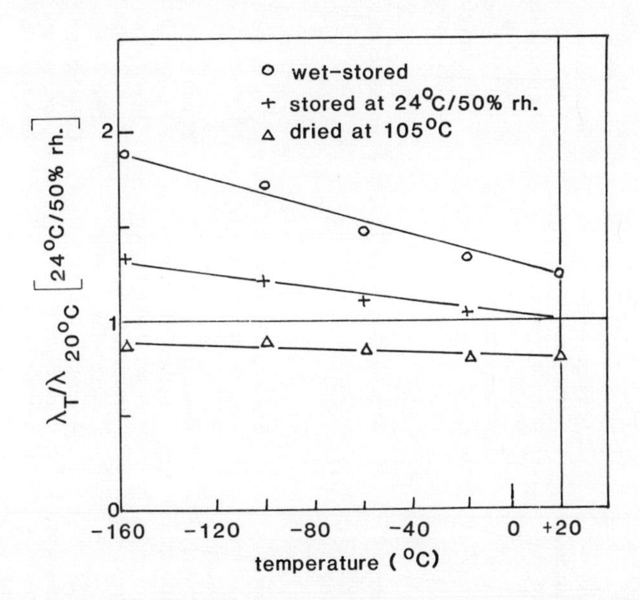

Figure 4.50 Average values of the thermal conductivity of concrete as a function of temperature and moisture content

Table 4.12 Effect of various factors on cryogenic properties of concrete

Property	Effect of influencing factor			
affected	Water/cement ratio	Moisture content	Age	Temperature
Relative compressive strength	Increases with increase in w/c; more pronounced with decreasing temperature.	Small increase in strength (25%) at 45–50% RH; may be proportional to m/c in temperature range 0°C to −120°C; can be 2.5 to 3.0 × ambient strength at −120°C when saturated	Possibly significant in 'young' concretes (< 200 days, say) being more important the younger the concrete; depends on curing history	Substantial increase as temperature falls, particularly in range from −10°C to −100°C; peaks at around −120°C (saturated concretes)
Relative tensile strength	May be assumed to relate to compressive strength and thus follows similar trends			Substantial increase in range 0°C to −25°C; peaks at around −70°C
Relative modulus of elasticity	For saturated concretes $E_R = 0.8\sqrt{\sigma_R}$ where E_R is modulus relative to ambient modulus and σ_R is corresponding relative compressive strength (temperature range −10°C to −100°C)			Similar pattern to compressive strength
Thermal contraction	Down to −60°C marked influence on expansion during freezing; uniform effect below −60°C	Significant in range 85–100% RH during freezing (material expands down to −60/−70°C); uniform contraction below −70°C	Possibly significant in 'young' concretes (< 200 days) in freezing range	Very important down to −70°C (depending on m/c and w/c)
Thermal conductivity	Aggregate type more important	Consider as for normal temperature	Aggregate type more important	
Specific heat	Minor effect	Important on freezing	Minor effect	Important around freezing point for moist concretes

temperatures (Sehnal et al, 1983) based on compressive strength tests. There is no reason to believe that other property values will be any less affected.

Table 4.12 summarises the effect of water/cement ratio, moisture content, age, and temperature on some properties of concrete at low temperatures (Marshall, 1983).

References

Ansquer, P. and Ledoigt, B. (1980) L'instrumentation d'une plate-forme béton et un aperçu de son comportement–un exemple: Frigg TCP 2. *Proc. Int. Symp. Behaviour of Offshore Concrete Structures, CNEXO*, Brest

Bajpai, A.C., Mustoe, L.R. and Walker, D. (1977) *Advanced Engineering Mathematics*. Wiley, London.

Bamforth, P.B. (1985) Concrete used for storage of cryogenic liquids with special reference to temperature effects. *Proc. Symp. Design, Construction and Maintenance of Concrete Storage Structures, Concrete Society*. London.

Berge, S. (1985) Basic fatigue properties of welded joints. *Fatigue Handbook: Offshore Steel Structures*. (ed. A. Almar-Naess) Tapir, Trondheim.

Biczók, I. (1964) *Concrete Corrosion & Concrete Protection, Akadémiai Kiadó* Budapest.

Blundell, R., Dimond, C. and Browne, R.D. (1976) *The properties of concrete subjected to elevated temperatures, Tech. Note 9 CIRIA-UEG*. London.

Boineau, A. (1980) Béton à la mer, mesures de la tenure en eau et de la qualité par auscultation dynamique. *Proc. Int. Symp. Behaviour of Offshore Concrete Structures, CNEXO*. Brest.

Bonnell, D.G.R. and Harper, F.C. (1951) *The thermal expansion of concrete, Nat. Bldg. Studies, Tech. Paper No. 7* HMSO, London.

Boresi, A.P., Sidebottom, O.M., Seely, F.B. and Smith, J.O. (1978) *Advanced Mechanics of Materials (3rd edn.)*. Wiley, New York.

Browne, R.D. (1968) Properties of concrete in reactor vessels. *Proc. Conf. Prestressed Concrete Pressure Vessels, Inst. Civil Engineers*. London.

Browne, R.D. and Baker, A.F. (1979) *The performance of concrete in a marine environment, Developments in Concrete Technology–1*. (ed. F.D. Lydon) Applied Science Publ.

Buenfeld, N.R. and Newman, J.B. (1984) Permeability of concrete in a marine environment. *Mag. Concrete Research*, **36(127)**, June: 67–80.

Buenfeld, N.R. and Newman, J.B. (1986) Permeability of marine concrete. *Marine Concrete, Proc. Int. Conf. Concrete in the Marine Environment, Concrete Society*. London.

Bureau of Reclamation (1940) *Thermal properties of concrete, Boulder Canyon Project Final Report, Bull. No. 1, Part VII*. US Dept. of Interior.

Butler, J.E. (1982) The influence of pore pressure upon concrete. *Mag. Concrete Research*, **34(118)**, March: 3–17.

Buyukozturk, O. and Zisman, J.G. (1982) Behaviour of concrete in biaxial cyclic compression. *Proc. Int. Conf. Behaviour of Offshore Structures, Vol. 2, Hemisphere*. Washington.

Cabrera, J.G. (1986) *The use of pulverised fuel ash to produce durable concrete, Improvement of Concrete Durability*. Thos. Telford, London.

Campbell-Allen, D. and Thorne, C.P. (1963) The thermal conductivity of concrete. *Mag. of Concrete Research*, **15(43)**.

CEB-FIP (1978) *Model Code for Concrete Structures, Comité Euro-International du Béton*. Paris.

Clayton, N. (1986) Concrete strength loss from water pressurisation. *Marine Concrete, Proc. Int. Conf. Concrete in the Marine Environment, Concrete Society*. London.

Concrete Society (1985) Permeability testing of site concrete–a review of methods and experience. *Permeability of Concrete and its Control, Concrete Society*. London.

Conjeaud, M.L. (1980) Mechanism of sea water attack on cement mortar. *Performance of Concrete in Marine Environment, SP-65, ACI*. Detroit.

Conjeaud, M., Guyot, R., Ranc, R. and Varizat, A. (1980) *Experimentations réalisées à la Société Lafarge sur la resistance des ciments à l'attaque chimique de l'eau de mer. Coll. Int. sur la Tenue des Ouvrages en Béton en Mer, CNEXO*.

Cornelissen, H.A.W. (1984) *Fatigue failure of concrete in tension, Heron* **29(4).**

Cornelissen, H.A.W. and Reinhardt, H.W. (1982) Fatigue of plain concrete in uniaxial tension and in alternating tension-compression loading. *IABSE Reports Vol. 37, Proc. IABSE Coll. Fatigue of Steel and Concrete Structures.* ETH-Hönggerberg, Zürich.

Cornelissen, H.A.W. and Reinhardt, H.W. (1984) Uniaxial tensile fatigue failure of concrete under constant-amplitude and programme loading. *Mag. of Concrete Research* **36(129)**, Dec: 216–226.

Cornelissen, H.A.W. and Siemes, A.J.M. (1985) Plain concrete under sustained tensile or tensile and compressive fatigue loading. *Proc. Int. Conf. Behaviour of Offshore Structures.* Elsevier, Amsterdam.

CUR-commissie B23 (1981) *Duurzaamheid maritieme constructies (Durability of maritime structures), Rapport 100, Commissie voor uitveering van Research.* The Netherlands.

CUR-VB (1984) *Report 112, Vermoeiing van beton, deel 1: drukspanningen* (*Fatigue of concrete, Part 1: compressive stresses*) Institut TNO, Betonvereniging, Zoetermeer.

Department of Energy (1977) *Offshore installations: Guidance on design and construction.* HMSO, London.

Diamond, S. (1976) Cement paste microstructure–an overview at several levels. *Proc. Conf. Hydraulic Cement Pastes: their Structure and Properties, Cement and Concrete Assocn.* Slough.

Doucet, Y.J., Thebault, J. and Trinh, L.J. (1986) Fatigue assessment of concrete structures based on in-service responses. *Proc. 18th Annual Offshore Tech. Conf.* Houston.

Dougill, J.W. (1968) Some effects of thermal volume changes on the properties and behaviour of concrete *Proc. Int. Conf. Structure of Concrete, Cement & Concrete Assocn.* London.

Duderstadt, J.J. and Martin, W.R. (1979) *Transport Theory.* Wiley, New York.

Duhoux, L. (1980) *Etat des ouvrages de l'usine maremotrice de la Rance après quinze années de service, Coll. Int. sur la Tenue des Ouvrages en Béton en Mer, CNEXO.* Brest.

Elwell, D. and Pointon, A.J. (1978) *Physics for Engineers and Scientists.* Ellis Horwood, Chichester.

Encyclopedia *of World Geography* (1974) Octopus Books, London.

Fédération Internationale de la Précontrainte (FIP) (1982) *Cryogenic Behaviour of Materials for Prestressed Concrete, FIP.* Slough

Fjeld, S., Furnes, O., Hansvold, C., Røland, B., Blaker, B., Morley, C. (1978) Special problems in structural analysis and design of offshore concrete platforms. *Proc. 10th Annual Offshore Technology Conf.* Houston.

Fjeld, S. and Røland, B. (1982) Experiences with eleven offshore concrete structures. *Proc. 14th Annual Offshore Tech. Conf.* Houston.

Fjell, (1977) *Check list, FIP Notes* **67**. March/April: 7.

Fookes, P.G., Simm, J.D. and Barr, J.M. (1986) Marine concrete performance in different climatic environments. *Marine Concrete–Proc. Int. Conf. on Concrete in the Marine Environement, The Concrete Society.* London.

Ford, H. and Alexander, J.M. (1977) *Advanced Mechanics of Materials, 2nd. ed.* Ellis Horwood, Chichester.

Gerwick, B.C. and Berner, D. (1984) Concrete in vessel building: realities, limits and potentialities. *Proc. FIP/CPCI Symp. Calgary, Vol. I, Canadian Prestressed Conc. Inst.* Ottawa.

Geyer, J.F. and Stahl, B. (1986) Simplified fatigue design procedure for offshore structures. *Proc. 18th. Annual Offshore Tech. Conf.* Houston.

Gibson, A.H. (1952) *Hydraulics and its Applications, 5th ed.* Constable, London.

Hawkins, N.M. (1976) Fatigue considerations for concrete ships and offshore structures. *Proc. Conf. on Concrete Ships & Floating Structures, Univ. of California.* Berkeley.

Haynes, H.H. (1980) Permeability of concrete in sea water. *Performance of Concrete in Marine Environment, SP-65, American Conc. Inst.* Detroit

Helmuth, R.A. (1960) Capillary size restrictions on ice formation in hardened Portland Cement pastes. *Paper VI-S2, Proc. 4th. Int. Symp. Chemistry of Cement.* Washington.

Helmuth, R.A. (1961) Dimensional changes of hardened Portland cement paste caused by temperature changes. *Proc. Highway Research Board,* **40**.

Higdon, A., Ohlsen, E.H., Stiles, W.B., Weese, J.A., Riley, W.F. (1978) *Mechanics of Materials (3rd. ed.)* John Wiley, New York.

Hoff, G.C. (1986) The service record of concrete offshore platforms in the North Sea. *Marine Concrete–Proc. Int. Conf. Concrete in the Marine Environment, The Concrete Society.* London.

Hofφy, A. and Hafskjold, P.S. (1983) *A survey of up to 60 years old concrete ships in subarctic environment, FIP Notes,* **2**: 16–21.

Holmen, J.O. (1982) Fatigue of concrete by constant and variable amplitude loading. *Fatigue of Concrete Structures, SP-75, American Concrete Institute.* Detroit.

Hornemann, U., Rothenhäusler, H., Senf, H., Kalthoff, J.F., Winkler, S. (1984) Experimental investigation of wave and fracture propagation in glass slabs loaded by steel cylinders at high impact velocities. *Proc. 3rd. Conf. Mechanical Properties at High Rates of Strain, Inst. of Physics.* Oxford.

Hsu, T.T.C. (1981) Fatigue of plain concrete. *ACI Journal.* July-Aug: 292–305.

Hughes, B.P. and Watson, A.J. (1978) Compressive strength and ultimate strain of concrete under impact loading. *Mag. of Concrete Research,* **30(105)**, Dec: 189–199.

Hughes, D.C. (1985) Pore structure and permeability of hardened cement paste. *Mag. of Concrete Research,* **37(133)**, Dec: 227–233.

Hunter, S.C. (1983) *Mechanics of Continuous Media, 2nd ed.* Ellis Horwood, Chichester.

Idorn, G.M. (1967) *Durability of Concrete Structures in Denmark.* Danish Technical Press, Copenhagen.

Ilantzis, A. (1980) *Comportement mécanique du béton impregne sous pression hydrostatique, Coll. Int. Tenue des Ouvrages en Béton en Mer, CNEXO.* Brest.

Int. Sub-Committee on Concrete for Large Dams (ISCCOLD) (1958) Collection of data on the thermal properties of concrete for dams. *Proc. 6th Congress on Large Dams, Vol. IV.* New York.

Jackson, R. (1977) *Transport in Porous Catalysts.* Elsevier, Amsterdam.

Jakob, M. (1949) *Heat Transfer, Vol. I.* John Wiley, New York.

Jungwirth, D. (1984) *Long-term behaviour of prestressed concrete structures–experience and conclusions, FIP Notes,* **2**: 8–12, 25.

Kallaby, J. and Price, J.B. Evaluation of fatigue considerations in the design of framed offshore structures. *Proc. 8th. Annual Offshore Tech. Conf.* Houston.

Kalthoff, J.F., Shockey, D.A. and Homma, H. (1984) Short pulse fracture mechanics. *Proc. 3rd Conf. Mechanical Properties at High Rates of Strain, Inst. of Physics.* Oxford.

Kaplan, S.A. (1980) Factors affecting the relationship between rate of loading and measured compressive strength of concrete. *Mag. of Concrete Research,* **32(111)**, June: 79–88.

Kapur, K.C. and Lamberson, L.R. (1977) *Reliability in Engineering Design.* John Wiley, New York.

Keckes, S. (1980) *Protecting the marine environment, Ambio* **12(2)**.

King, C.A.M. (1975) *Introduction to Physical and Biological Oceanography.* Edward Arnold, London.

Kronen, H. and Andersen, J.H. (1983) Properties of cryogenic concrete. *Nordic Concrete Research,* **No. 2**, *Nordic Concrete Federation.* Oslo, Dec.

Lea, F.M. (1970) *The Chemistry of Cement and Concrete, 3rd ed.* Edward Arnold, London.

Van Leeuwen, J. and Siemes, A.J.M. (1979) *Miner's rule with respect to plain concrete, Heron,* **24(1)**.

Van Leeuwen, J. and Siemes, A.J.M. (1979) Miner's rule with respect to plain concrete. *Proc. 2nd. Int. Conf. Behaviour of Offshore Structures, BHRA, Fluid Engineering.* Cranfield.

Lenschow, R. and Seguin, M. (1982) Chargement dynamique aléatoire à long terme de structures en béton. *Proc. 2nd. Coll. Int. Tenue des Ouvrages en Béton en Mer, CNEXO.* Brest.

Linsley, R.K., Kohler, M.A. and Paulhus, J.S. (1975) *Hydrology for Engineers, 2nd. ed.* McGraw-Hill, New York.

Lotsberg, I. and Almar-Naess, A. (1985) *Fatigue life calculations, Fatigue Handbook: Offshore Steel Structures.* (ed. A. Almar-Naess) Tapir, Trondheim.

Mainstone, R.J. (1975) Properties of materials at high rates of straining or loading. *Building Rsch. Estab. Current Paper CP 62/75, reprinted from Matériaux et Construction,* **8(44)**: 102–116.

Marin, J. (1962) *Mechanical Behaviour of Engineering Materials.* Prentice-Hall, London.

Marshall, A.L. (1972) The thermal properties of concrete. *Building Science,* 7: 167—174.

Marshall, A.L. (1982) Cryogenic concrete. *Cryogenics,* Nov: 555–565.

Marshall, A.L. (1983) Cryogenic concrete and the marine storage of LNG. *Proc. 2nd. Int. Conf. Cryogenic Concrete, The Concrete Society.* London.

Marshall, A.L. (1985) Behaviour of concrete at Arctic temperatures. *Proc. 8th. Int. Conf. Port and Ocean Engineering under Arctic Conditions, Danish Hyd. Inst.* Copenhagen.

Marshall, A.L. and Drake, S.R. (1987) Thermal and dynamic response of concrete at low temperatures. *Proc. 9th Int. Conf. Port & Ocean Engineering under Arctic Conditions, Univ. of Alaska.* Fairbanks.

Marshall, P.W. and Luyties, W.H. (1982) Allowable stresses for fatigue design. *Proc. 3rd. Int. Conf. Behaviour of Offshore Structures, Hemisphere.* Washington.

Mehta, P.K. (1980) Durability of concrete in marine environment—a review. *Performance of Concrete in Marine Environment, SP-65, American Concrete Institute.* Detroit.

Mihashi, H. and Wittman, F.H. (1981) Probabilistic concept to describe the influence of rate of loading on strength of concrete. *Trans., 6th Int. Conf. on Structural Mechanics in Reactor Technology.* North Holland Publ. (for the EEC).

Mills, R.H. (1984) The influence of stress on the permeability of concrete. *Proc. FIP/CPCI Symposia, Calgary, Vol. I, Canadian Prestressed Concrete Inst.* Ottawa.

Mitchell, L.J. (1953) Thermal expansion tests on aggregates, neat cements and concretes. *Proc. American Soc. for Testing Materials,* **53**: 963–977.

Morley, C.T. (1979) *Theory of pore pressure in reinforced concrete cylinders, FIP Notes* **79**, March/April: 7–15.

Neal, J.A. and Kesler, C.E. (1968) The fatigue of plain concrete, The Structure of Concrete and its Behaviour under Load. *Proc. Int. Conf. 1965, Cement & Concrete Assocn.* London.

Nelson, E. L., Carasquillo, R.L. and Fowler, D.W. (1987) Fatigue of high strength concrete subjected to biaxial cyclic compression. *Proc. Symp. Utilisation of High Strength Concrete.* Tapir, Trondheim.

Neville, A.M. *Properties of concrete.* Pitman, London.

Nishiyama, M., Mugurama, H, and Watanabe, F. (1987) On the low-cycle fatigue behaviours of concrete and concrete members under submerged condition. *Proc. Symp. Utilisation of High Strength Concrete.* Tapir, Trondheim.

Nolte, K.G. and Hansford, J.E. (1976) Closed-form expressions for determining the fatigue damage of structure due to ocean waves. *Proc. 8th. Annual Offshore Tech. Conf.* Houston.

Normand, R. (1986) Review of the performance of concrete coastal structures in the Gulf area. *Marine Concrete—Proc. Int. Conf. on Concrete in the Marine Environment, The Concrete Society.* London.

Nyame, B.K. and Illston, J.M. (1981) Relationships between permeability and pore structure of hardened cement paste. *Mag. of Concrete Research,* **33(116)**, Sept: 139–146.

Okada, T. and Iguro, M. (1978) Bending behaviour of prestressed concrete beams under low temperature. *J. of Japan Prestressed Concrete Eng. Assocn.* **20**, *Special Issue for 8th FIP Congress*: 17–35.

Partington, J.R. (1954) *General and Inorganic Chemistry.* Macmillan, London.

Pomeroy, C.D. (1986) *Requirements for durable concrete, Improvement of Concrete Durability.* Thos. Telford, London.

Powers, T.C. (1953) Moisture effects in concrete, Nat. Rsch. Council—Division of Building Research. *Proc. Conf. Bldg. Rsch.* Ottawa.

Radjy, F. and Richards, C.W. (1969) Internal friction and dynamic modulus transitions in hardened cement paste at low temperatures. *Matériaux et Construction,* **2(7)**: 17–22.

Raithby, K.D. (1979) Behaviour of concrete under fatigue loading. *Developments in Concrete Technology-1* (ed. F.D. Lydon) Applied Sciences Publ.

Raithby, K.D. and Whiffin, A.C. (1968) *Failure of plain concrete under fatigue loading—a review of current knowledge, RRL Report LR 231. Road Rsch. Lab.* Crowthorne.

Rao, S.R. (1972) *Surface Phenomena.* Hutchinson, London.

Regourd, M. (1980) Physico-chemical studies of cement pastes, mortars and concretes exposed to sea water. *Performance of Concrete in Marine Environment, SP-65, American Concrete Institute.* Detroit.

Røren, E,M.Q., Sollie, T. and Carlin, B. (1985) Case histories of structural damages—lessons learned. *Proc. Conf. on Behaviour of Offshore Structures.* Elsevier, Amsterdam.

Rostasy, F.S., Weiss, R. and Wiedemann, G. (1980) Changes of pore structure in cement mortars due to temperature. *Cement and Concrete Rsch.* **10(2)**: 157–164.

Rostasy, F.S. and Wiedemann G. (1981) Strength, deformation and thermal strains of concrete at cryogenic conditions. *Proc. First Int. Conf. Cryogenic Concrete, The Concrete Society.* London.

Sehnal, Z.A., Kronen, H. and Marshall, A.L. (1983) Factors influencing the low temperature strength of concrete. *Proc. Second Int. Conf. on Cryogenic Concrete, The Concrete Society.* London.

Siemes, A.J.M. (1982) *Fatigue of plain concrete in uniaxial compression, IABSE Reports Vol. 37 Coll. Fatigue of Steel & Concrete Structures* IABSE/ETH-Hönggerberg, Zürich.

Siemes, A.J.M., (1982) Miner's rule with respect to plain concrete. *Fatigue of Concrete Structures, SP-75, American Concrete Inst.* Detroit.

Skinner, B.J. (1966) Thermal expansion. *Section 6, Handbook of Physical Constants, Geol. Soc. of America,Memoir 97.*

Søding, H. (1974) Calculation of long-term extreme loads and fatigue loads of marine structures. *The Dynamics of Marine Vehicles and Structures in Waves, Inst. Mech. Eng.* London.

Somerville G. and Taylor, H.P.J. (1978) Concrete properties. *Proc. 8th. FIP Congress, Part 1, Cement and Concrete Assocn.* Slough.

Sparkes, F.N. (1954) The control of concrete quality: a review of the present position. *Proc. Symp. on Mix Design and Quality Control of Concrete, Cement and Concrete Assocn.* London.

Sparks, P.R. (1982) The influence of rate of loading and material variability on the fatigue characteristics of concrete. *Fatigue of Concrete Structures, SP-75 American Concrete Inst.* Detroit.

Sparks, P.R. and Menzies, J.B. (1973) The effect of rate of loading upon the static and fatigue strengths of plain concrete in compression. *Mag. Concrete Rsch.* **25(83)**, June: 73–80.

Streeter, V.L. and Wylie, E.B. (1979) *Fluid Mechanics, 7th ed.* McGraw-Hill/Kogakusha, Tokyo.

Swanson, S.R. (1976) The use of random processes in fatigue engineering. *Proc. Conf. on Fatigue Testing and Design, Vol. I, Soc. of Environmental Engineers.* London.

Taylor Woodrow Research Labs. (1980) Marine durability survey of the Tongue Sands Tower. *Concrete in the Oceans Tech. Report No. 5, Cement & Concrete Assocn.* Slough.

Tepfers, R. (1982) Fatigue of plain concrete subject to stress reversals. *Fatigue of Concrete Structures, SP-75, American Concrete Inst.* Detroit.

Terzaghi, K. and Peck, R.B. (1960) *Soil Mechanics in Engineering Practice.* John Wiley, New York.

Tibbetts, D.C. (1971) Performance of concrete in sea-water. *Proc. Thorvaldson Symp. on Performance on Concrete.* Univ. of Toronto Press, Toronto.

Traina, L.A. and Jeragh, A.A. (1982) Fatigue of plain concrete subjected to biaxial-cyclical loading. *Fatigue of Concrete Structures, SP-75, American Concrete Inst.* Detroit.

Turekian, K.K. (1972) *Chemistry of the Earth.* Holt, Rinehart & Winston, New York.

Valenta, O. (1969) Durability of concrete. *Proc. 5th. Int. Symp. on Chemistry of Cement, Part III, Properties of cement paste and concrete, Cement Assocn. of Japan.* Tokyo.

Venuat, M. (1980) *Comportement des ciments et des bétons à la mer, Coll. Int. Tenue des OUvrages en Béton en Mer, CNEXO.* Brest.

Verbeck, G.J. and Helmuth, R.H. (1969) Structure and physical properties of cement paste. *Proc. 5th Int. Symp. on Chemistry of Cement, Vol. III, Cement Assocn. of Japan.* Tokyo.

Veritas, Det norske (1980) *Experiences from in-service inspection and monitoring of 11 North Sea structures, Coll. Int. Tenue des Ouvrages en Béton en Mer, CNEXO.* Brest.

Vughts, J.H. and Kinra, R.K. (1976) Probabilistic fatigue analysis of fixed offshore structures. *Proc. 8th Annual Offshore Tech. Conf.* Houston.

Vuorinen, J. (1985) Applications of diffusion theory to permeability tests on concrete. *Mag. of Concrete Rsch.*, **37(132)**, Sept: 145–161.

Walker, M.J. (1985) The general problems for reinforced concrete in the Gulf region and the use of 'design life' and 'life cycle costing' to justify improvements. *Proc. First Int. Conf. Deterioration and Repair of Reinforced Concrete in the Arabian Gulf, the Bahrain Soc. of Engineers,* Manama, Bahrain.

Watson, A.J. and Oyeka, C.C. (1981) Oil permeability of hardened cement pastes and concrete. *Mag. Concrete Rsch.* **33(115)**, June: 85–95.

Watson, A.J. and Sanderson, A.J. (1984) Cratering and subcrater fractures produced by explosive shock on concrete. *Proc. 3rd. Conf. Mechanical Properties at High Rates of Strain, Inst. of Physics.* Oxford.

Wiebenga, J.G. (1980) Durability of concrete structures along the North Sea coast of the Netherlands. *Performance of Concrete in Marine Environment, SP-65, American Concrete Inst.* Detroit.

Zielinski, J. (1984) Fracture of concrete under impact loading. *Proc. Int. Conf. Structural Impact and Crashworthiness.* (ed. J. Morton) Elsevier, London.

5 Behaviour of reinforced and prestressed concrete

5.1 Introduction

The sequence of the preceding chapter will be followed in this as far as possible so that durability, fatigue, impact and thermal response will be considered in turn. However, an important difference is the incorporation of the behaviour of steel. Consequently phenomena such as corrosion and bond must be included, resulting in some shift in emphasis. Differential movements too acquire fresh significance due to the conjunction of dissimilar materials. Moreover, concrete itself has to be viewed as a protective coating as well as a structural material.

The picture is thus more complex, even confused. However, the underlying physical processes are, if not entirely, at least fairly well understood so that behaviour can be evaluated, although not necessarily always quantified. This should make it easier to avoid potential fault and to ensure the elimination of possible deterioration or at the very least to delay it to beyond the point at which it ceases to be of economic significance.

5.2 Durability

Even a cursory glance at the literature indicates that the principal source of concrete degradation is the corrosion of reinforcement. In essence, rusting steel expands causing the covering concrete to crack and spall. This provides easier access for the agents of corrosion thereby accelerating the process of deterioration until eventually the protection disappears and, unless unchecked, ultimately the steel as well. The two primary requirements for a sound structure then are:

(a) both general and detail design must minimise, if not eliminate entirely, the risk of damage;
(b) the quality of materials and workmanship must be of adequate standard.

Clearly these place obligations upon both designer and builder. However, there is also an obligation upon the purchaser of the structure. Improvement in initial quality and probable endurance imply greater capital investment (through a higher purchase price). If maintenance is to be reduced and the need for remedial work averted, appropriate measures have to be taken at the outset and they have a cost, albeit marginal in many cases, which the purchaser must be prepared to meet.

5.3 Corrosion

Corrosion is a complex and variable process so that, on the one hand there are reinforced concrete structures in good condition after 100 years exposure to various environments (Wilkins and Lawrence, 1985) while, on the other hand, there has been at least one case of noticeable deterioration within a year of construction (Sneddon, 1985). Why is there such divergent behaviour? Obviously there will be differences in quality of concrete, quality of workmanship, and nature of exposure but it is necessary to consider more fundamental processes in order to arrive at a better understanding of why these factors are significant.

5.3.1 *Corrosion as an electro-chemical process*

Four states of corrosion of steel in concrete have been defined (Arup, 1985):

(a) the passive state;
(b) pitting corrosion;
(c) general corrosion;
(d) active, low-potential corrosion.

These relate to corrosion as an electro-chemical process which requires a potential difference between two connected electrodes in an electrolyte, as in a car battery. In concrete the electrodes may be neighbouring points on the same reinforcing bar, or be separate bars or groups of bars which, for whatever reason, have a potential difference between them.

5.3.1.1 *Permeability and moisture content* The electrolyte is the liquid in the concrete and so obviously will be affected by its chemical composition; its electrical conductivity will be governed also by the length and cross-sectional area of the path between the electrodes as in any electrical conductor. Liquid contained in the pores of another material (such as concrete) may have the same (electrical) current intensity or carrying capacity as the 'free' liquid but the more dispersed it is, the less total 'current' it will carry through a given cross-section of the containing material. Hence one might reasonably expect the behaviour of a particular electrolyte, or the rate at which it reacts, to be affected by the permeability of the concrete since that controls the availability and access of the electrolyte. However, if the electrodes (and particularly the cathode) are large enough, then the limiting effect of permeability will be countered to some extent by the expanded conducting path.

Moisture content must be a factor to be considered so that corrosion is not to be expected in dry concrete (which in any case would require a completely dry surrounding atmosphere—a rare occurrence) but may be anticipated in saturated concrete. This does not necessarily occur however, or the process may be so slow as to be quite insignificant. Fairly complex phenomena are involved which may be affected principally by chlorides, oxygen and carbon dioxide. The net result is that, generally, the worst cases of corrosion damage occur in moist concrete to which all three substances have access; this generally means in the splash zone in marine structures, or in structures close to the sea which are affected by other specific conditions, as in the Arabian Gulf.

5.3.1.2 *Electrolytic potential* It is important to bear in mind the electro-chemical nature of corrosion in considering what happens. Different substances have different electric potentials, giving rise to an electromotive force between connected pairs of electrodes in the presence of a conducting fluid (the electrolyte). It is customary to measure such potentials relative to a standard electrode such as the standard hydrogen electrode (she) and standard calomel electrode (sce). The electro-chemical series is a grading system for substances according to their potential difference from the standard, some being relatively electro-positive and others electro-negative.

When two metallic plates in a cell are connected externally to complete the circuit, the electrolyte is split into positive and negative ions which then react separately with the two electrodes. By convention the flow of electric current is *deemed* to be a drift of positively charged particles whereas the reality is that it is an electron drift in the opposite direction, from negative to positive electrode. Now during electrolysis, because of the nature of the reactions at the electrodes, the *actual* electron drift occurs from the more electro-positive to the lesser. Consequently, to accord with convention, in the electrolytic cell the anode (positive electrode) is the more electro-negative with the cathode (negative electrode) then being the more electro-positive. *Conventional* current flow is from anode to cathode, of course.

The electrode potential e (volts) is given by the Nernst equation

$$e = \frac{RT}{yF} \ln \frac{p}{P} \tag{5.1}$$

where p is the osmotic pressure of the ion, P is the solution pressure of the metal, R is the gas constant in electrical units, T is the absolute temperature, F is 1 faraday (= 96 500 coulombs) and y is the valency of the ion (Partington, 1954).

From this it can be shown that

$$e = (RT/yF)\ln a + e_0 \tag{5.2}$$

where a is the activity and e_0 is the standard electrode potential. In dilute solutions the activity a is the same as the concentration. The potential difference $(e - e_0)$ is an important factor in the corrosion process so that there is clearly some dependence on temperature and concentration. However, for this aspect at any rate, the difference between behaviour in say Arctic and tropical waters is not a major one: about 10% for a temperature of 35°C as compared with 5°C.

5.3.1.3 *Oxidation and reduction* Electrolysis means that there is dissociation in the electrolyte, with negatively charged ions moving to the anode (anions) and positively charged ions (cations) to the cathode. Separate actions then occur at both anode and cathode which depend on the chemical compositions of electrolyte and electrodes, with metal ions being released from the anode. In general, oxidation takes place at the anode and reduction at the cathode. Here, oxidation has the general meaning of *removal* of electrons from the element or compound which is oxidised, while reduction means *gain* of electrons (Mackay and Mackay, 1986). Reduction can result from removal of oxygen or electronegative elements *or* gain of hydrogen or electropositive elements. The potentials tabulated to produce the Electrochemical (or Electromotive) Series referred to above are known as redox potentials and the change in the redox system at either electrode may be expressed as

$$\text{Ox} + ne^- = \text{Red} \tag{5.3}$$

where Ox and Red represent the oxidised and reduced forms respectively and ne^- is the number of electrons exchanged. Thus for iron

$$Fe^{3+} + e^- = Fe^{2+} \tag{5.4}$$

and

$$Fe^{2+} + 2e^- = Fe$$

The potential for the first of these, the oxidation of ferrous (Fe^{2+}) to ferric (Fe^{3+}) ions (or reduction of ferric to ferrous ions) is +0.77 v while the potential for the second, the oxidation of iron to ferrous ions, is −0.41 v. The two conversions therefore occupy very different places in the Series, which may be regarded as indicating the relative tendencies of various substances to gain or lose electrons. The Series is, of course, a reflection of their performance in relation to hydrogen as a reducing agent, with the negative sign indicating 'better'.

Where the oxidised and reduced forms in a redox system are not at unit activities, the Nernst equation becomes

$$e = e_0 + \frac{RT}{nF}\ln\frac{[\text{Ox}]}{[\text{Red}]} \tag{5.5}$$

with the symbols as for eq. (5.2), n the number of electrons as in eq. (5.3) and [ ] signifying activity or ion concentration. At 25°C and substituting for the constants, eq. (5.5) becomes

$$e = e_0 + \frac{0.059}{n} \log \frac{[\mathrm{Ox}]}{[\mathrm{Red}]} \tag{5.6}$$

One further term requires some explanation. Dissociation of an electrolyte means that it separates into positive and negative ions, as has been mentioned already. These then shunt along until there is a reaction (or reactions) at the cathode and at the anode resulting in bonding and electron drift. However, it appears that in some solids there is a disparity in sizes between ions. Where, for example, the cation is very small and the anion large, there is a distortion of charge and a dipole is induced in the anion and it is thus *polarised*. When polarisation occurs at an electrode it introduces an additional surface resistance so that the reaction is slowed or impeded: the reduction in efficiency of a cell when a plate becomes coated with hydrogen bubbles is due to polarisation for example, and the electrolyte has to be modified to avoid it otherwise mechanical action would be needed to remove the bubbles.

5.3.2 *Corrosion reactions*

Corrosion of steel in concrete has been postulated as in Figure 5.1 (Beeby, 1978). As has already been indicated, anode and cathode may be on different bars or may occur near each other on the same bar, due to the presence of mill scale perhaps, or differences in quality, or impurities in the steel. In the region

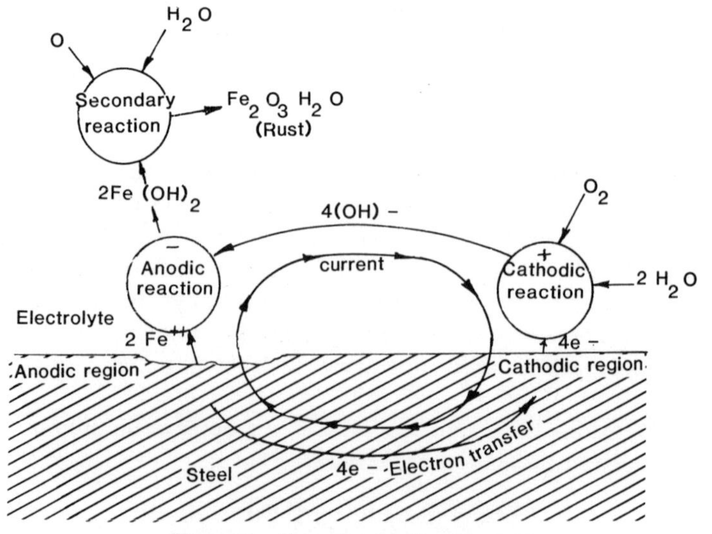

Figure 5.1 Corrosion reaction (idealised)

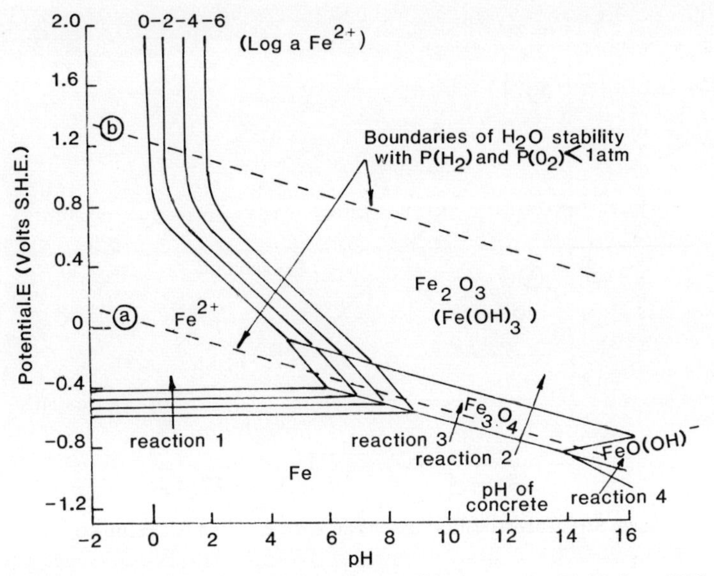

Figure 5.2 Potential-pH equilibrium diagram for the systems Fe-H_2O at 25°C

of the anode, four corrosion reactions prove to be the primary ones (Wilkins and Lawrence, 1980). They are:

$$Fe \rightarrow Fe^{2+} + 2e' \tag{a}$$

$$Fe + 3H_2O \rightarrow Fe(OH)_3 + 3H^+ + 3e' \tag{b}$$

$$3Fe + 4H_2O \rightarrow Fe_3O_4 + 8H^+ + 8e' \tag{c}$$

$$Fe + 2H_2O \rightarrow FeO(OH)' + 3H^+ + 2e' \tag{d}$$

Which takes place depends on both the potential and the pH of the electrolyte, as illustrated on the so-called Pourbaix diagram of Figure 5.2 (Wilkins and Lawrence, 1980). Bearing in mind that a pH value of 7 represents neutrality, with higher values indicating increasing alkalinity, it is clear from the simplified version of Figure 5.3 (Wilkins and Lawrence, 1980) that certain conditions must exist before corrosion can take place. (It should be noted that $pH = -\log a_H$ where a_H is the hydrogen ion activity, often taken as the hydrogen ion concentration.) The pore solution of Portland cement concretes reaches pH values of 13 to 14 within a few hours of hydration and they are maintained for very long periods in the absence of leaching effects or carbonation (Page, 1985). Cement composition and water cement ratio have some influence but pH values over 13 are still normal even in the presence of cement substitutes such as pozzolanas and slags.

5.3.2.1 *Passivation* Some explanation of the term 'passivity' or 'passivation' is warranted at this juncture. In highly alkaline concrete the anodic steel

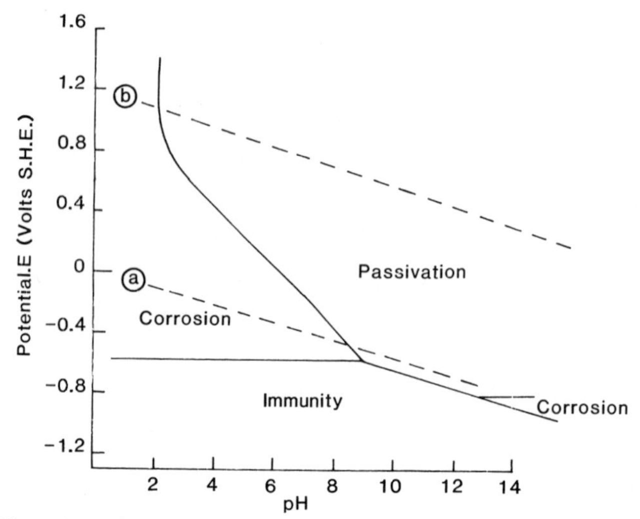

Figure 5.3 Theoretical conditions for corrosion and passivation of iron

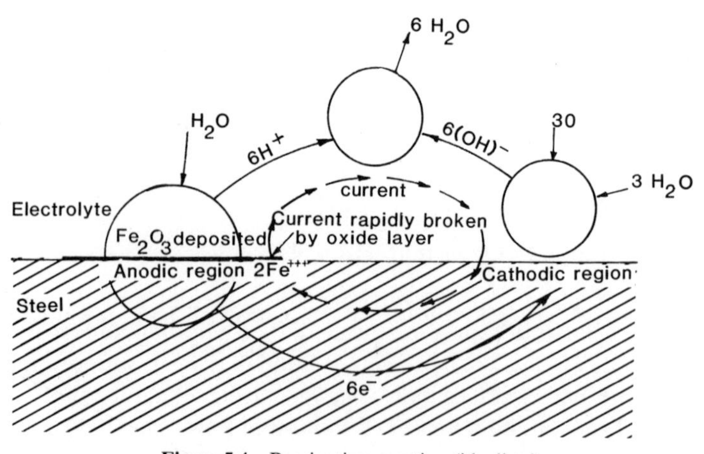

Figure 5.4 Passivating reaction (idealised)

surface becomes coated with a very thin protective layer of ferric oxide which is stable over a wide range of potentials and so acts as a protective coating—corrosion in fact continues but is so slow as to be virtually imperceptible: of the order of 1 μm/year (Page, 1985). The idealised reaction is shown in Figure 5.4 (Beeby, 1978). This anodic polarisation means that a higher potential difference or current density is needed for corrosion to take place. However, it is affected by the presence of chloride ions, and may be related to their concentration relative to the hydroxyl ions released at the cathode (Page, 1985). The precise mechanism appears to be uncertain but it amounts to an

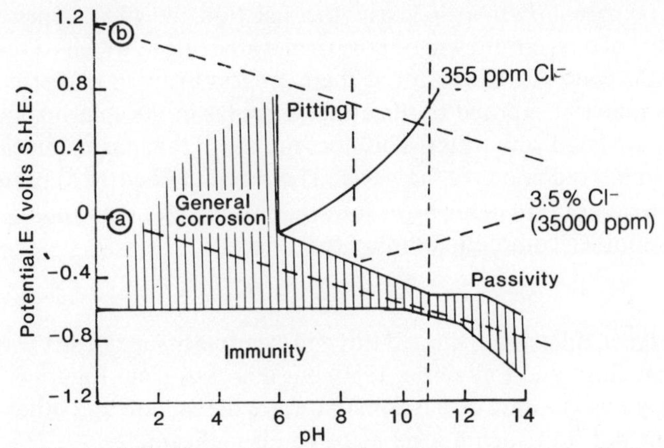

Figure 5.5 Influence of chloride on corrosion and passivation of iron in well-stirred solutions

effective perforation of the ferric oxide coating. Figure 5.5 (Wilkins and Lawrence, 1980) illustrates the outcome.

5.3.2.2 *Diffusion of chloride ions* The concentration of chloride ions varies with cement type and appears to be higher for higher C_3A content in OPC but this last seems to have a less significant effect on the effective diffusivity of chloride ions. A ranking for some different cements is given in Table 5.1 and it is clear that sulphate resisting cement is by far the poorest performer, with cement replacement by blast furnance slag being the best (Holden et al, 1985). This suggests that corrosion (and hence corrosion damage) might be least with blast furnace slag concrete, which seems to be borne out by the performance of the Dutch structures referred to in the preceding chapter. Sulphate resisting cements are perhaps best avoided.

The metal ions released at the anode give rise to the formation of the corrosion product for which hydroxyl ions released at the cathode are also needed:

$$\tfrac{1}{2}O_2 + H_2O + 2e^- = 2OH^- \tag{5.8}$$

Table 5.1 Effective diffusivity of chloride ions at 25°C in various cement pastes of w/c 0.5 (Holden et al, 1985).

Type of cement	Diffusivity ($\times 10^8$) $cm^2 s^{-1}$
OPC—A	3.14
OPC—B	4.47
OPC—B/30% PFA	1.47
OPC—B/65% BFS	0.41
SRPC	10.00

The C_3A content of OPC-A, OPC-B, and SRPC was 7.7, 14.3 and 1.9% respectively.

5.3.2.3 *Oxygen diffusion* Clearly the reaction, eq. (5.8) depends on the availability of oxygen and will be governed by the rate at which oxygen diffuses through the concrete. Corrosion is therefore less likely in immersed concrete than in a material exposed to the atmosphere (as in the splash zone) because the pores are filled with water. This does not mean that corrosion cannot take place in immersed concrete, however. The reaction of eq. (5.8) partners (a) of eq. (5.7). Reactions (b) to (d), however, release hydrogen ions and these may go towards another common cathodic reaction:

$$2H^+ + 2e^- = H_2 \tag{5.9}$$

Where oxygen diffusion is limited this will eventually supersede the reaction of eq. (5.8) (Wilkins and Lawrence, 1980) but it is clear from Figure 5.2, without embarking on elaborate discussion, that rates of reaction and other necessary conditions put it at such a level as to be insignificant.

5.3.3 *Carbonation*

An important factor in preventing or limiting the corrosion of steel in concrete is obviously the alkalinity of the concrete. So long as that can be maintained at a sufficently high level then the risk is minimised. Unfortunately, concrete exposed to the atmosphere is subject to carbonation in which the (slaked) lime in the concrete is converted to calcium carbonate so reducing the pH value of the medium (carbon dioxide also dissolves in water to produce the weak carbonic acid).

On the other hand there may be accompanying pore-blocking which reduces permeability although the extent must depend on which form of $CaCO_3$ is precipitated—it can exist in three polymorphic forms, calcite, vaterite and aragonite, the last of which increases the solid volume in converting $Ca(OH)_2$ by 3% while the first increases it by 11% (Lea, 1970). The chemical reaction is

$$Ca(OH)_2 + CO_2 = CaCO_3 + H_2O \tag{5.10}$$

and it has been pointed out (Venuat and Alexandre, 1968/9) that the speed of reaction will depend on the rate of removal of the water formed. In other words, carbonation depends on the existence of a drying atmosphere and is impeded in the presence of water. On the other hand dry CO_2 does not react with dry $Ca(OH)_2$ so the presence of moisture is essential to the carbonation process. In short the 'optimum' moisture content for carbonation is intermediate between zero and saturation. This results in a curve such as that of Verbeck in Figure 5.6 (Venuat and Alexandre, 1968/9). (It should be noted too that carbonation also occurs in cement, such as high alumina cement, which contains no calcium hydroxide—carbon dioxide attacks and decomposes all the hydration products of cement (Lea, 1970) although it is not clear how that affects corrosion.

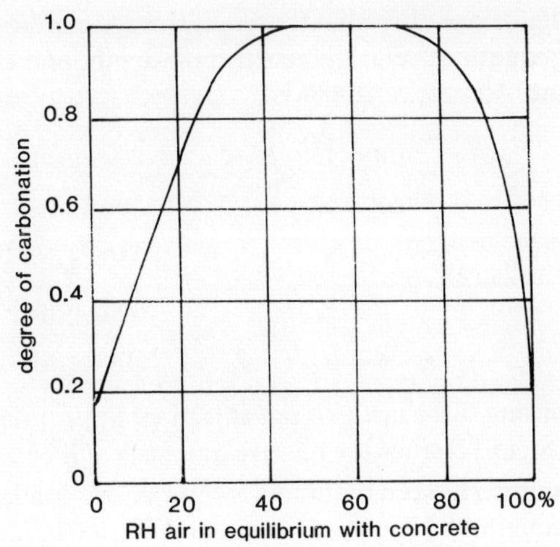

Figure 5.6 Degree of carbonation as a function of relative humidity

Fortunately, carbonation can be a very slow process, so that it may not reach the depth of the reinforcement within the life of the structure. Various relationships have been proposed for depth of penetration, the common feature being that depth is proportional to the square root of time in years. They are generally of simple form, one based on German work being (Browne et al, 1985)

$$d = \sqrt{2Dt} \tag{5.11}$$

where d is the depth of penetration (mm), t is the time in years and D is the diffusion coefficient (mm^2/year). It is suggested that a high value of D would be 15 and a moderate value, 5. A similar relationship, based on work in Finland, is quoted elsewhere (Allen and Forrester, 1985):

$$d = 10B\sqrt{t} \tag{5.12}$$

where the constant B depends on quality of concrete and storage condition.

Table 5.2 Variation in factor B with storage conditions for several concrete grades. (Factor B is used in the prediction of the depth of carbonation).

Concrete quality	Storage	B
Low Strength	Outdoors (moist)	0.6
	Indoors	1.0
Middle Strength	Outdoors (moist)	0.2
	Indoors	0.5
High Strength	Outdoors (moist)	0.1
	Indoors	0.2

Assuming, reasonably, that the worst storage condition would be 50%, a set of French equations relating penetration depth, and time to cement content (Venuat and Alexandre, 1968) is

$$
\begin{aligned}
d &= 15\sqrt{t} + 2 && \text{at } 200\,\text{kg/m}^3 \\
d &= 12\sqrt{t} + 2 && 300\,\text{kg/m}^3 \\
d &= 5\sqrt{t} + 2 && 350\,\text{kg/m}^3 \\
d &= 5\sqrt{t} && 400\,\text{kg/m}^3 \\
d &= 5\sqrt{t} - 2 && 500\,\text{kg/m}^3
\end{aligned}
\tag{5.13}
$$

In comparing these last two sets it is more appropriate to take the indoors values for eq. (5.12) since the relative humidity will be closer to the 50% of eqs (5.13). It can then be seen that there is some correspondence, since $d = 5\sqrt{t}$ for 'Middle strength, Indoors' and for a content of 400 kg/m^3. This would result in a D value of 12.5 for eq. (5.11), which is somewhat higher than one might expect from the guidance given. That is, the first relationship underestimates the depth of penetration in comparison with the other two. However, if one adopts a life of 100 years the carbonation layer is still only 50 mm thick on a worst estimate.

5.3.4 *Chloride concentration*

Carbonation of concrete does not in itself corrode steel, of course, it provides the environment in which corrosion can take place. If it can be prevented or delayed, then corrosion becomes less likely, and clearly sufficient cover and cement content help in that direction: for example, a minimum of 50 mm, and 400 kg/m^3 is indicated by the figures above. The other factors over which control is necessary are the access of chloride ions governing the anodic reaction, the access of oxygen governing the cathodic reaction, and the availability of moisture. These are interrelated undoubtedly, in that increasing moisture content increases chloride content but limits oxygen diffusion (as well as carbon dioxide).

Figure 5.7 (Tuutti, 1980) indicates how surface concentration of chloride ions, cover, diffusivity, and threshold concentration affect the initiation of corrosion. The threshold concentration indicated is that of chloride ions and varies with cement type as suggested previously but, quoting other sources, the ratio of chloride to hydroxyl ions required for (local) corrosion is constant at 0.61. (A figure similar to Figure 5.7 for carbonation is also given in the reference but is omitted here to avoid confusion with the relationships already given. One point worth emphasising, however, is that increased CO_2 in the atmosphere does increase the rate of carbonation—the normal concentration is about 0.032%.) Assuming that penetration of chloride ions is a diffusive process a method of estimating the distribution has been developed (Nagano

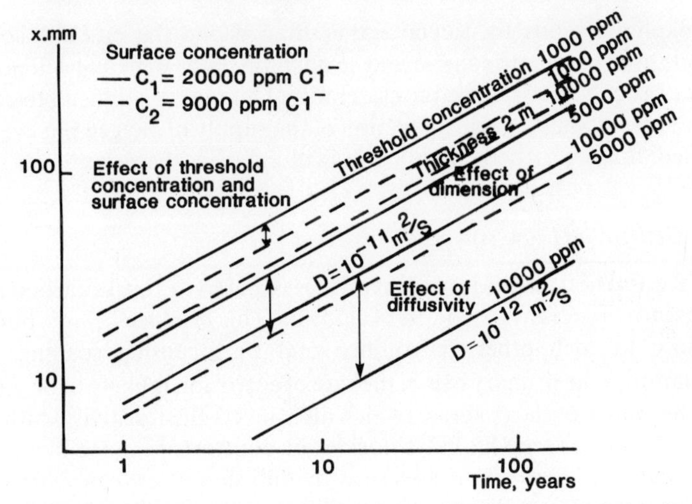

Figure 5.7 Calculated initiation periods in connection with chloride penetration (D permeability of material)

and Naito, 1986) based on the well known solution of the diffusion equation (to be found in standard mathematical texts):

$$C_c(x,t) = C_0\left(1 - \operatorname{erf}\frac{x}{2\sqrt{D_c t}}\right) \tag{5.14}$$

where C_c is the percentage chloride ion concentration in concrete by weight at depth x cm. and time t seconds, D_c is the coefficient of diffusion (cm²/s) of apparent chloride concentration in the concrete, and C_0 is the chloride supply concentration i.e. in the water at the immersed concrete surface; erf is the error function.

The allowable chloride concentration suggested is 0.06% so that, as long as the level at the steel is less, there is no need for concern. For the case studies to which the method was applied in the reference, for a 50 N/mm² concrete D_c was 0.50×10^{-8} cm²/s and for a 24 N/mm² concrete it was 2.2×10^{-8} cm²/s.

Four states of corrosion were alluded to at the outset of this discussion and now warrant some explanation. The passive state has already been considered leaving the other three to be dealt with (following Arup, 1985).

5.3.5 *Pitting*

Pitting results from local breakdown of the passive film where it is weak or there is a high chloride ion concentration. Adjacent passive steel acts as the cathode (for which a relatively large area may be needed) and the anode dissolves away to produce a pit. Polarisation of the cathode reduces the corrosion potential preventing the development of new pits. High resistivity of the concrete also constrains the action, as low resistivity encourages it and high alkalinity tends to deepen rather than widen the pit. The corrosion products, moreover, are soluble and so spalling of the concrete will not occur until the process has advanced considerably. This makes it difficult to detect, of course. Development is also a function of the supply of oxygen but even if this is limited, a large cathode/anode area will intensify the process.

5.3.6 *General and low-potential corrosion*

General corrosion results from general loss of passivity due to carbonation or the presence of excessive amounts of chloride. This produces a large number of pits close to each other generating solid rust, causing spalling. Where carbonation is the primary cause, the rate of corrosion is likely to be governed by the humidity of the concrete (which also affects the resistivity) rather than the supply of oxygen. Excess chloride, in contrast, is more likely in wet concrete and so corrosion is governed by diffusion of oxygen. This suggests that tidal concrete may be more susceptible to chloride damage, and concrete immediately above the tidal range more susceptible carbonation effects.

Active, low-potential corrosion is very slow, occurring where passivity cannot be maintained due to lack of oxygen but in strongly alkaline concrete. Pitting cannot occur because the potential is too low and the corrosion rate is so low as to be insignificant.

5.3.7 *Influence of cracking*

It is to be expected that the corrosion pattern will be affected by cracking in the surrounding concrete. An extensive discussion is provided elsewhere, (Beeby, 1978) so it may be curtailed here. The preferred view is to regard a crack as offering a path to the interior of the concrete for depassivating and corroding agents. Due to bond slip and internal cracking around deformed bars in particular, there will also be seepage for a distance along a bar on both sides of a given crack, so corrosion will occur much earlier than would arise with penetration from exposed surfaces. The pattern is illustrated in Figure 5.8 (Beeby, 1978) while Figure 5.9 from the same source shows how cracks might be expected to develop.

Crack width is not constant, being wider at the free surface and tapering inwards along the crack until it reaches the steel (or not as the case may be). In relation to corrosion, its significance is that it brings forward the onset of the action, after which the rate will be governed by the processes already described. However, the crack may become blocked or partially blocked by chemicals such as brucite and aragonite as described in the preceding chapter, as indeed may the pores (Fidjestøl and Nilsen, 1980). This protects against corrosion but is likely to depend on the nature of loading: static loading will aid, while cyclic or other dynamic loading can impede protection since the pumping action of water in the crack may carry away the products as well as

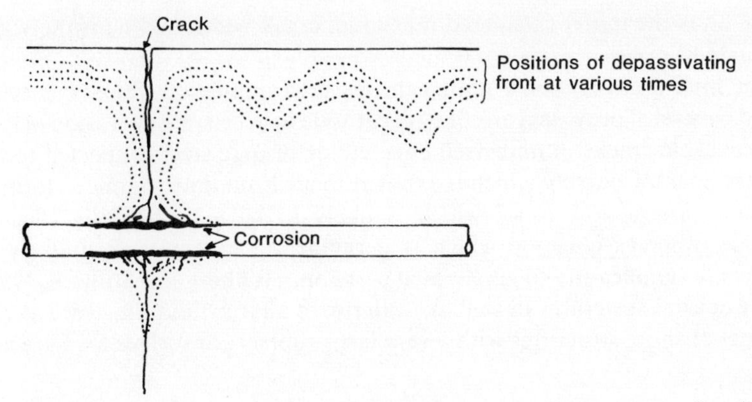

Figure 5.8 Schematic picture of advance of depassivating front in region of crack

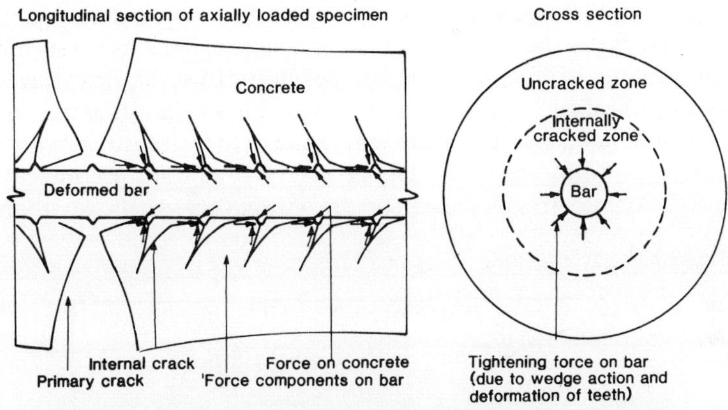

Figure 5.9 Schematic of conditions close to a deformed reinforcing bar

ensuring that discontinuity is maintained (this is returned to later). Crack width *per se* is unlikely to be directly related to corrosion rate. Dynamic loading in itself seems unlikely to affect corrosion (Nilsen and Espeliol, 1985) but there is little doubt that cracking does, particularly if cracks are aligned along bars, as distinct from being transverse to them (Stillwell, 1983). However, in the short term at least, trouble develops only if the crack reaches the bar but, if it is sufficiently wide (0.1 to 0.2 mm), macro-cell corrosion occurs rather than the micro-cell epitomised by Figure 5.1 (Okada and Miyagawa, 1980). Crack widths are increased by repeated loading, however, (Lovegrove and El Din, 1980), so that the risk of corrosion may be increased. A suggested relationship is

$$w_n = w_0(0.382 - 0.0227 \log n) \log n \tag{5.15}$$

where w_0 is the initial calculated maximum crack width and w_n is the width after n load cycles.

One final point is worth noting in this brief discussion. 'Passive steel in sound concrete [provides] an effective cathode to accelerate corrosion of steel at vulnerable cracks in immersed concrete or of bare steel connected to the reinforcement. Concrete which is exposed to air from time to time... forms a "galvanic interaction zone" which is likely to be a particularly effective cathode, whereas concrete which is permanently immersed is unlikely to contribute significantly to accelerated corrsion.' (Wilkins and Stillwell, 1986). Hence coastal structures in shallow water with a large tidal range are at risk whereas offshore structures with a very large submerged surface area are not.

5.3.8 *Prestressed concrete*

Thus far the discussion has been concerned primarily with reinforced concrete and, in general, the arguments can also be applied to prestressed concrete, certainly in so far as the concrete element is concerned: the steel may differ in speed of response with variation in composition. However, there have been cases where so-called stress corrosion has occurred in prestressed concrete, although it is argued that it is in fact generally due to hydrogen embrittlement (Burdekin and Rothwell, 1981). The necessary conditions are indicated in Figure 5.10. Anodic stress corrosion occurs only in very specific environments

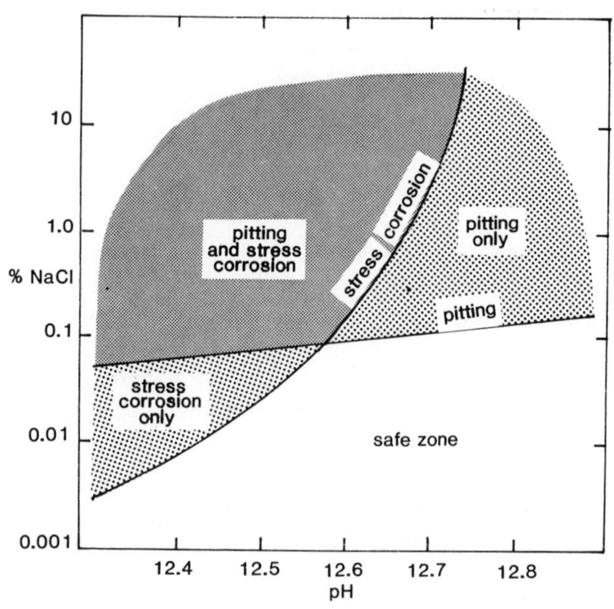

Figure 5.10 Comparison of conditions for pitting corrosion and stress corrosion cracking

and under particular electrochemical conditions such as in concrete containing aggressive anions like nitrate. The mechanism of hydrogen embrittlement, on the other hand, involves absorption by the steel of atomic hydrogen from the corrosion process (or from any other source). It leads to rapid low ductility failure when the overall fracture stress is exceeded being 'particularly trouble some in high strength steels, at high stress levels, and the rate of failure and to some extent the possibility of failure depends on the stress level.' (Burdekin and Rothwell, 1981). The environmental range is much wider than for anodic stress corrosion which requires hydrogen ions and, frequently, substances such as sulphide which obstruct the re-combination of hydrogen atoms to form molecules.

The most likely conditions for hydrogen evolution (which requires substantial reduction in pH levels) are in ungrouted or badly grouted tendons. Even there, however, the likelihood appears to be that tendons in voids will be covered with a protecting film of grout. Nevertheless the possibility of local damage should not be discounted. To guard against delayed fracture, storage and installation procedures should ensure that unprotected steel does not corrode prior to concreting and/or grouting—hydrogen introduced in such a case can diffuse subsequently to high stress areas.

Specification of cold drawn steels substantially reduces the probability of hydrogen embrittlement under embedded conditions so that the risk to tendons is minimised (although not eliminated). Anchorages, on the other hand, cannot be safeguarded quite so easily, with the principal sources of trouble being due to shrinkage of the filling material and penetration of it by aggressive matter (Eberwijn and de Waal, 1985). Preventive measures include use of primers in anchorage recesses, reinforcement, grouting, waterproofing, capping of the ends of tendons, and so on. In short it is imperative that seawater and oxygen penetration be prevented.

5.3.9 *SRB (sulphate-reducing bacteria)*

One further source of corrosion warrants attention. The destructive effect of hydrogen sulphide in sewers is well known (Lea, 1970). However, quite apart from any direct chemical effects on the concrete itself, it is to be expected that its alkalinity will be reduced, thereby increasing the likelihood of steel corrosion. The hydrogen sulphide is produced by anaerobic bacteria which feed on available sulphates in the absence of oxygen. This can also happen under certain conditions in offshore structures and the resulting problems can be grouped under two headings as shown at Table 5.3 (Hamilton and Sanders, 1984).

Sulphate reducing bacteria (SRB) are said to be

(a) capable of causing cathodic depolarisation by utilisation of the electrochemically produced hydrogen,

Table 5.3 Problems associated with particular types of structures

Production systems	Type of problem	Offshore structures	Type of problem
Reservoir	souring, plugging	Marine macrofouling	external corrosion of underlying structure
Oil-carrying lines (risers, pipelines)	internal corrosion	Natural sediments	
Water-carrying lines (injection systems)	internal corrosion	Deposited materials	
Storage	H_2S gas, corrosion, souring		

(b) instrumental in the continuous production of sulphide and the deposition of sulphide films.

'Experience offshore has shown that significant concentrations of hydrogen sulphide are generated by SRB in stagnant water in offshore installations, in particular in those platforms with oil/water storage systems.' (Wilkinson, 1984).

Stagnant water conditions result from the consumption of oxygen by the growth of aerobic bacteria in large volumes of water in, for example, water-filled legs and storage cells. This generates carbon dioxide and lowers the pH value as before but, more significantly perhaps, it produces the conditions for anaerobic bacteria (SRB) to develop. They in turn produce H_2S and lower the pH value. Furthermore, if the sulphide reaches oxygen then it can be oxidised to produce sulphuric acid.

Quantification is difficult but steady state corrosion conditions are quoted as being between 0.5 and 0.75 mm/year (Wilkins & Lawrence, 1980) so that the problem has to be taken seriously.

Apart from the stagnant water conditions referred to, other sources cited are marine fouling, bacterial activity in sediments, dumped drilling cuttings, and discharged production water (produced naturally in the reservoir together with oil and gas) (Wilkinson, 1984). Clearly the existence of currents in the water around the structure will help to reduce some of these external effects, while recirculation in storage cells will help internally as will the use of added chemicals—in stagnant conditions the concentration of H_2S doubles every 10 to 15 hours. Chlorination and aeration, however, are obviously undesirable and extra precautions against corrosion may be required.

While these effects may be peculiar to the production of hydrocarbons, it is as well to remember that certain geographical areas result in exposure to H_2S, as indicated in the preceding chapter. Structures situated near sewer outfalls may also be at some risk and one might hazard a guess that shallow waters in volcanic areas will have heavier concentrations of sulphur.

5.3.10 *Interpretation of measurements: a cautionary note*

It is clear that proper assessment of conditions in a given concrete depends on sophisticated analysis. While *in situ* measurements of potential differences may be possible, and resistivities can be of help, there are many traps for the unwary in interpreting results. Two brief quotations might serve in this regard '…very careful interpretation of data is required if electrochemical measurements are to be correctly interpreted, even qualitatively' (Wilkins and Lawrence, 1980). '…potentials taken blindly in a splash zone could be interpreted as active rebar where it was really just a polarisation effect' (Browne et al, 1985).

To counter this need for elaborate techniques it is tempting to look for easy-to-apply empirical relationships. They are not too difficult to derive, given careful selection (Beeby, 1980), but it is emphasised that 'the data do not permit the isolation of the critical variables in an unambiguous way.' From this particular study it emerges that it is also possible to be over-conservative, although that is less undesirable than being unsafe.

The moral may be that the results of exposure tests can assist in studying the effect of varying 'engineering' parameters such as cover to reinforcement, bar diameter, water/cement ratio, and so on. However, such effects can only be evaluated properly by examining the specific conditions of test in relation to the basic physico-chemical processes involved, such as have been described above. Unfortunately the requisite information is too often not available, especially for long-term tests now at an advanced stage. Considerable judgement is needed in such cases.

5.3.11 *Temperature dependence*

Brief reference has been made to temperature effects and some elaboration may be useful. It is well-known that the speed of most chemical reactions doubles with an increase in temperature of about ten degrees. However, the temperature dependence of *electro*-chemical processes is somewhat less, as has been indicated above (as in eq. (5.2)), so that the actual corrosion reactions themselves may be only marginally affected. On the other hand, the carbonation reaction may be accelerated significantly although again that is more complicated than first appears and will depend on the controlling factors. If evaporation of water is the most important, then it depends on temperature and vapour pressure differences and so will be governed by atmospheric conditions i.e. whether the atmospheric sink is cold or hot, wet or dry, and so on. If diffusion of CO_2 is the most important factor, then it too will be dependent on *absolute* temperature which means that the effect will be marginal. In any case, carbonation presumably affects the time to the *onset* of corrosion and then provides the conditions for corrosion to take place, rather than affecting the corrosion reaction itself (although it might perhaps alter the composition of the electrolyte).

Consequently one would conclude that temperature may be less important than might at first appear. (To return to the cell or car battery analogy, one would not normally expect temperature variation to make a vast difference to battery output although a drop in temperature does affect the load to be met, as a result of higher oil viscosity in cold weather starting, for example.) Corrosion does not vanish in Arctic conditions (Fotinos and Hsu, 1984). In contrast, bacteria may be more prolific in warmer waters, and sulphate attack and other chemical effects more 'productive'.

5.4 Bond

5.4.1 *Nature of bond*

Bond is the most important characteristic of the joint action of steel and concrete in reinforced and prestressed concrete. Without such interaction or load transfer there would obviously be little point in putting the two materials together. Consequently, the possibility must be considered that bond might be affected by the conditions of marine use and particularly by corrosion, since it is the primary modifier of the steel surface. For that it is necessary first to examine the nature of the phenomenon although it is not altogether easy to define.

Whilst it is common practice to distinguish between the bond of smooth (or plain) bars and of deformed bars, it may be best viewed as a three-stage process in which bar type determines which is dominant. There is a degree of interdependence between the stages and they also may coincide in any one location or occur simultaneously in different parts of a given length of bar or strand.

It is perhaps as well to begin by defining what bond is *not*, particularly since it is generally modelled in relation to engineering rather than chemical concepts. For example, it is not one of the predominant forms by which molecules and compounds are formed. That is, it is neither ionic, covalent nor metallic bonding (Mackay and Mackay, 1986). There may, however, be the weaker form known as Van der Waals bonding which occurs when two or more 'inert' atoms or molecules are brought close together (Elwell and Pointon, 1978). (In theory, two clean molecularly flat surfaces would adhere strongly—if they could be produced.) This is the principle on which many adhesives function, including among their important properties certain which are characteristic of cement paste (Nurse, 1968):

(a) an initial fluid state, enabling the adherends to be wetted easily;
(b) surface activity to wet the adherend and displace air or contaminants;
(c) setting or hardening to achieve the desired strength, or in the case of tacky adhesion, viscosity.

Unfortunately, relative to steel, the paste does not possess a fourth main characteristic, that of chemical similarity. Nevertheless there must be some adhesion otherwise, for example, steel surfaces could not be passivated by coating with a cement film. Adhesion is improved with cleanliness of surface but perfection in this is impossible to achieve in practical conditions since the surfaces of metals normally become contaminated instantaneously due to the chemisorption of oxygen to form an oxide film (Rao, 1972).

5.4.2 *Stages of bond*

The initial stage of bond then is adhesion. The second stage is friction as a result of shrinkage of the matrix around the steel, setting up radial compression in it and loop tensile stresses in the concrete as indicated in Figure 5.11(b) (Beeby, 1978). This interfacial frictional load or 'grip' (Wilby, 1983) will resist or transfer longitudinal load.

These first two types, or stages, of bond occur without the application of 'external' load. The third and final stage, due to mechanical interlock, will be manifested under load, the interlock ranging from that between hydrated cement 'grains' and surface roughnesses in the steel, to the matrix bearing stresses set up by ribs or other deformations in the steel as indicated in Figure 5.9, the idealised view of which is given in Figure 5.11. The three primary stages of bond, then, are

(a) adhesion;
(b) friction;
(c) mechanical interlock.

While this sort of model is not new and is somewhat elementary, it is useful in postulating what the likely effects are of corrosion. That the detailed picture is more complicated is demonstrated by pull-out tests which indicate that de-bonding is in a thin layer (predominantly of $Ca(OH)_2$ crystals out of which C—S—H crystals have grown) on 'smooth' wires or bars. On rougher surfaces relatively smooth slip lines develop in a porous matrix layer surrounding the surface layer where micro-cracking and the grinding of crushed 'cement-stone' are a feature. This crushing and grinding is said to lead to some radial stress release. Ribbed bars have zones of crushed material in front of the ribs, complicating the simplified model of Figure 5.9, (Stroeven and de Wind, 1982).

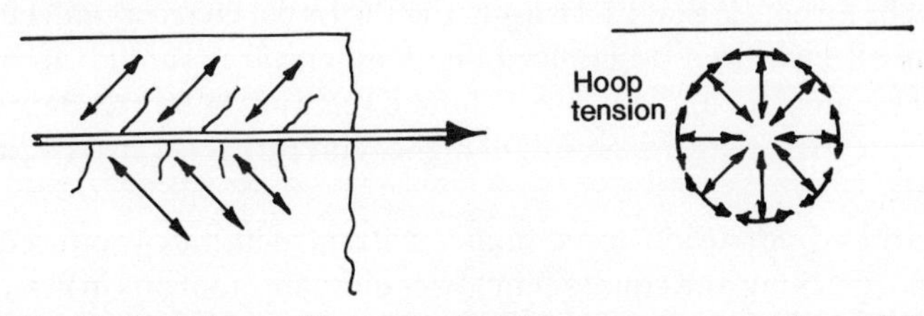

Figure 5.11 Idealised picture of forces developed in concrete surrounding a deformed bar

Other aspects of bond which of course should not be forgotten include (Wilby, 1983):

dilatancy—	the resistance to slip resulting from the wedging action of loosened particles;
tangential friction—	due to changes in direction in the region of a crack;
wedging action—	resulting from Poisson's ratio effects in a bar along which there is a change of stress. It may be appreciable in prestressed concrete, insignificant in reinforced concrete.

5.4.3 *Bond and corrosion*

Since anodic corrosion results in local and/or general corrosion, creating pits and generating oxides of iron, it can be expected to affect bond. The corrosion products are 2 to 3 times the volume of originating metal (Beeby, 1978) and this expansion is what causes bursting or spalling of the surrounding concrete. There must then be two countervailing effects on bond or at least upon its first two stages.

Substitution of rust for metal implies a breakdown in adhesion so that depassivation will coincide with loss of adhesion. Expansion as a result of precipitation of oxides of iron implies an increase in hoop-stress and radial compression so that friction bond could increase. Similarly if corrosion occurs on the bearing face of a deformed bar, the bearing force or stress must increase although the nature of the interface changes. Thus one can visualise a layer of loose material building up between steel and sound matrix, facilitating relative longitudinal movement between them. At the same time, the layer is coming under increasing constraint due to its expansion making relative movement more difficult.

The model of Figure 5.9 and the normal radial stresses would result eventually in bond failures of the type shown in Figure 5.12 (Taylor, 1983). This of course is the same pattern as longitudinal corrosion cracking (which increases the radial stresses). It is quite conceivable then, that bond strength might increase initially with corrosion although it will show an eventual decline as the probability increases of slip planes developing in the friable surrounding layer of rust. Lateral cracking will also affect the outcome and it should not be forgotten that the stress in the steel must increase under constant load as its effective area is reduced. This latter effect should become less significant for greater bar diameters although there might be a tendency for the corrosion rate to increase as a result of higher currents generated by the larger electrodes.

In accelerated corrosion tests, higher current densities produced earlier longitudinal cracking and consequent lower ultimate moments in flexural tests (Lin, 1980). While the exact figures are a function of the specific tests, the

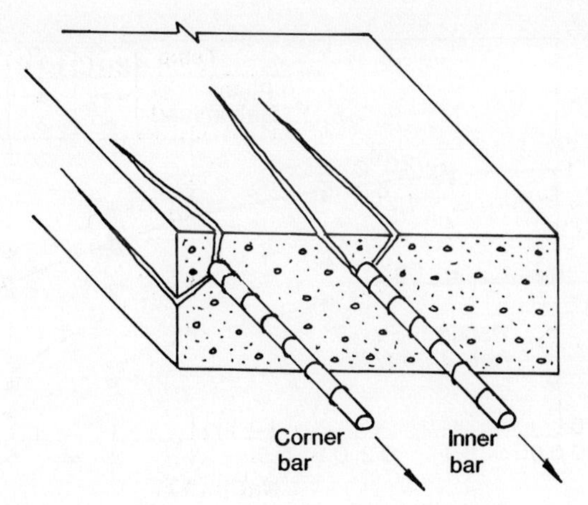

Figure 5.12 Failure mechanism: pull-out of deformed bar

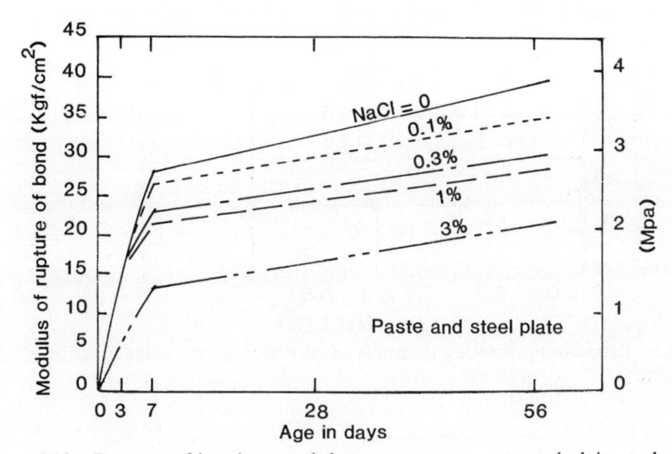

Figure 5.13 Progress of bond strength between cement paste and plain steel plate

reductions were substantial: up to 50%. Furthermore, the average ultimate flexural bond strength was reduced by about one third.

Incorporating sodium chloride in the concrete at mixing significantly reduces adhesive bond (Sakamoto and Iwasaki, 1982) as Figure 5.13 makes clear. Figure 5.14 shows also how ultimate bond decreases with increased salt content and temperature although galvanised bars are less affected by temperature. These results are for young concretes and, while there are no results presented for older concretes, there is clearly a dramatic increase in corrosion products with increased salt content and age—Figure 5.15. It is obviously most undesirable to use sea water for making concrete to be used in conjunction with reinforcement.

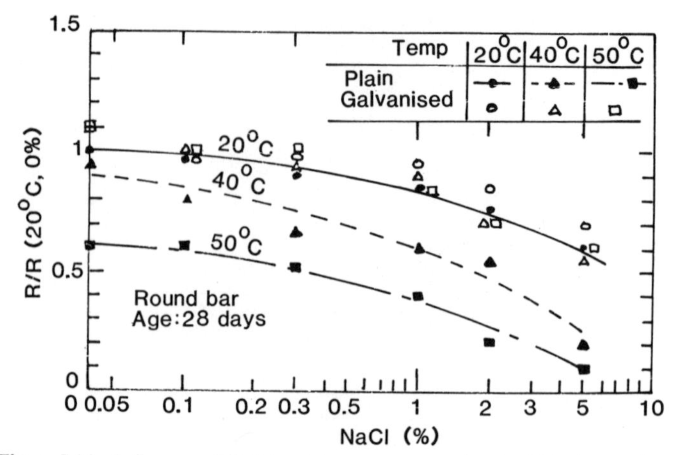

Figure 5.14 Influence of NaCl and temperature on bond of black round bar

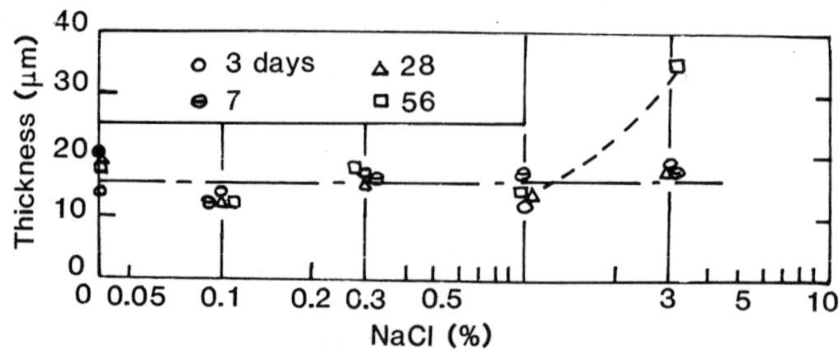

Figure 5.15 Relationship between thickness of corrosion products layer and NaCl content of cement paste

Finally concerning bond, it has been shown that there is a substantial linear increase in bond strength in uncorroded plain bars with increase in lateral stress or confining pressure (Robins and Standish, 1982). This tends to support the argument above, at least to the extent that, even if frictional bond is not actually increased at some stage, it will not be reduced by as much as one would expect from simple slip *without* lateral load. It is also evident that deformed bars, while having a substantially higher pull-out strength, do not have as great a relative increase with pressure since the mode of failure of the surrounding concrete is limited ultimately by shearing at higher loads.

Apart from the general effects of deterioration, the increased risk of bond failure through corrosion has to be taken seriously. It has been well demonstrated that failure of bond of deformed bars can be very dangerous since there is no prior warning of ductile deformation (Tepfers, 1979).

5.5 Fatigue

In reinforced and prestressed concrete load cycling has three main aspects: behaviour of the concrete, behaviour of the steel, and interaction between them. Concrete was discussed in the preceding chapter, along with some general fatigue considerations, now it is appropriate to turn to the steel prior to examining the composite.

5.5.1 *Reinforcing steel*

European, British, and American specifications for reinforcement have been summarised, accompanied by a comprehensive list of relevant standards for both concrete and reinforcement, primarily by the British Standards Institution and the American Society for Testing Materials (Rygol, 1983). Whilst some standards and codes of practice are revised periodically, or even replaced, the list provides a useful starting point although here we shall confine reference for the moment to the European code (CEB-FIP, 1978).

5.5.1.1 *Categories of steel* Three modes of categorising steel are indicated:

(1) according to method of production
  —hot rolled steel (natural steel);
  —cold worked steel (either by torsion and/or tension, or by cold drawing and/or rolling);
  —special steel (e.g. hardened and tempered steel).
(2) according to surface properties
  —plain smooth bars or wires (including welded mesh);
  —high bond bars or wires (including welded mesh).
(3) according to weldability
  —Class (a) (not weldable): 'steel that cannot be welded with any practical method with acceptable results';
  —Class (b) (weldable in certain conditions): 'steel which can be welded with acceptable results, by using special methods or normal methods accompanied by special safety precautions';
  —Class (c) (weldable): 'steel which can be welded with the usual methods with acceptable results.'

The hot rolled (weldable) steels are classified according to their chemical composition, type of welding and diameter as indicated in Table 5.4.

Characteristic strengths have been specified for various types of reinforcement (BSI, 1972) which give some idea of the range of properties to be expected (Table 5.5).

5.5.1.2 *Characteristics* Detailed consideration of the characteristics of the various steels is the province of the metallurgist but a few general remarks are germane here.

Table 5.4 Classification of weldable steels

Class	Welding by continuous weld	Tack welding
(b)	$C \leqslant 0.35\%$	$6 \leqslant \phi \leqslant 16: C \leqslant 0.28\%$ $16 < \phi \leqslant 25: C \leqslant 0.26\%$ $\phi > 25: C \leqslant 0.24\%$
	$C_{rep} \leqslant 0.60\%$	$C_{rep} \leqslant 0.54\%$
(c)	$C \leqslant 0.24\%$	$6 \leqslant \phi \leqslant 16: C \leqslant 0.22\%$ $16 < \phi \leqslant 25: C \leqslant 0.20\%$ $\phi > 25: C \leqslant 0.18\%$
	$C_{rep} \leqslant 0.52\%$	$C_{rep} \leqslant 0.48\%$

C_{rep} is the equivalent carbon defined by $C_{rep} = C + Mn/6 + (Cr + Mo + V/5) + (Ni + Cu/15)$.
ϕ = diameter (mm).

Table 5.5 Characteristic strengths of various types of reinforcement

Designation	Nominal sizes mm	Specified Characteristic Strength N/mm²
Hot rolled mild steel	All sizes	250
Hot rolled high yield	All sizes	410
Cold worked high yield	$\leqslant 16$	460
	> 16	425
Hard drawn steel wire	$\leqslant 12$	485

There is a trade-off in steel production between increased strength (and higher yield) and loss of ductility. Major influencing factors are carbon content, heat treatment and work-hardening. This has been known for some time, of course. For example, for carbon contents of less than 1%, an empirical relationship between strength and carbon content has been proposed, modified here for SI units (Withey and Washa, 1954):

$$\sigma_t = 6.895\ (45 + C) \tag{5.16}$$

where σ_t is the tensile strength (N/mm²)
and C is the carbon content in hundredths of one percent i.e. for 1% carbon content $C = 100$

The yield point is between 50 and 60% of the tensile strength, depending upon the carbon content, but for around 0.20% the characteristic strength would be in the region of 225–270 N/mm² which may be compared with the first value in Table 5.5. To obtain a strength of 410 N/mm² on this reckoning would require a carbon content of over 0.60%. Figure 5.16 reproduces some appropriate stress/strain curves for various carbon contents from which the loss of ductility is evident (Withey and Washa, 1954).

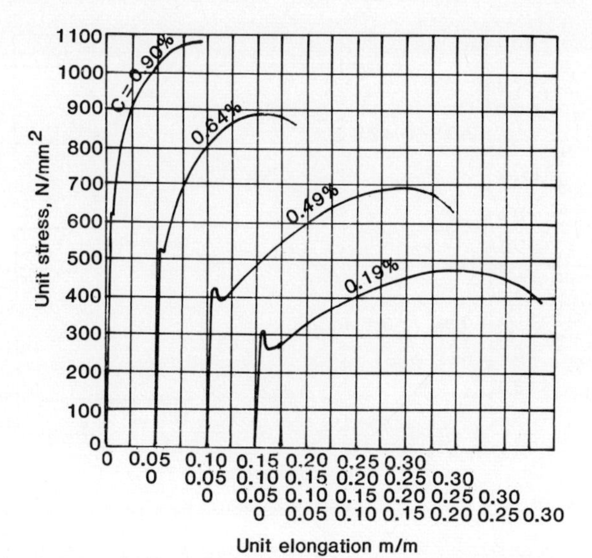

Figure 5.16 Tension stress-strain diagrams of hot-rolled steel bars

Cold working also results in higher yield points and strengths, and increases further, the higher the carbon content, but again at the expense of ductility.

5.5.1.3 *High strength reinforcement* In the present context of fatigue there are penalties for the higher strength reinforcements, particularly for bent, ribbed bars. A survey of a number of sources indicates increasing loss of fatigue strength in the sequence hot-rolled polished bars/hot-rolled cold-drawn/cold deformed ≈ hot-rolled high yield/hot-rolled spiral-ribbed/hot-rolled

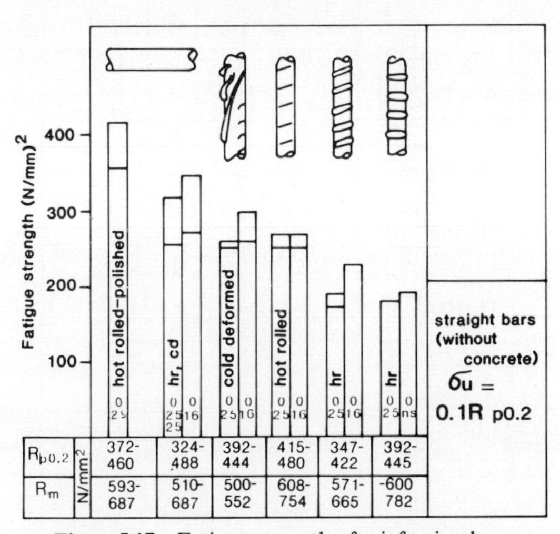

Figure 5.17 Fatigue strength of reinforcing bars

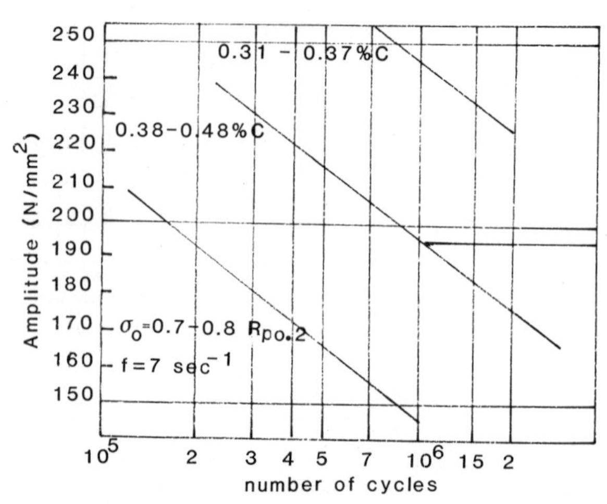

Figure 5.18 Fatigue behaviour of hot-rolled ribbed bars, bend samples in concrete

circumferential-ribbed, as indicated in Figure 5.17 (Nürnberger, 1982). Increased carbon content also seems to reduce the fatigue strength, as illustrated by Figure 5.18 (loc cit), while bending can result in considerable reduction, particularly as the radius of the bend decreases, Figure 5.19 (loc cit) (the data are for bent bars embedded in concrete).

A number of researchers have noted that failure initiates at the root of ribs on the 'side' of the bar of greatest tensile stress (Bennett and Joynes, 1979). In contrast to such failure in bending, it has been observed that failure in axial

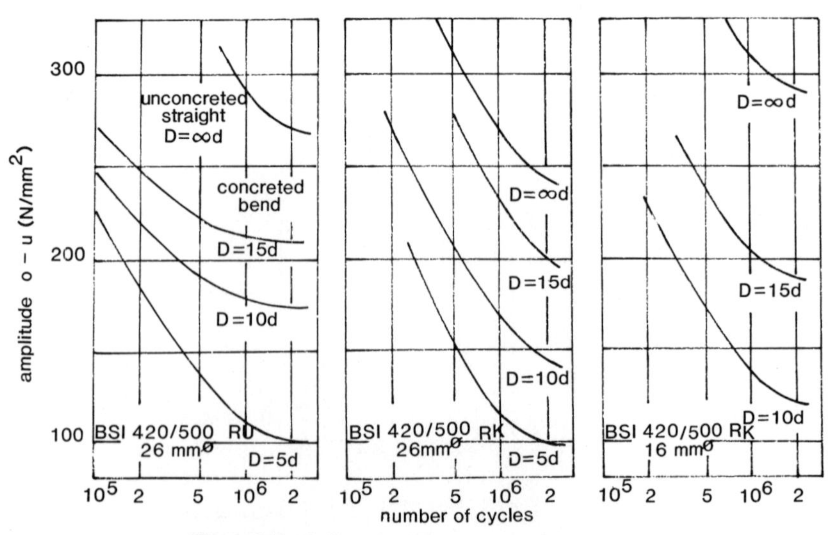

Figure 5.19 Influence of bending ratio on fatigue

tests initiates at surface imperfections (Tilly and Moss, 1982). It has been further suggested that the concept of a fatigue limit at about 2×10^6 cycles for reinforcement is not supported by the evidence. However, in the necessary long endurance tests, there is conceivably a degree of corrosion which has been found to eliminate the limit in structural steels (Bardal, 1985). Since galvanising of reinforcement seems to improve fatigue performance (Roper, 1982), it follows that corrosion must diminish it. This will be the case for pitting corrosion in particular, since pits will themselves act as crack initiators similarly to other flaws.

5.5.1.4 *Influence of bar bending* The view has been expressed that fatigue strength depends rather on features such as connections and bends than on the response of straight bars (Morf, 1982). Moreover, 'reinforced concrete structures are likely to be insensitive to fatigue failure because of the great number of bars in one structural element.' In other words, simultaneous fatigue failure in more than one bar is extremely unlikely in normal circumstances and the stress levels in practice would mean that the concrete is more likely to fail first (immersion in sea-water, however, gives rise to different circumstances, as will be considered later).

On the other hand, stirrups are more susceptible to fatigue failure due to the sharpness of the bends required. Whilst failure of an individual stirrup in itself is not necessarily critical, it increases the likelihood of shear failure of reinforced concrete elements, as has indeed been demonstrated in tests, referred to a little later on (Horii and Ueda, 1977). To counter this, headed bars have been developed, such as forged or fabricated T-bars, providing better ties and greater shear resistance (Gerwick, 1986).

5.5.1.5 *Bar size* Fatigue strength reduces with increasing bar size (Tilly and Moss, 1982; Morf, 1982), stresses to produce a given endurance being about 30% lower for diameters of 32 and 40 mm than for 16 mm bars. Wöhler lines give relationships for stress range σ_r and cycles to failure N of the general form

$$\sigma_r^n N = K \tag{5.17a}$$

with n for 16 mm bars being 9. K values are shown in Table 5.6 (Tilly and Moss, 1982).

Based on tests on American steel, other suggested relationships for straight bars are (Waagard, 1977):

$$\sigma_{rl} = 165 - 0.33 S_{\min}\ \mathrm{N/mm^2} \tag{5.17b}$$

and

$$\log N = 6.5 - 2.3\frac{\sigma_r}{\sigma_y} - 0.002\, S_{\min} \tag{5.17c}$$

where σ_{rl} is the stress range at the fatigue limit and σ_y is the yield stress. N is the fatigue life for a stress range above the fatigue limit.

Table 5.6 Values of K factor used to determine fatigue strength of steel bars

Type of loading	$K \times 10^{27}$ 16 mm ϕ	32 and 40 mm ϕ
axial	0.75	0.11
bending	3.09	0.31

5.5.1.6 *Strain softening* Of potential significance to bars in cyclically-loaded cracked concrete is the Bauschinger effect (Marin, 1962; Ford and Alexander, 1977). The beneficial influence of strain- or work-hardening on yield is well-known. However, if after, say, yield in tension, the load is reversed to yield in compression, the resulting compressive yield stress is lower than the tensile yield stress and there is a subsequent **reduction** in tensile yield strength. Similarly, if compressive yielding takes place first, the tensile yield stress is lower than the compressive, and so on. In other words, reverse loading beyond the yield point results in strain-**softening** rather than strain-hardening. Consequently where reverse loading occurs, as from waves, fatigue failure might occur earlier than would otherwise be expected. This point will be returned to below.

5.5.2 *Prestressing steel*

Prestressing steels also are categorised by the European Model Code (CEB-FIP, 1978):

(1) according to treatment
  (a) heat treatment—patenting
    —hardened and tempered steel
  (b) mechanical treatment—cold rolled or drawn steel
    —cold worked by torsion or tension
(2) according to type of product
  —wires and bars
  —strands or cables
(3) according to their shape
  —plain round bars or wires (straight or crimped)
  —bars or wires which are not round and/or not smooth

They do not seem to have an endurance limit (Naaman, 1982) and the fatigue strength depends on type, whichever method of categorisation is employed. Type of anchorage and degree of bond contribute, as is to be expected: bending at an anchorage can cause problems. A general relationship has been proposed for lives up to two million cycles and is applicable to various steel types since strength is expressed as the ratio of stress range S to ultimate strength σ_{tu} (Naaman, 1982):

$$\frac{S}{\sigma_{tu}} = -0.123 \log N + 0.87 \tag{5.18}$$

This fits the results of a number of investigators.

Cold drawing and heat treatment can be expected to have effects analogous to those on reinforcing and structural steels but with considerable improvement for wire and strand which, for example, have higher allowable stresses than bars in design codes (note that this concerns *absolute* values as distinct from the *relative* value of eq. (5.18)).

Equation 5.18 makes use of stress range, in common with many fatigue relationships. However, an alternative has been suggested as providing a better basis for data correlation because 'it treats the fatigue limit as an open, variable property of the material' (Warner, 1982). Figure 5.20 shows test data for 5 mm diameter crimped wires, 12.5 mm 7-wire strand, plus data for 7/16 inch (11 mm) tested some time previously. The resulting curve is expressed by

$$\log N = \frac{1.169}{R} + 5.227 - 0.031\,R \tag{5.19}$$

R is the stress increment defined by

$$R = S_{\max} - S_{\lim} \tag{5.20}$$

where $S_{\max}$ is the maximum stress in the cycle and $S_{\lim}$ is the 'fatigue limit' corresponding to the minimum stress level $S_{\min}$ in the cycle. $S_{\lim}$ values (for different minimum stress levels) are shown in Table 5.7.

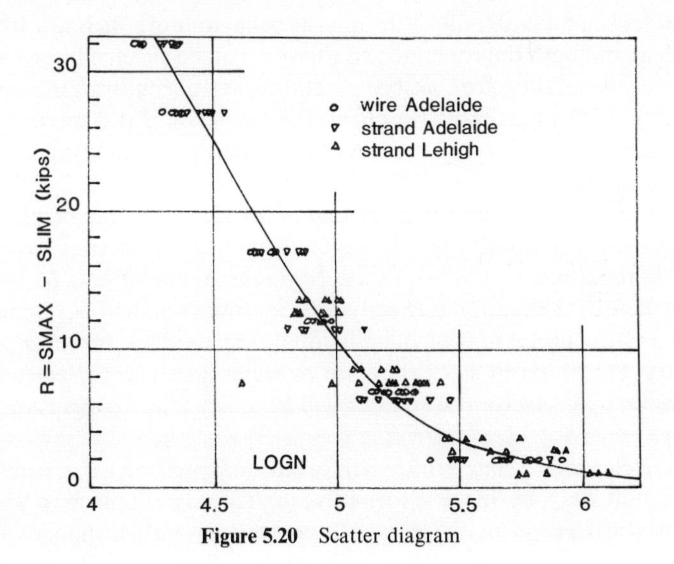

Figure 5.20 Scatter diagram

Table 5.7 Fatigue limit values for several wires at two different minimum stress levels

Test	Minimum stress level %	Fatigue limit %
Wire: Adelaide	40	53
Strand: Adelaide	40	53
Strand: Lehigh	40	56.5
Wire: Adelaide	60	73
Strand: Adelaide	60	73
Strand: Lehigh	60	73

5.5.3 *Composite behaviour of steel and concrete*

Where two materials as different as steel and concrete are used compositely, their behaviour will be governed by the structural design, their response to load, and their behaviour in the environment to which they are exposed. Thus the distribution of load between them will be determined by design and material response—for example, load on an over-reinforced beam will be limited by the strength of the concrete and on an under-reinforced beam, by the strength of the steel. The modes of collapse also are likely to differ. However, the materials' properties may be modified by the environment resulting in a re-distribution of load between them. The pattern of loading itself may affect them, as it does in long-term cyclic loading. Consequently it is advisable to establish, if possible, how their composite use corresponds to assessments based on their separate behaviour, particularly under conditions similar to those of actual use.

This means in the present context, for example, that reinforced and prestressed concrete should be fatigue-tested in sea water at realistic frequencies since it is already clear that the behaviour of concrete is frequency-dependent and both the concrete and the steel can be affected by sea water. As a result, a substantial effort has been made in several countries in recent years to investigate this particular problem. The data-base is still relatively sparse but enough has been learned to formulate a number of conclusions which can be summarised here.

The spur has been provided primarily by development of oil and gas resources under the North Sea, in which a number of countries have a direct and/or indirect interest. Work in the Netherlands was discussed in the last chapter, here consideration is mainly of programmes in the UK, Norway, and France. Instead of elaborating on individual contributions, however, a general summary will be given since they are to some degree complementary and confirmatory. A selection of sources will be given at the conclusion.

5.5.3.1 *Role of cracking* Both reinforced and prestressed concretes were tested but it may be more informative to consider them as representing different stress levels in the concrete component rather than as different materials. Certainly the stress level appears to have a significant bearing on the response of cracked concrete, or a preferable subdivision might be into cracked and un-cracked concrete, with the cracked material warranting more extended discussion. In effect, uncracked concrete was discussed in the preceding chapter so that all the factors considered there will apply with due allowances being made for possible steel corrosion. In practice this will only be relevant to structures or parts of structures where tension is not allowed or does not occur, that is in prestressed, partially prestressed, or mass concrete structures. Where cracking does occur, performance changes significantly. Cracking may stem from various causes: construction cracks as a result of initial or drying shrinkage, settlement, heat of hydration; load cracks during normal service as a result of flexure, stress reversal, torsion, shear or following exceptional loading, under 100-year storm conditions, for instance. Construction cracks can be avoided by appropriate mix design and control, placing, and construction procedures; load cracking is inevitable unless designed against as indicated. The remainder of this discussion will concern cracked concrete.

The fatigue life of cracked reinforced concrete is reduced by immersion in sea-water with failure tending to occur in steel corroded at a crack in the concrete. Pitting corrosion at the crack and surface corrosion along the bar take place as suggested in Figure 5.8. Furthermore, it seems reasonable to expect that corrosion will occur in all bars at a crack so that the remarks made earlier about the likelihood of only one bar failing at a particular section do not apply here. Crack blocking proceeds as discussed previously but its efficacy depends on stress levels in the concrete. In reinforced concrete specimens (and some prestressed) the deposits were sufficiently strong to resist normal loads but in other prestressed concrete specimens they were washed out, presumably as a result of crushing—the 'transition' stress level was somewhere between 6 and 12 N/mm^2. Whether or not failure occurred in reinforced or prestressed specimens under test is irrelevant, it is the stress level which appears to be the significant factor, with crack blocking at low stress levels and crack widening at higher stress levels.

5.5.3.2 *Crack blocking and cyclic stiffening* Crack blocking produces increased stiffness of flexural elements since the tests under cyclic loading between fixed maximum and minimum stresses displayed a marked reduction in deflection range. Load deflection curves took the form of Figure 5.21: the maximum deflection shows a gradual increase with time but the minimum deflection has a much more rapid increase. That is, the deflected beam does not return to its previous position as the open crack fills up (producing progressive closure from the narrow, inner end outwards). Rapid cycling does not allow sufficient time for the process to take place thereby contributing towards an apparent frequency effect at low cycling rates. In air there can be a similar but much smaller effect due to particles of fractured concrete lodging in the cracks.

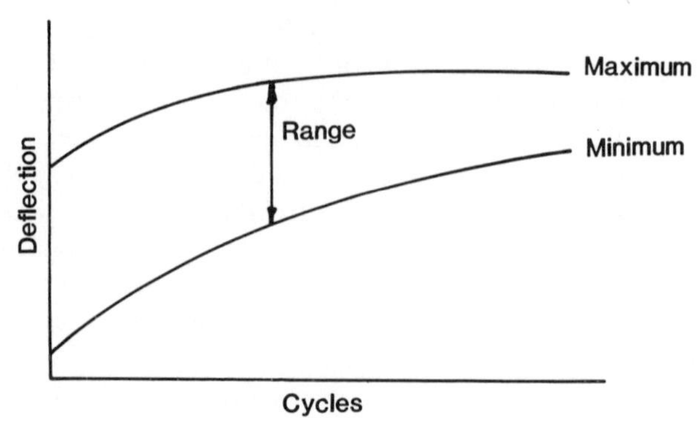

Figure 5.21 Variation in deflection range for beams tested in seawater

Reduction in sea-water-immersed deflection range was as much as 80% in the tests in question (it is important to emphasise that this phenomenon does not occur in freshwater since the crack-blocking is the result of a chemical reaction with *sea* water. Nor should such crack blocking be confused with autogenous healing.)

This cyclic stiffening means that the steel strain range is reduced with the consequence that the effective stress range also is reduced. Hence if the S–N curves are plotted on the basis of the initial stress range they will give a misleading impression because the *actual* stress range for a given fatigue life will be lower, in fact. Examples are given in Figures 5.22(a) and (b) (Paterson and Dill, 1982). Against that it should be noted that crack-blocking does not prevent corrosion and so the reduced steel strain presumably is being applied to a reduced steel section. The adjusted or effective stress ranges have been calculated in direct ratio to reduction in deflection but the stress in the steel at a crack where corrosion is advanced may be higher than is to be anticipated by this method of calculation: the complexity of the problem is self-evident.

There is limited evidence to suggest that external water pressure has no significant effect on fatigue life, at least up to 1.4 N/mm^2, equivalent to a depth of about 140 m. How or whether such a pressure might affect pumping of water in the cracks and consequent erosion of the concrete is not clear. A low frequency effect, at least partly due to crack blocking, has already been alluded to but a contributory factor may be the stress range. Normally fatigue life is extended with a lower stress range. However, this also means there is more time available for corrosion with its consequent reduction in steel section. The UK work showed that, for initial stress ranges of 200 N/mm^2 or less, fatigue failures initiated at corrosion sites instead of at the so-called mechanical initiation sites at rib intersections, typical of shorter endurance tests. In general, fatigue strength in sea water is lower than it is in air.

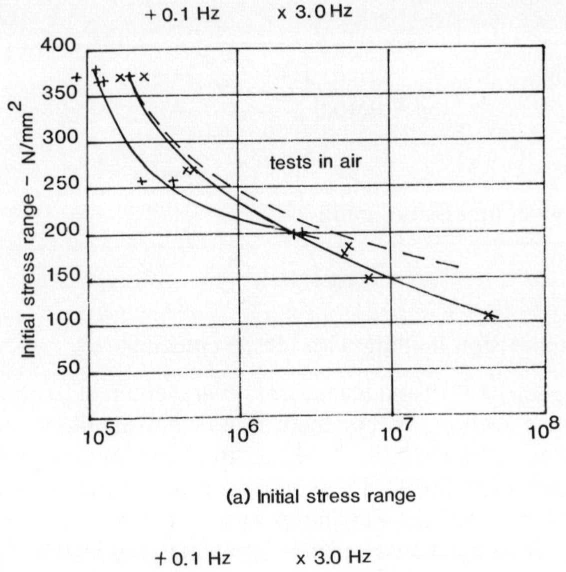

(a) Initial stress range

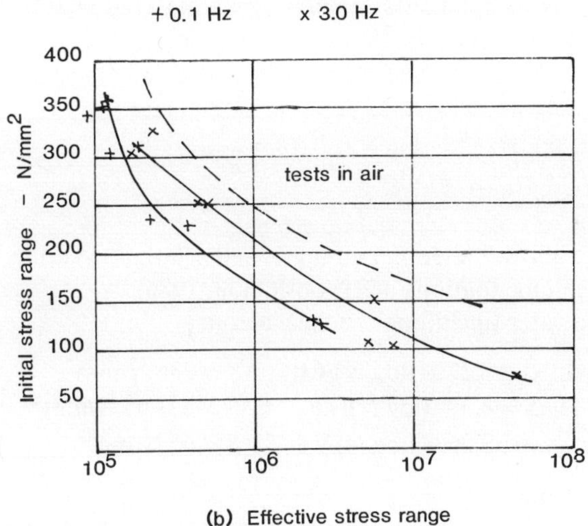

(b) Effective stress range

Figure 5.22 S–N curves

5.5.4 *Design recommendations*

Design recommendations have been produced from both the Norwegian and the UK work. The Norwegian curves represent a linear-elastic and a non-linear approach according to whichever set of regulations is being adopted (the classification society Det norske Veritas or the Norwegian Petroleum Directorate, respectively) (Waagaard, 1986):

Linear

$$\log N = 10.0\frac{\left(1-\dfrac{\sigma_{\max}}{\alpha\cdot f'_c/\gamma_m}\right)}{\left(1-\dfrac{\sigma_{\min}}{\alpha\cdot f'_c/\gamma_m}\right)} \tag{5.21}$$

for compression–compression loading.

$$\log N = 8.0\left(1-\frac{\sigma_{\max}}{\alpha\cdot f'_c/\gamma_m}\right) \tag{5.22}$$

for tension-compression loading with tensile cracking.

$\alpha = 1.19 - (\varepsilon_{\max}/\varepsilon_{\min}) \times 0.19$ and is a factor taking account of the flexural strain gradient across a section. Where there is no gradient (as in purely axial loading) $\varepsilon_{\max} = \varepsilon_{\min}$ and $\alpha = 1$ ($\varepsilon_{\max}$ and $\varepsilon_{\min}$ are the maximum and minimum strains in the outermost fibres). These equations are for the range $3.5 < \log N < 6.2$. $\sigma_{\max}$ and $\sigma_{\min}$ are the maximum and minimum stress levels in the concrete and f'_c is its cylinder strength. γ_m is a material factor = 1.25.

The non-linear equation is

$$\log N = 7.0\frac{1-\dfrac{\sigma_{\max}}{\sigma_c}}{1-\dfrac{\sigma_{\min}}{\sigma_{\max}}} \tag{5.23}$$

where σ_c is the allowable design strength under fatigue loading.

The UK recommendations are (Booth et al, 1986):
concrete in seawater (including the splash zone)

$$\log S = \begin{cases} 3.27 - (\log N)/6.0 & S > 235\ \text{N/mm}^2 \\ 4.3 - (\log N)/2.8 & 235 > S > 65\ \text{N/mm}^2 \\ 3.26 - (\log N)/4.8 & S < 65\ \text{N/mm}^2 \end{cases} \tag{5.24}$$

S is the stress range.

Fully submerged or cathodically protected beams, where corrosion is less likely, will be less critical but there is insufficient data to permit them to be treated less conservatively. These curves make no allowance for crack blocking.

For concrete in air (straight reinforcing bars embedded in a dry salt-free environment) the recommendations are:

$$\begin{aligned} \log S &= 3.27 - (\log N)/6.0 \qquad (400 < S < 120\ \text{N/mm}^2) \\ \log S &= 2.73 - (\log N)/11.0 \qquad S < 120\ \text{N/mm}^2 \end{aligned} \tag{5.25}$$

The UK work cautions against reliance on crack blocking in practical structures since it is not yet clear to what extent the reaction takes place.

Details of the French, Norwegian and British tests may be found in the following selection of sources: Trinh, 1980; Peyronnet et al, 1977; Trinh and Peyronnet, 1978; Waagaard, 1977, 1986, 1987; Arthur et al, 1982; Banerjee et al, 1985; Booth et al, 1986; Paterson, 1980; Paterson and Dill, 1982; Paterson et al, 1982.

5.5.5 *Fatigue and shear*

Flexural fatigue is not necessarily the most critical aspect of cyclic loading. It has already been indicated that stirrups are particularly susceptible to fatigue failure and, indeed, fatigue strength in shear has been shown to be considerably smaller (perhaps $< 10\%$) than in flexure at low frequencies (Horii and Ueda, 1977). Cracked concrete shows significantly reduced stiffness under cyclic shear loadings which also result in rapid degradation of bond (Gerwick and Venuti, 1980). The major factors appear to be reinforcement ratio, concrete strength, and stirrup quantity. In tests on beams with reinforcement percentages of 0.7, 1.11 and 1.59 shear failure occurred in concretes of strength $23.5\,N/mm^2$ but not for $32.4\,N/mm^2$. There were three modes of failure: rupture of main reinforcement in the bending moment span, likewise in the shear span, and stirrup failure in the shear span, with the governing factor being the tensile stress in the stirrups. For a stirrup ratio over half the main reinforcement ratio, failure occurred in bending. For lower ratios, bending failure resulted from using higher strength concrete while shear failure followed from the lower strength concrete. Reversed bending flexural fatigue strength was also lower than that for uni-directional bending (Horii and Ueda, 1977).

Fatigue tests directed towards bridges rather than offshore structures also suggest that stress *range* is a more reliable parameter to use in assessing fatigue in shear than maximum or minimum stress, since there is much less scatter in results (Frey, 1982). Span/depth ratios may affect the results, however (Okamura and Ueda, 1982), and it appears that stirrup failure is not necessarily only at bends: 'fracture is generally along the main diagonal crack making the beam to fail'.

In larger structures, or those with larger structural elements such as shear walls, it is worth while making use of vertical prestress to reduce the diagonal tension, particularly when they are submerged (Gerwick and Venuti, 1980).

Reference has already been made to degradation of stiffness and bond in cracked concrete under cyclic shearing action. The fatigue strength of plain and prestressed concrete under pure torsion appears to be about the same as under compression or flexure (Okada and Kojima, 1982): that is, the fatigue strength is around 55% of the static strength at ten million cycles of loading. However, it is somewhat less for normal- (and fibre-)reinforced concretes depending upon the reinforcement ratio, being as low as 38% for a ratio of 1.0%, 53% for a ratio of 1.5%, and 51% for a ratio of 2.0%. The first two of these were under-reinforced leading to failure being governed by response of the steel reinforcement and the progress of cracking. Two separate categories of torsional fatigue failure in reinforced concrete are suggested:

(a) failure at low cycles with a high maximum load ratio—similar to static failure in its mechanism;
(b) failure at high cycles with a low maximum load ratio—the mechanism passes from the uncracked state gradually through the cracked state up to failure, with load being passed from the concrete to the steel. Cracking is diagonal and there is a certain amount of dowel action by the reinforcement.

Cracking has already been discussed but with emphasis on its relation to corrosion and blocking and their effect on the load response of concrete rather than with how cracking itself affects load response. Bond slip will not be considered directly here since it is a familiar concept discussed at length elsewhere and particularly with reference to the calculation of crack widths (Beeby, 1978). However, cracking has other implications, two of which have a definite bearing on fatigue behaviour, especially in shear, as does the progressive nature of bond slip.

5.5.6 *Tension stiffening*

At a crack, load is carried entirely by the steel which, in seawater, may be of reduced section. Consequently strain-hardening and/or strain-softening may take place—the Bauschinger effect already mentioned (ultimately, of course, fatigue failure may occur). *Between* cracks, however, there is likely to be a degree of tension stiffening, where the tensile capacity of the concrete means that the strain in the steel is reduced with the net result that the *average* force in the reinforcement is less than it is at the crack. In other words, even after cracking, the concrete continues to share the tensile load (although obviously not across the crack). The idealised representation of this is given in Figure 5.23 (Beeby, 1978) where ε_l is the strain carried by the steel alone under load P while ε_m is the average strain under the same load. $\Delta\varepsilon$, the reduction in strain, is then the tension stiffening effect for that load.

A bond effectiveness index λ has been introduced (Jakobsen, 1982):

$$\lambda = \frac{\varepsilon_0}{\varepsilon_m} \tag{5.26}$$

where ε_0 is the maximum steel strain (in a crack).

It has been shown that ε_m and ε_0 are related and that λ may be represented in terms of ε_m, the reinforcement ratio μ and the tensile strength of the concrete f_t:

$$\varepsilon_m = \varepsilon_0\left[1-\left(\frac{f_t}{\mu E_r \varepsilon_0}\right)^2\right] \tag{5.27}$$

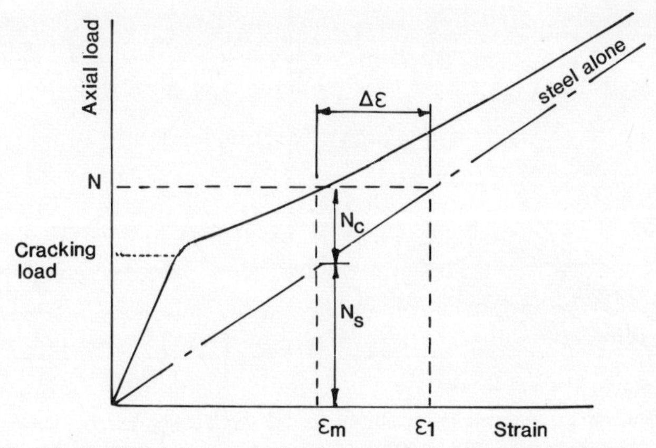

Figure 5.23 Idealized load-strain relation for a reinforced member subject to pure tension

$$\lambda = \tfrac{1}{2}\left[1 + \sqrt{1 + k\left[\frac{2f_t}{\mu E_r \varepsilon_m} g(\alpha)\right]^2}\,\right] \tag{5.28}$$

E_r is the modulus of elasticity of the reinforcement and k and $g(\alpha)$ are, respectively, factors to allow for the over-estimation of bond and skew angles between reinforcement and cracks. Figure 5.24 (Jakobsen, 1982) shows the variation of λ with increasing strain for different concrete qualities and steel ratios under monotonic loading. λ may be interpreted as the ratio of the effective secant modulus of the steel to its actual secant modulus and $\lambda - 1$ $(= \Delta\varepsilon/\varepsilon_m)$ indicates the contribution of tensioned concrete to the overall stiffness. As load (and hence strain) is increased, the diminishing value of λ

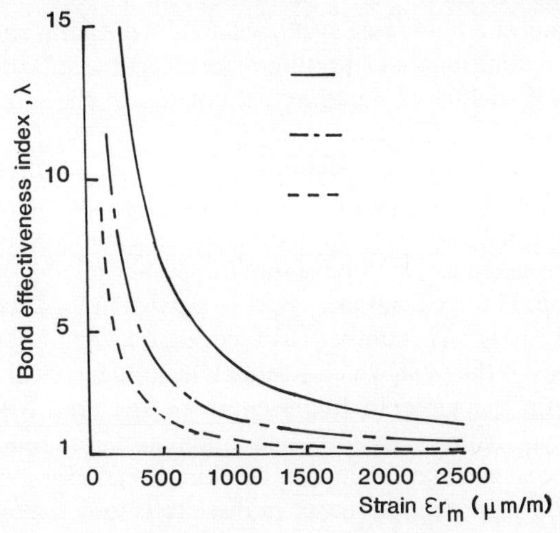

Figure 5.24 Bond effectiveness index versus mean strain, monotonic loading

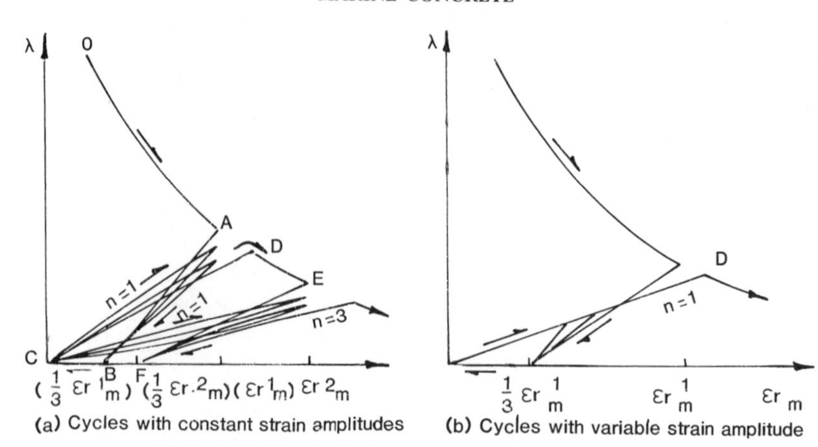

Figure 5.25 Bond effectiveness under general load cycling

signifies increased degradation of bond and cracking. Higher reinforcement ratios also mean that the concrete is taking a smaller share of the tension load as is to be expected (irrespective of whether it is direct or flexural). Moreover, at a given strain a certain amount of degradation, deterioration, or damage will have occurred. Consequently if the concrete is unloaded and then re-loaded, by the same amount, it will return to near the same point on the λ–strain curve. When the load is increased beyond the previous maximum, the curve will be followed and subsequent unloading and re-loading will have the same pattern (assuming that the cycles are few).

This is represented in Figure 5.25 (Jakobsen, 1982) where OADE is the appropriate curve from Figure 5.24. Concrete loaded to a strain ε_{ml} will follow it to A and then, on unloading, from A to B (where $\varepsilon_m \approx \frac{1}{3}\varepsilon_{ml}$) and then to C, although in practice there will be some residual strain. Provided ε_{ml} is not exceeded, loading and unloading cycles will follow the paths shown with slight degradation. Loading beyond ε_{ml} resumes the λ–ε curve at D until further later cycling follows C-E-F-C. (*Un*loading cannot follow the λ–ε curve because cracking has occurred and concrete cannot be *un*-cracked.)

Cyclic torsion tests on box sections suggest that the hypothesis is not unreasonable. That the secant stiffness is reduced is confirmed by the typical hysteresis curves, with and without axial load, of Figures 5.26 and 5.27. (The secant stiffness is taken as the slope of the straight line joining the two apices of a hysteresis loop. The cycle numbers refer to particular load sequences during the tests, thus 13.10 is cycle number 10 of sequence 13: details are omitted here for brevity since it is the *trend* which is more relevant.) It is clear that the energy dissipated within the material (represented by the area within the loops) diminishes as cycling proceeds, although it increases each time the amplitude of the load cycle is increased, as further cracking and deterioration take place. Load cycling during a given sequence in these tests took the form of applying constant angles of twist, thereby maintaining constant shear strain with the

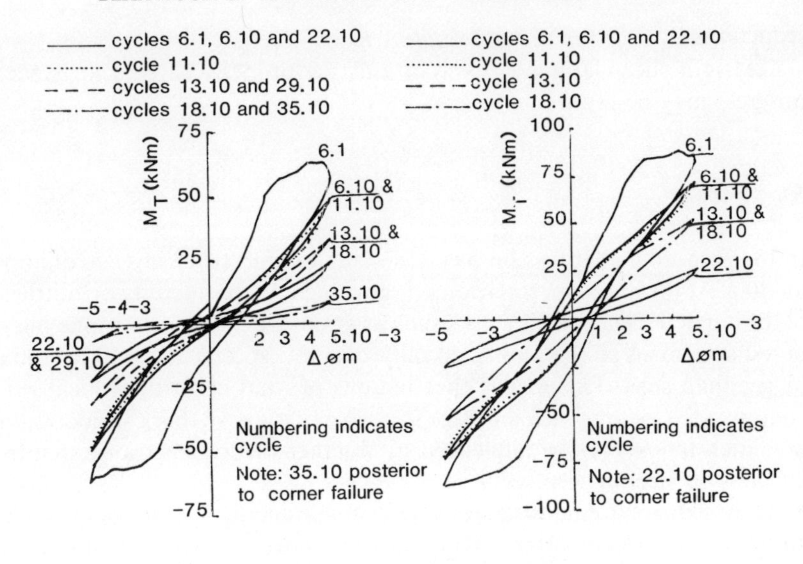

Figure 5.26 Development of hysteresis curves (after Jakobsen, 1982)

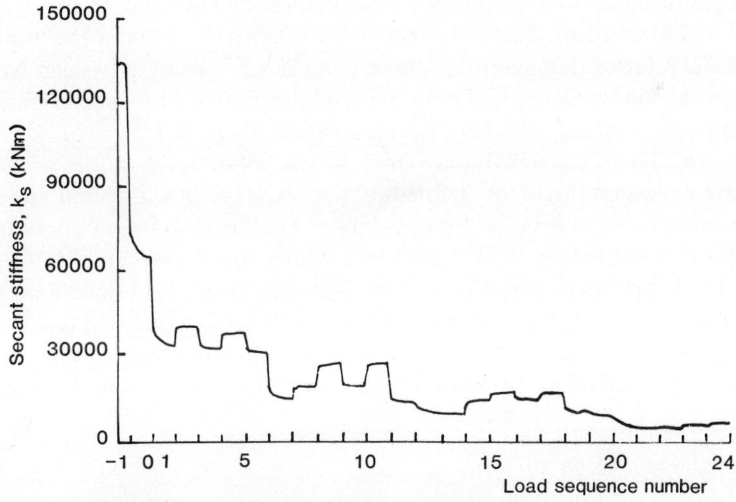

Figure 5.27 Development of secant stiffness (Jakobsen, 1982)

torque reducing progressively. The dramatic fall in secant stiffness during the early stages is illustrated by Figure 5.27.

While the actual amount and rate of occurrence of damage is test-specific, it is evident that cyclic shear loading can result in considerable cracking and bond slip within relatively few cycles. In fact, in this particular case, stiffness reductions by a factor of 8 to 30 prior to failure were observed.

Clearly if shear fatigue is likely it must be properly catered for since its progress may be rapid and severe.

5.6 Impact

In the preceding chapter allusion was made to differences in scale of impact loading. At the micro level response is governed by the ductility or brittleness of the impacted material. Factors such as grain size help to determine the size of test specimens so that, not unusually, concrete specimens are substantially bigger than steel. This means that notions of what constitutes localised or concentrated loading may differ for the two materials, so that consideration of any inter-action may be influenced by whether the load is transferred from concrete to steel or vice-versa.

At the other extreme, at the macro level, the energy absorption of a whole structure may be involved as in collision between a ship or aircraft and structure. Here, in particular, the deformability of the impactor becomes highly significant: if it is smaller than for the impacted object, then the impact is 'hard'; if it is larger, the impact is 'soft'. (Many automobiles are now designed with collapse or energy-absorbing zones to give a 'soft' impact to protect the occupants in the event of collision.) While local damage occurs, much energy may be absorbed by elastic deformation of the structure (and the impactor) with the relative damage through collapse of each being governed by their structural configuration and their material properties. Momentum transfer is an important factor, whether into momentum or strain energy or deflection.

Intermediate between the extremes, at the 'meso' level, comes a range of situations which might be defined as the response of individual structural elements (such as beams, columns, slabs, domes, and so on) to concentrated loads (from falling objects, for example). A substantial amount of testing is at this level, for fairly obvious reasons—testing prototype structures under conditions up to collapse is expensive and only really appropriate if they are due for demolition! Structural elements can be rendered capable of laboratory control and provide a somewhat larger scale than direct materials testing.

5.6.1 *Load classification*

The FIP Commission on Concrete Ships (Buijs, 1978) produced a load classification which has clear relevance to fixed structures also and introduces some of the factors which have to be considered (Table 5.8).

Explosion is a specialised topic beyond the scope of this exercise. Precautions to be adopted against it depend upon the nature of the structure: hydrocarbon storage or leakage involves certain risks but any overload is likely to be due to shock wave rather than the concentrated load from the seat

Table 5.8 Classification of impact loads

	load due to	duration (sec)	load cycles	momentum	force	area	nature of load
	explosion	10^{-4}	1	large	large	small-great	shock
impact	collision	1	2	very large	very large	small	shock
	vibration	10^{-2}	10^{11}	small	small	small	alternating
impact	wave slam	1–10	10^{9}–10^{8}	small-large	small-large	small-great	alternating
elements	ship motion	10	10^{8}	small-large	small-large	small-very great	alternating
no	live load	10^{9}	10^{3}	zero	very large	very great	static
impact	dead weight	10^{9}	10	zero	very large	very great	static

of a chemical explosion. The equivalent maximum uniformly distributed overpressure from detonation of the gas-air mixture caused by liquefied methane spillage or leakage is about 2 atmospheres (Buijs, 1978). In contrast the pressure range at the centre of a chemical explosion is 100–300 kN/mm^2 with temperatures about 3500–4000°C (Henrych, 1979). It is questionable whether non-military structures need be concerned with concentrated loads of such high intensity (or perhaps even higher).

Another useful suggested classification of loads (Table 5.9) sets out various sources of impact and forms of loading accordingly (Perry and Brown, 1982).

5.6.2 *Load categories*

The foregoing discussion and tables lead to the conclusion that it might be helpful to categorise impact or impulse loading in three ways:

(a) uniformly distributed loads—corresponding to explosion shock wave;
(b) concentrated loads—corresponding to dropped objects, projectiles, etc. (explosion may also be included as an extreme case);
(c) localised-cum-general loads—corresponding to collision.

Explosion will not be considered further here so discussion will concern categories (b) and (c) which form useful sub-divisions for examining some of the literature. They fall fairly conveniently into the two broad areas of experimental work and structural analysis respectively, distinguishing between concentrated loads and localised loads. A concentrated load can be regarded as relating to impact at a point or small area which produces stress waves in the impacted member and damage in the immediate vicinity. A localised load on the other hand results in a chain or sequence of damage and/or failure of structural member(s). The difference is one of degree as much as anything since structural failure or collapse will begin with damage at the first point of contact of two colliding structures.

5.6.3 *Concentrated loads*

New drillpipe for oilwells can weigh from about 7.2 to 37.6 kg/m and may be in lengths of nearly 14 m (Rabia, 1985). The potential damage to the roof of an underwater oil storage cell can readily be appreciated should a length fall end-on into the water from a height (free fall from 30 m gives a water-entry velocity of just over 24 m/s). Its momentum would be equivalent to that of a 50 kg projectile travelling at 250 m/s. Not all falling or wind-blown objects are laden with the same destructive potential, of course, but clearly the possible consequences of such hazards require careful examination.

5.6.3.1 *Types of damage* A certain terminology has developed in the study of these problems, illustrated in Figure 5.28 (Kennedy, 1975). At low velocities

Table 5.9 Classification of load according to cause

Class	Description of loading	hard or soft	Examples	Typical velocities m/s	Relative incidence of extreme loading	Principal cause of concern
1	(a) static (b) quasi-static	h h/s	machinery vehicles ships	0 0–10	rare frequent at < 2 m/s	buildings bridge parapets and piers, marine structures
2	Dropped objects	h/s	containers, steelwork, pipes, tornado-borne debris	5–40	frequent	buildings during construction, offshore marine structures, sea bed capsules
3	Aircraft	a) s b) h	overall parts	200–300	rare	nuclear installations
4	Ballistic	h	bullets missiles shells	1000	frequent in certain areas	military and certain civilian structures
5	Nuclear		beyond present scope			

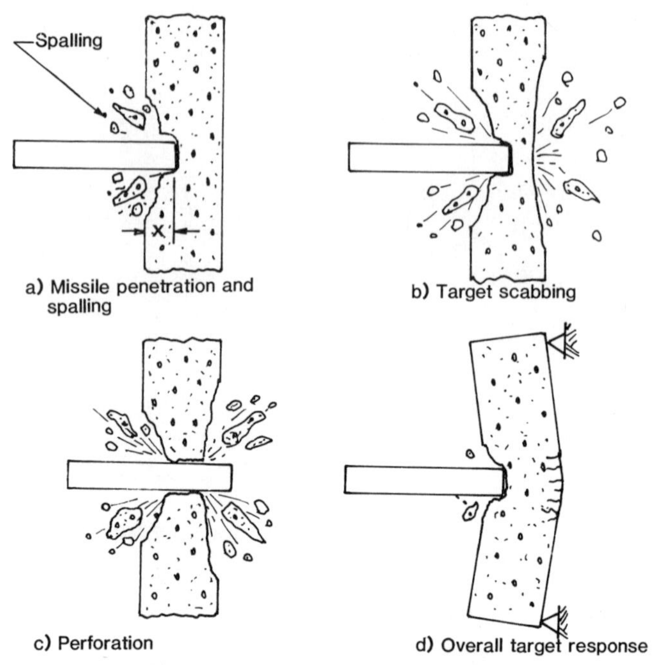

Figure 5.28 Missile impact phenomena

a missile would simply bounce off the wall without causing more than superficial damage. Higher velocities would produce increasing *cratering or spalling* on the impacted surface up to the stage where the velocity is sufficiently high for the missile to *penetrate* beyond the crater into the wall without rebound. *Perforation* occurs when the velocity is so high that penetration occurs right through the concrete, preceded by cracking and *scabbing* on the rear face (i.e. breaking-out of particles of concrete). If the velocity is high enough the missile will pass through without sticking. This presumes hard impact of course, such as by a steel missile on a concrete target. Soft impact as in wood on concrete would result in crushing of the wood without significant damage to the concrete. An intermediate stage (prior to perforation), the formation of a *shear plug*, has also been proposed (Perry, 1986). Figure 5.28(d) shows so-called 'plastic' impact where there is sufficient energy input to cause penetration and deformation of the 'target' (and possibly adjacent parts of the structure). Scabbing and plastic impact imply progressive debonding of any reinforcement on the rear face. There is presumably also resisting dowel action by reinforcement orthogonal to the path of the missile.

The problem is complex so it is worthwhile looking at some of its fundamental aspects prior to considering empirical or theoretical relationships which have been proposed. This may be helpful in eliciting critical factors which are likely to be influential.

5.6.3.2 *Contact stresses and stress waves* Contact stresses are obviously important in relation to concentrated loads and the theory for elastic bodies (due to Hertz) is set out in various texts (see Timoshenko and Goodier, 1951 and Boresi et al, 1978 for example). It is too elaborate to repeat here but, in essence, it shows that when two curved bodies in contact are placed under load, the stresses around the point or area of contact are a function of the load, the elastic properties of the bodies, their curvatures, and the angle between them. The solution is considerably simplified when one of the bodies has a flat face and is effectively of infinite extent and the other is axi-symmetrical. The criterion for failure of a ductile material is its shear strength and for a brittle material, its tensile strength.

Under dynamic loading the propagation and reflection of stress waves are important factors. Remote from the point of contact between colliding elastic bodies, there is a dilatation (or longitudinal) wave in the direction of travel and a distortion (or transverse) wave which cause longitudinal and lateral deformation respectively. They travel at different speeds and are expressed in the following equations (Timoshenko and Goodier, 1951):

Dilatation
$$\frac{\partial^2 u}{\partial t^2} = c_1^2 \cdot \frac{\partial^2 u}{\partial x^2} \tag{5.29}$$

$$c_1 = \sqrt{\frac{E(1-\nu)}{(1+\nu)(1-2\nu)\rho}} \approx 1.05\sqrt{\frac{E}{\rho}} \quad \text{for concrete} \tag{5.30}$$

Distortion
$$\frac{\partial^2 v}{\partial t^2} = c_2^2 \cdot \frac{\partial^2 v}{\partial x^2} \tag{5.31}$$

$$c_2 = c_1\sqrt{\frac{1-2\nu}{2(1-\nu)}} \approx 0.65\sqrt{\frac{E}{\rho}} \quad \text{for concrete} \tag{5.32}$$

u is the particle displacement in direction x at time t and v is the corresponding displacement in direction y. c_1 and c_2 are the respective wave velocities and E and ν are the elastic modulus and Poisson's ratio.

Solution of eqs. (5.29) and (5.31) depends on the specific problem under consideration but it is clear that particle displacement (and hence strain) must be a function of $\sqrt{E/\rho}$. The implication here then is that any relationship purporting to estimate impact damage must incorporate some function of E.

5.6.3.3 *Material failure* The failure criterion for brittle materials is the maximum tensile stress. According to Hertz theory, for a sphere striking a flat-faced semi-infinite solid (i.e. a 'half-space') this is the radial stress σ_r at the edge of the contact circle (Engel, 1976).

$$\sigma_r = 0.05585\,(1-2\nu_2)(E_r^4 V^2 \rho)^{1/5} \tag{5.33}$$

where V is the velocity of the sphere which has mass density ρ. The suffices 1 and 2 refer to sphere and impacted body respectively. E_r is the so-called 'reduced modulus' given by

$$E_r = \left(\frac{1 - v_1^2}{\pi E_1} + \frac{1 - v_2^2}{\pi E_2} \right)^{-1} \quad (5.34)$$

Apparently ring cracks have been noted in brittle materials such as glass and plexiglas. It seems not unreasonable then, to expect that there be an analogous effect in concrete, reinforcing the view that postulated impact assessments must incorporate (non-linear) functions of E, V and ρ *and* relative velocity.

Presumably the 'ripples' of stress occur throughout the depth of the concrete (associated with the distortion wave), diminishing with depth as the wave decays. Consequently both cratering and scabbing (to a lesser extent) are likely to be associated with it. In addition the compressive dilatation wave will be reflected as a tensile wave from the far face and this will assist both actions as well as the de-bonding of concrete from surface reinforcement. Any reinforcement orthogonal to the direction of impact can be expected to resist transverse stresses, provided bond is not impaired. Increasing depth of concrete is likely to reduce some of the potential for scabbing, debonding and cratering as energy is dissipated in the initial and reflected waves. Consequently a further parameter to be included could be some inverse function of concrete thickness.

This is supported by considering punching shear since, at its simplest, the stress must be inversely proportional to impactor perimeter multiplied by effective concrete thickness (i.e. the potential area of concrete exposed to punching shear). Dowel action of reinforcement must also add to the resistance.

5.6.3.4 *Damage assessment and estimation* In summary then, assessment of impact damage should be some function of load, velocity, elastic properties of impactor and concrete, their surface shapes, thickness of concrete, and amount of reinforcement. These are all fairly easily quantifiable; rather more difficult is the quantification of crack growth, although some investigators (e.g. Anderson et al, 1984, Watson and Sanderson, 1984) have sought correlations with the volume of cratered or scabbed concrete which may provide a measure of fracture: crack propagation must dissipate some proportion of the energy pulse. Concrete strength might also be regarded as an appropriate indicator although in the derivation of empirical relationships it may not be clear to what extent compressive strength is used as a yardstick for elastic modulus or tensile strength or vice versa: within limits the three may be interrelated.

Various formulae have been proposed from time to time. Figure 5.29 (Kennedy, 1976) compares penetration depths from several of these, while Figure 5.30 (from the same source) compares penetration thicknesses. It can be seen that there is some scatter and it was concluded that the NDRC formula was the most appropriate for local effects from hard impact. Subsequent tests

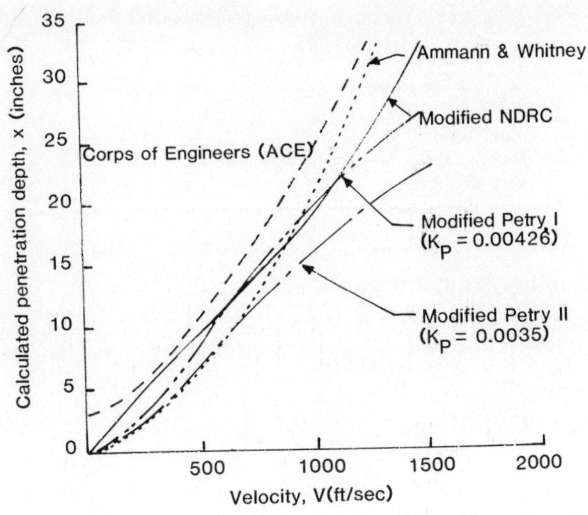

Figure 5.29 Comparison of concrete penetration depths calculated by various formulae for the the case of a typical missile

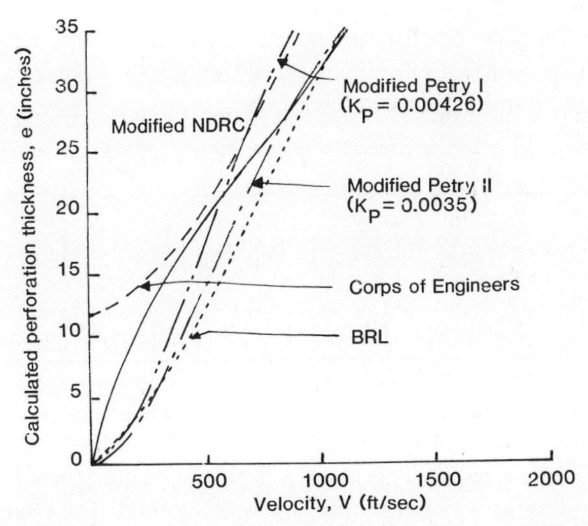

Figure 5.30 Comparison of concrete perforation thickness calculated by various formulae for the case of a typical missile

(Woodfin and Slitter, 1981) have reaffirmed its suitability and a methodology has been proposed for its use in conjunction with probability rather than a safety factor (Haldar, 1981). Consequently it seems worth quoting here in the form given in the latter reference, even though the system of units may now be somewhat anachronistic. The so-called Modified NDRC (National Defense

Table 5.10 Variation in missile shape factor, N

Shape	N
flat nose	0.72
blunt nose	0.84
average bullet-nose (spherical end)	1.00
very sharp nose	1.14

Research Committee) equations are:

$$\frac{x}{d}=\left[\frac{4KNW}{d}\left(\frac{V}{1000d}\right)^{1.8}\right]^{1/2} \quad \text{when} \quad \frac{x}{d}\leqslant 2.0$$

and

$$\frac{x}{d}=1+\left[\frac{KNW}{d}\left(\frac{V}{1000d}\right)^{1.8}\right] \qquad \frac{x}{d}>2.0 \tag{5.35}$$

x = total penetration depth (inches); d = missile diameter (inches); W = missile wt. (pounds); $K=180/\sqrt{\sigma_c'}$; σ_c' = concrete ultimate strength (pounds/square inch); V = striking velocity of missile (feet/second); N = missile shape factor, according to shape of missile (Table 5.10).

Two criteria may be applied to use of this formula, the prevention of scabbing or the prevention of perforation. If the slab thicknesses required to achieve this are x_s and x_p respectively, then:

$$\left.\begin{aligned}\frac{x_s}{d}&=7.91\left(\frac{x}{d}\right)-5.06\left(\frac{x}{d}\right)^2 \quad \left(\frac{x}{d}\right)<0.65\\ \frac{x_s}{d}&=2.12+1.36\left(\frac{x}{d}\right) \quad 0.65\leqslant\left(\frac{x}{d}\right)\leqslant 11.7\end{aligned}\right\} \tag{5.36}$$

$$\left.\begin{aligned}\frac{x_p}{d}&=3.19\left(\frac{x}{d}\right)-0.718\left(\frac{x}{d}\right)^2 \quad \left(\frac{x}{d}\right)<1.35\\ \frac{x_p}{d}&=1.32+1.24\left(\frac{x}{d}\right) \quad 1.35\leqslant\left(\frac{x}{d}\right)\leqslant 13.5\end{aligned}\right\} \tag{5.37}$$

If the target thickness is T (inches) and F is a safety factor then $x_s=T/F$ or $x_p=T/F$ according to whichever criterion is being used. The recommended value of F is 1.2.

In preference to using such a safety factor, it has been suggested (Haldar, 1981) that the NDRC equations be considered the upper bound estimate for penetration. The equivalent lower bound estimate recommended is to use half that value of penetration depth. These upper and lower bounds will then have a spectrum of equations between them, each providing an appropriate velocity; that is, a set of damage probabilities can be generated according to the distribution selected between the bounds. The overall damage probability for a given impact can then be obtained. (It should be remembered here that generally the designer would either select a specific case for impact—the deterministic approach—or would look for a specific probability of occurrence.)

An extensive French programme (Berriaud et al, 1978) produced a formula of the form

$$\frac{\rho V_c^2}{\sigma_c} = k\left(\frac{M}{\rho\phi^2 T}\right)^a\left(\frac{\phi}{T}\right)^b \tag{5.38}$$

where V_c is the minimum velocity for perforation to be just achieved by a projectile mass M, diameter ϕ; k, a and b are constants. Experimental work resulted in

$$V_c^2 = 1.7\sigma_c\rho^{1/3}\left(\frac{\phi T^2}{M}\right)^{4/3} \tag{5.39}$$

for reinforcement contents of 150–300 kg/m^3, compressive strengths of 30–50 N/mm^2 and values of the 'characteristic parameter' $M/\phi T^2$ of 2000 to 30 000. The velocity range was 50 to 200 m/s.

It is asserted that the ratio of target thickness to missile diameter (T/ϕ) helps to determine the form of damage most likely to occur (Perry, 1986). As it decreases from, say, 6 to 1, damage changes from predominantly penetration and spalling (cratering) to likely scabbing and shear plug formation. For offshore structures the scabbing limit is considered to be the most important and a non-dimensional number N_1 is introduced:

$$N_1 = \frac{M^{0.5}V}{\phi^{0.5}T(1 + T/\phi)} \cdot \frac{E^{0.5}}{\sigma_\tau} \tag{5.40}$$

σ_τ is the concrete shear strength. 'The hypothesis is that when N_1 is greater than some critical value, the nominal shear stress exceeds the nominal shear strength and inclined cracking, shear plug formation and scabbing resulting from shear plug movement are likely to occur.' (Perry, 1986). Critical values of N_1 are not given, however.

Other techniques have been applied to the solution of problems involving concentrated loads and impact. Yield-line theory for prestressed slabs subjected to concentrated loads (Perry and Brown, 1982); the possibility of using the method of characteristics (Brown and Perry, 1979); finite difference techniques have each been tried. One of the most powerful techniques seems likely to be finite elements. Figure 5.31 (de Rouvray et al. 1984) shows a typical example and suggests considerable promise. However, like the other mathematical modelling techniques, it does depend on sound understanding of the physical phenomena involved and the availability of applicable data. To what extent these criteria can be satisfied is not clear so it seems probable that empirical formulae will be needed for some time to come.

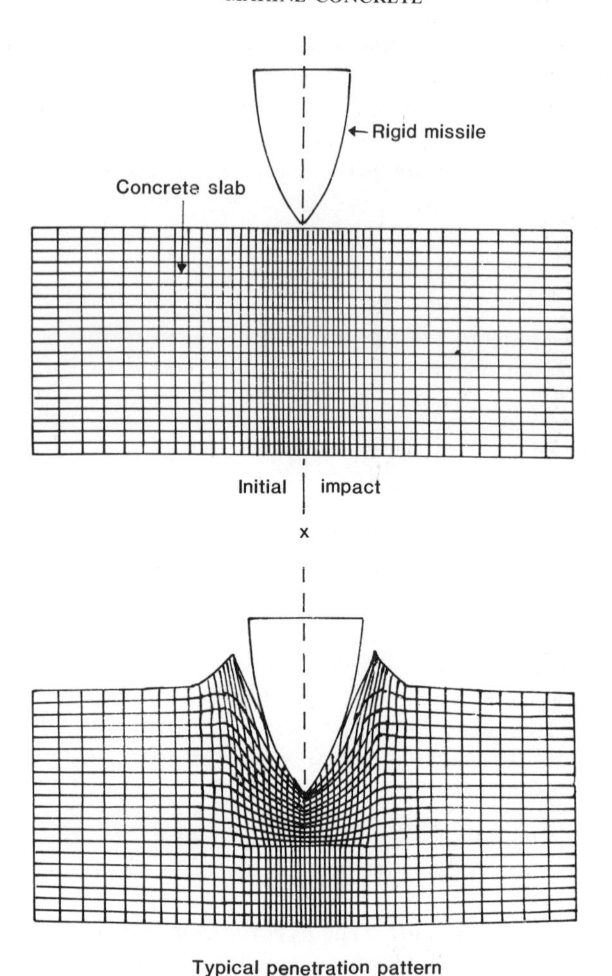

Figure 5.31 Missile impact and penetration in a concrete slab

It will be noted that eqs (5.38) and (5.40) employ the *ratio* of missile 'width' to slab depth. Earlier reference to punching shear implied a vertical-sided shear plug whereas in reality the plug is more likely to resemble a flat-topped volcano in profile (Braestrup and Nielsen, 1983). Similar failures occur in concrete shells (Kavyrchine and Ashtari, 1982 and 1983). This is not the place to consider plastic analysis, it is sufficient to say that it leads to the proposal of a non-dimensional load parameter for static loading

$$\frac{\tau}{\sigma_c} = \frac{P}{\pi(\phi + 2T)\sigma_c} \qquad (5.41)$$

$$= \frac{P}{\pi T(\phi/T + 2)\sigma_c}$$

in which form there are certain similarities to eq. (5.40). (τ is the average shear stress on a control cylinder at a distance T from the loaded area). Consequently it appears that ϕ/T is a factor which also should be incorporated at some stage.

Equations (5.35) and (5.39) indicate that resistance to impact is not particularly sensitive to concrete quality, at least as expressed by compressive strength. Plastic analysis, however, shows that punching shear is very sensitive to the ratio of tensile to compressive strength, with the failure surface tending to become more conical (less flat) as it increases (Braestrup and Nielsen, 1983). Improved tensile strength therefore seems to be a desirable aim. Indirect confirmation of this comes from the improvement in impact performance noted from the incorporation of fibres in the concrete (Anderson et al, 1984; Hughes, 1981) although such a statement over-simplifies the case.

While much of this discussion has concerned slabs it is worth while emphasising that other shapes of element have been tested also, with the 'specimens' being quite substantial in some instances: H–frames with the bar of the H having a clear span of 0.4 m (Watson and Ang, 1984); domes of 1.5 m span (Perry and Dinic, 1985; Kufuor and Perry, 1984); tubes 3 m long × 0.5 m diameter (Brakel and Oostlander, 1979); beams 8 m long with one end dropped through 3.75 m (Ammann, 1984; Ammann et al, 1981).

5.6.4 *General loads (collision)*

Collision may involve hard and/or soft impact. Disintegrating rotating machinery, such as a turbine, can result in hard missiles travelling at appreciable velocities and tests have been conducted accordingly, with parts having masses up to 2100 kg projected at velocities up to 132 m/s (Woodfin and Sliter, 1981). The mean low-flying velocity of a Phantom jet fighter is about the same and, on collision with a concrete structure, would give a linear increase in load from zero to about 24 MN in 0.05 s (Zorn et al, 1981) from a total mass of 17 000 kg. The larger but slower-flying Boeing 707 would have a greater contact area and so cause less severe damage to the structure than the Phantom (Zimmerman et al, 1981). For crashing automobiles, at low speeds crushing of the vehicle body dominates generation of impact forces on a structure but at high speeds the impact force signature will have a 'spike' due to impact of the engine (Chiapetta and Costello, 1981).

These observations, from different papers presented at the same meeting, illustrate how collision forces are governed, to a considerable extent, by the nature of the colliding body. Consequently, for offshore concrete structures in particular, the effects on the structure are likely to be very different due to a typical supply boat, an oil tanker, falling weights, or a crashing helicopter.

Direct impact from a helicopter on the concrete part of the structure is unlikely (as distinct from impact on the superstructure) but collision from a marine vessel is an 'everyday' possibility (albeit normally a mild impact) so it is hardly surprising that the question has received some attention. Clearly much depends on the type of vessel involved and, aside from navy ships, it has been suggested that they be classified thus (Incecik and Samoulidis, 1982):

1. ships with two or more longitudinal bulkheads (tankers, LNG carriers, containerships, the larger passenger ferries);
2. ships without internal longitudinal bulkheads (dry cargo ships, bulk cargo carriers, Ro-Ro ships, etc.).

This appears to be slanted more towards ship to ship collisions but it forms a useful starting point for ship to platform impacts since the classifications may be indicative of the energy–absorbing and penetration capacities of the larger vessels.

5.6.4.1 *Collision categories* A more general classification scheme has been offered along with comment on the suitability of predictive techniques available to the designer (Dallard and Miles, 1984). There are three prerequisites to categorisation:

(a) estimation of the amount of energy to be stored or absorbed during impact, usually by rigid body calculations relating to conservation of energy, conservation of momentum, etc;
(b) estimation of the load (P)/deflection (δ) characteristics of the colliding or impacting bodies; initially quasi-static calculations will provide the elastic stiffness characteristics of the structure(s) and the collapse loads can be estimated on the basis of plastic flow, plastic hinge, yield line formation, buckling, or whatever;
(c) estimation of peak forces and event durations by equating the energy estimates of (a) to the area under the P–δ curve(s) of (b).

The collapse characteristics of the five suggested categories are shown in Figure 5.32 (Dallard and Miles, 1984).

1. Elastic structures
   The impact energy to be stored or absorbed is less than the capacity of the structure to store energy elastically. In practice some localised damage will normally occur, thereby limiting the peak load generated. Dynamic response may result in excitation of natural frequencies. Commonplace elastic structures—billiard balls.
2. Local plastic structures
   Localised impact damage of steel structures takes the form of 'local plasticity' where plastic flow occurs right at the point of impact. Example structures—metal shock limiters on packages requiring impact protection are designed to flow plastically i.e. collapse.
3. Global collapse structures (stable)
   Collapse extends through large parts of the impacted body. The impact energy to be absorbed or stored is significantly greater than the elastic strain energy storage potential. 'Once a collapse mechanism has been

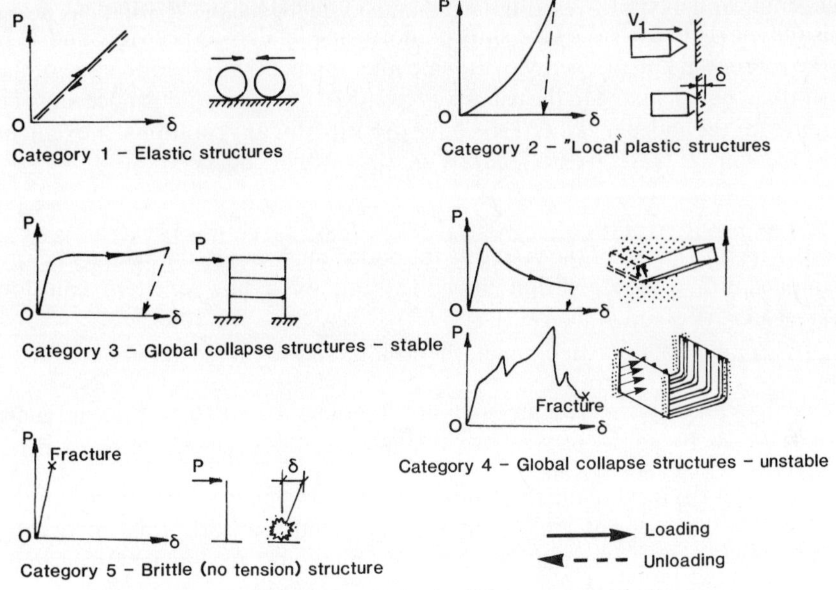

Figure 5.32 Collapse characteristics for different categories of structure

established, the plastic part of the P–δ curve is either perfectly plastic or hardens monotonically.' Examples—encastré beams and plates and non-buckling frameworks as in steel building frames.

4. Global collapse structures (unstable)
   Similar to category 3 *but* the P–δ curve does not rise monotonically after the establishment of a collapse mechanism, rather there are abrupt changes of stiffness and the curve may decline after the point of initial collapse. Examples—sheet metal fabrications and welded plate structures such as ships and automotive structures.
5. Brittle materials
   Energy storage or absorption is severely limited by a predisposition towards brittle failure which occurs before plastic distortions can take place.

Useful as such a classification is, a particular structure is not necessarily uniquely categorised. For example a concrete oil storage platform may belong to category 1 for loads from a supply boat, category 2 for a falling drilling pipe and category 3 for an out-of-control or mis-directed oil tanker while the tanker itself would be a category 4 structure.

5.6.4.2 *Energy considerations* For a collision between a ship and a fixed platform, the energy conservation equation is (Søreide et al, 1982):

$$E_k = E_s + E_p + E_f + E_r \tag{5.42}$$

E_k is the kinetic energy of the striking vessel immediately before impact, $E_{s,p,f}$ is the energy absorbed by the ship, platform and fenders respectively and E_r is the rotational kinetic energy of the ship after impact. The kinetic energy of the ship can be calculated in the usual way (eq. (5.43)) but some allowance must be made for the added mass, as it will have to be also for any rotational movement of the ship. If there are no fenders, then $E_f = 0$, of course.

$$E_k = \tfrac{1}{2}(m + m')v_1^2 \tag{5.43}$$

where m is the mass of the vessel, m' the added mass and v_1 the velocity before impact. For a side collision $m' = 0.4m$ and for a bow or stern collision $m' = 0.1\,m$ (Røland et al, 1982).

Assuming fully plastic impact, the residual energy after impact is

$$E_p = \frac{m^2 v_2^2}{2(m + m_c)} = \tfrac{1}{2}mv_2^2 \tag{5.44}$$

for local shaft response

where m_c is the mass of the shaft and v_2 the vessel velocity after impact. For the vessel to be stopped, ship and structure must absorb this energy by deformation. Assuming that the shaft is infinitely stiff, the impact force can be calculated from the area under the P–δ curve of the vessel.

For small vessels (say 5000T), the ratio of duration of impact and natural period of local bending of the shaft is about 30. Typically, the kinetic energy of the shaft/deck frame at the end of the impact is about 4% of the impact energy. This would appear to justify treating the local impact statically. For large vessels, however (say 150 000T), the response is global rather than local and dynamic effects must then be considered (Røland et al, 1982). The natural periods of local vibrations in typical tubular off-shore structures are (Sørenson, 1976):

$$\left.\begin{aligned} T_h &\approx 0.1\text{–}0.5 \\ T_m &\approx 10\text{–}20 \\ T_b &\approx 50\text{–}500 \end{aligned}\right\} \times 10^{-3}\,\text{s}$$

where T_h is the natural period for vibrations across the thickness of a plate or shell, T_m is the natural period for membrane vibrations and T_h is the natural period of bending vibrations of the first mode in a circular cylindrical cell.

5.6.4.3 *Ship collision with concrete towers/shafts* The response of any given platform to a collision is a function of the initial design parameters adopted and the nature of the particular event which occurs: whether it is a bow or sideways collision from a large or small vessel. General conclusions are therefore very difficult to formulate. However, certain studies indicate that (Fjeld, 1979 and 1982), for typical offshore platforms with shafts < 20 m diameter:

(a) impact from supply ships with velocity significantly more than 2 m/s can easily be taken by walls 0.6–1.0 m thick;

(b) a significantly thicker wall will be required to withstand the impact from a 150 000T tanker; while the global frame can carry the load involved, local failure might result, depending on the circumstances.

Tests on concrete cylinders indicate the following pattern of failure (Brakel and Oostlander, 1979), illustrated in Figure 5.33:

(a) a longitudinal crack (1) develops on the inner surface just under the loaded area, extending and opening rapidly with increasing load;
(b) external cracks (2, 3, 4.....*h*) develop, the last crack (*n*) being about three times the wall thickness from the edge of the loaded area;
(c) further increase of load suddenly pushes (or punches) a cone-shaped part of the wall though the cylinder.

After crack (*n*) has developed there is no remaining rotation capacity in the wall so that a shear crack, angle ϕ, develops between (1) and (*n*) with increase in load. This increase has to be carried by a plate shaped part of the wall, fixed longitudinally and free transversely. Further structural equilibrium is not possible so the load punches through with the shear angle similar to that in tests on flat plates.

Based on these and other tests, the following formula has been proposed (Caldwell and Billington, 1981):

$$P_{br} = 2.92\, k_s k_t k_p h_n (a + h_n)\sqrt{\sigma'_c} \tag{5.45}$$

where P_{br} is the punching shear strength of a hollow concrete cylinder and σ'_c is the compressive strength of concrete from cylinder tests (N/mm^2). $k_{s,t,p}$ are factors allowing for effective depth of wall h_n, ratio of thickness to external diameter D, and prestress respectively.

$$\begin{aligned} k_s &= 0.65 + \frac{0.15}{0.12 + h_n} \\ k_t &= 0.5 + 50\,(t/D)^2 \\ k_p &= 1.0 + 0.02\sigma_c \end{aligned} \tag{5.46}$$

σ_c is the compressive strength in the tower walls (N/mm^2). For $Y/\sqrt{rt} > 3.0$ local yield, rather than more brittle punching shear failures, will occur. Y is the breadth of the loaded area and r is the mean radius of the tower. t, D, h_n and Y are all in metres.

For a collision spread over an area x by Y, the maximum area of punching shear failure $L \times B$ is given by

$$\begin{aligned} L &= x + k_L t \\ B &= Y + k_B t \end{aligned} \tag{5.47}$$

where k_L and k_B are damaged area length and height factors with average values of 5.76 and 2.59 respectively.

Finally, in concluding the discussion on impact, it should be noted that an

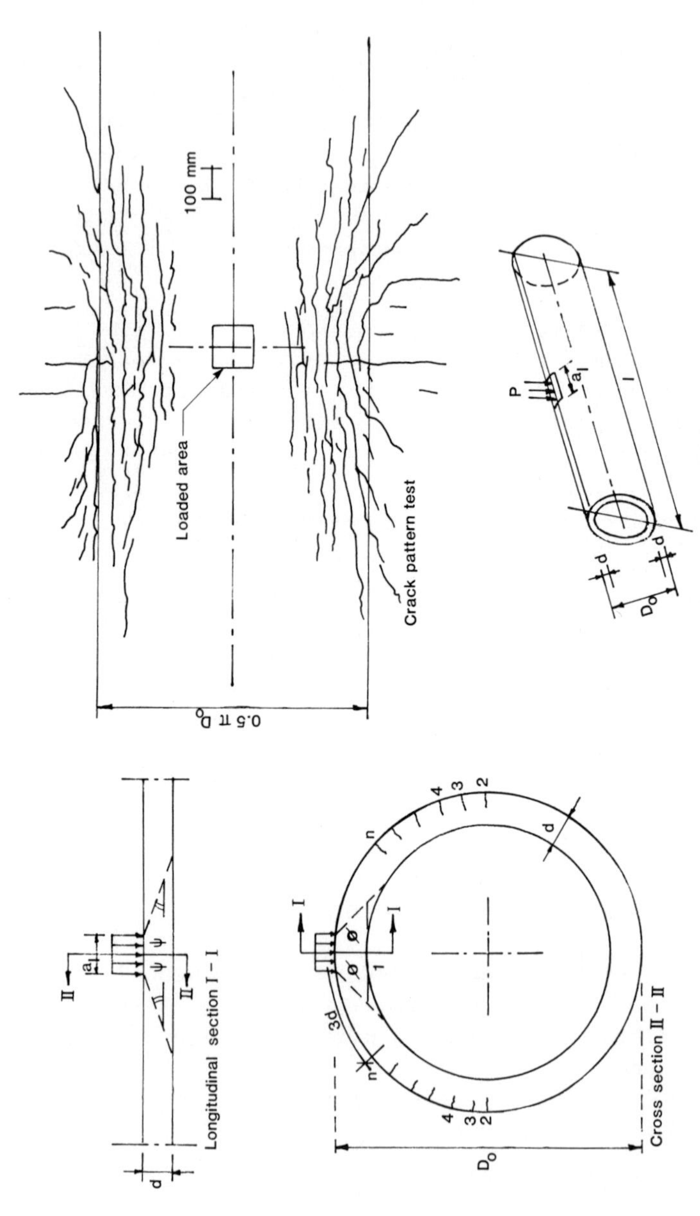

Figure 5.33 Concrete cylinder with concentrated load

alternative method has been proposed for predicting collapse of a concrete cylindrical column (Davies and Mavrides, 1981). This replaces the cylinder with a flat circular plate of half the column diameter to represent the area of the structure responding most rapidly to collision forces. For an infinite flat plate, this load zone has a diameter equal to about 13 times the plate thickness. Making use of these criteria, computer programs have been produced incorporating a complex constitutive model for concrete and an appropriate forcing function for the collision force. This is intended to provide more exact information on damage and distortion, limiting the work of divers in inspection by enabling assessment of the extent of damage.

5.7 Implosion

Explosion has been referred to briefly and its consequences can be catastrophic. However, structures subjected to external hydrostatic pressure can also *imp*lode i.e. collapse inwards if the pressure is sufficiently high. Shells of revolution minimise the use of material to resist uniform pressure loads so it is not surprising that oil storage and production platforms have cells and legs of predominantly cylindrical shape. Clearly then it is important to develop appropriate design procedures and the US Navy's Civil Engineering Laboratory has conducted tests accordingly over a number of years. Collapse loads and mechanisms have been established for both spheres and cylinders but attention here will be confined to the latter in view of their predominance.

Cylinders have been divided into the two familiar primary categories of *thin-* and *thick-walled*, with thin-walled being further sub-divided into *moderately long* and *long* cylinders (Haynes, 1979). Long cylinders are those not influenced by end closures, and behave as if they were infinitely long, while *short* thin cylinders can be incorporated into the thick-walled category. Division between the thin- and thick-walled categories depends on the criteria adopted. At its simplest, one may use Lamé's equations (Timoshenko and Goodier, 1951) for a cylinder of internal and external radii r_i and r_0 subjected to internal and external pressures p_i and p_0. Then the circumferential or hoop stress σ_{hr} at radius r is given by

$$\sigma_{hr} = -\frac{r_i^2 r_0^2 (p_0 - p_i)}{r_0^2 - r_i^2} \cdot \frac{1}{r^2} + \frac{p_i r_i^2 - p_0 r_0^2}{r_0^2 - r_i^2} \tag{5.48}$$

For zero internal pressure the stresses at the outer and inner fibres σ_{ho} and σ_{hi} then are

$$\sigma_{ho} = -\frac{r_0^2 + r_i^2}{r_0^2 - r_i^2} \cdot p_0 \tag{5.49}$$

and

$$\sigma_{hi} = -\frac{2r_0^2}{r_0^2 - r_i^2} \cdot p_0$$

Table 5.11 Variation of factors k_i, k_0 with R, ratio inner to outer radius: used in thin cylinder theory to calculate inner and outer fibre stresses and to estimate errors involved

R	0.90	0.95	0.97	0.99
k_0	1.05	1.025	1.015	1.005
k_i	0.95	0.975	0.985	0.995

For a thin-walled pipe thickness t with uniform hoop stress σ_h

$$\sigma_h = -\frac{p_0 r_0}{t} = -\frac{p_0 r_0}{r_0 - r_i} \tag{5.50}$$

If one then introduces an 'error factor' k such that $\sigma_h = k_0 \sigma_{h0}$ or $k_i \sigma_{hi}$, k can be calculated for various values of $R = r_i/r_0$ i.e. ratios of internal to external radius. From eqs (5.49) and (5.50)

$$k_0 = \frac{(1+R)}{(1+R^2)}$$

and

$$k_i = \tfrac{1}{2}(1+R) \tag{5.51}$$

Hence it can be seen that for a ratio of inner to outer radius of 0.90 (i.e. wall thickness t to outer diameter D_0 of 0.05) thin cylinder theory over-estimates the outer fibre stress by 5% and underestimates the inner fibre stress by the same amount. For $R = 0.99$ i.e. $t/D_0 = 0.005$ the corresponding stress errors are $\pm 0.5\%$. Division into categories can then be defined by the error which is acceptable.

In fact, defining thick-walled geometry for concrete cylinders as that in which material failure controls the implosion pressure requires t/D_0 to be about 0.06 (Highberg and Haynes, 1978): for spheres it is about 0.02. This takes account of adjustment of through-wall variations in hoop stress due to creep and so on.

Rather than attempt to summarise the theory and experimental results, which involves considerations of buckling and inelastic behaviour among other things, it is simpler here to reproduce the design charts which reflect the outcome. Figure 5.34 (Haynes, 1979) plots t/D_0 against L/D_0 for various values of implosion pressure p_{im} as a fraction of the uniaxial concrete compressive strength f'_c. L is the length of the cylinder. Hence, for a cylinder of particular configuration, the implosion pressure is easily found for concrete of a specific grade. No factor of safety is incorporated, nor is any allowance made for reinforcement which will depend, in any case, upon the extent to which it is tied against movement. The reader is advised to refer to the report and associated references for further details (see, for example, Haynes 1975, 1976; Highberg and Haynes, 1977; Leick and Bode, 1978).

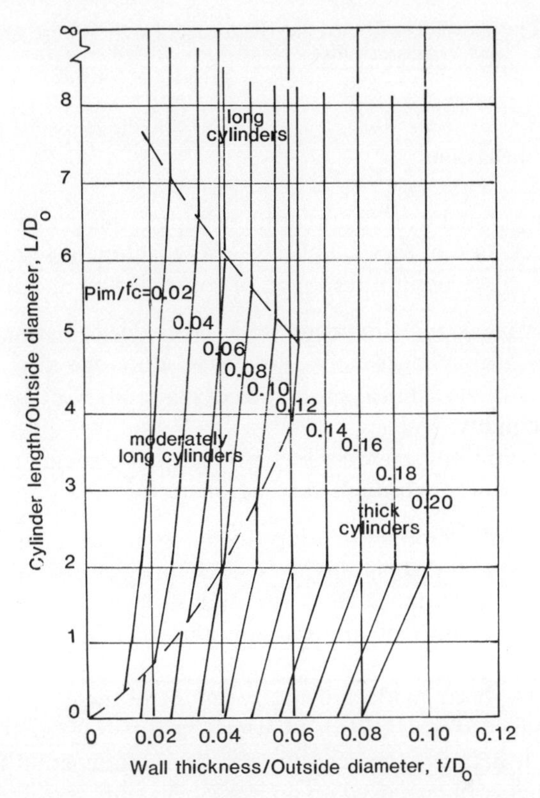

Figure 5.34 Design guide for predicting implosion of concrete cylinder structures

Resistance to high hydrostatic pressures mobilises the multi-axial strength of concrete and the concept has been carried further in the development of designs for seabed applications down to water depths of 1000 metres (Collard and Skillman, 1985). These utilise concrete sandwiched between steel skins, some of the criteria for which are discussed elsewhere (Montague and Goode, 1979). The most important stresses are the circumferential stresses in the steel skins and the inside of the concrete filler, and the radial interface stress between the inner skin and the filler. As pressure increases so does the radial displacement until the inner steel skin yields when there is a change in slope (reduction) of the pressure/displacement line. Thereafter the shell will fail when the concrete fails, possibly preceded by yielding of the outer skin, depending on the yield strength of the steel.

The performance of shells subjected to external pressure is affected by initial imperfections and the interaction between strength and instability. For steel-concrete-steel shells these factors do not appear to cause problems so long as the total wall thickness exceeds about $5\frac{1}{2}\%$ of the external diameter when the full theoretical failure pressure based upon strength alone should be attained. The subject will not be discussed further here since the object is primarily to introduce the concept.

5.8 Thermal behaviour

Following the parallel with the preceding chapter, this chapter will conclude with some discussion of thermal effects. Many factors have to be accounted for in considering structural response, of course (Iding and Bresler, 1978):

(a) stress-related instantaneous and creep deformation, as well as dimensional changes caused by temperature or shrinkage variations;
(b) changes in mechanical properties of materials with age, temperature and humidity; and
(c) degradation in local cross sections through cracking and crushing in concrete and/or through yielding of steel.

Being necessarily selective in order to put some limit on the discussion means that including some involves excluding others. That is not to downgrade their importance, however, and all factors have to be assessed but certain applications of, or uses for, the material demand particular attention.

Reference has been made previously to the storage of hot oil and of refrigerated liquefied gases, both of which give rise to particular thermal problems. As it happens they encompass the environmental thermal range so useful pointers can be picked up on likely behaviour in Arctic and tropical conditions.

5.8.1 *Thermal restraint*

Temperature change can alter the way in which a material behaves or the rate at which it responds to external influences, as we have seen for concrete. Fortunately, by and large steel is minimally affected within the range which is our present concern, so its properties and characteristics can be regarded as being sensibly constant. However, it differs from concrete in some respects and due account must be taken of those differences. Furthermore, even in an isotropic, homogeneous continuum, temperature change and heat flow involve temperature gradients which have consequential strain gradients. The resulting thermal stresses have to be sustained if structural integrity is to be maintained. Moreover, structural elements are connected to, and restrained by, other structural elements (here walls, shells and foundations are all considered 'elements') and such restraints induce stresses—generally absence of restraint means absence of stress caused by temperature or moisture change. It is the quantification of the restraint which can sometimes occasion difficulty.

At its simplest level, if a reinforced concrete element is subjected to a temperature change of T degrees, then there will be differential thermal movement between the steel and concrete unless their thermal coefficients (α) are identical. The consequent thermal strain is $(\alpha_s - \alpha_c)T$ where the suffices s and c denote steel and concrete respectively, and the stress in the steel will be $E_s(\alpha_s - \alpha_c)T$ and in the concrete, $E_c(\alpha_s - \alpha_c)T$. This is almost trivial but the addition of thermal gradients, differential restraint and thermal influence on concrete properties (including the effect on moisture gradients) soon complicates the issue. Given appropriate analytical and computational techniques and knowledge of the properties, however, thermal loads are calculable and can be added to those due to other imposed loads such as waves, hydrostatic pressure, plant, self-weight, and any additional dead and live loads. Load sharing between the concrete and any reinforcement then depends on their relative quantities and distribution but it is also affected by a fundamental difference between the two materials which is in turn governed by time through duration of loading (of whatever form).

5.8.2 *Thermal creep*

Within the envisaged limits of use, concrete creeps while steel does not. (Concrete is also affected by shrinkage but in the marine environment that can normally be ignored.) The statement is hardly revelatory but it is the combination of creep and thermal transients which is as yet imperfectly understood (Khoury et al, 1985) and therefore warrants some emphasis.

Figure 5.35 (Ross et al, 1965) shows a typical variation of specific creep (creep strain per unit stress) for concrete subjected to different durations of loading. Over the range 20–60°C at least, the lines are virtually straight and,

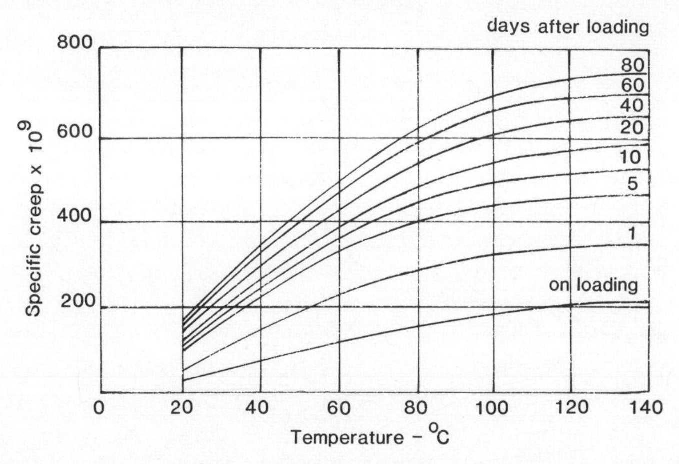

Figure 5.35 The variation of specific creep with temperature and time after loading for sealed concrete first loaded at 10 days

for durations of 5 days and more, pass through or very close to the origin. Since they are fairly evenly spaced (for the same temperature range) above a duration of 7 days or so, it seems reasonable to suppose that the **rate** of specific creep is proportional to temperature. Consequently the proposed two-dimensional constitutive equation, eq. (5.52), and based on more extensive considerations, seems quite acceptable (Richmond, 1977; Richmond et al, 1980):

$$\begin{aligned} \dot{\varepsilon}_x &= \sigma_x T\dot{\phi}(t) - \nu\sigma_y T\dot{\phi}(t) \\ &= T\dot{\phi}(t)(\sigma_x - \nu\sigma_y) \end{aligned} \tag{5.52}$$

where ε_x is the strain in the x-direction, T is the temperature °C, $\phi(t)$ is the specific creep per degree, σ_x and σ_y are the orthogonal stress components, ν is the creep Poisson's ratio and the dot denotes differentiation with respect to time. ($\phi(t)$ has been referred to as pseudo time or creep time because a creep specimen could be used as a clock with its output indicated as strain.)

5.8.2.1 *Transitional thermal creep* Such a relationship is useful for concrete under load at constant temperature but complications ensue when the temperature is increased. It has been shown (Illston and Sanders, 1973) that when loaded concrete has its temperature raised to a given level for the first time, there is a sharp increase in creep strain before the previous strain rate is resumed. This is not recoverable when the temperature is reduced nor does it manifest itself should the temperature be raised again to or below the previous elevated level. However, should it be raised beyond that, there is a

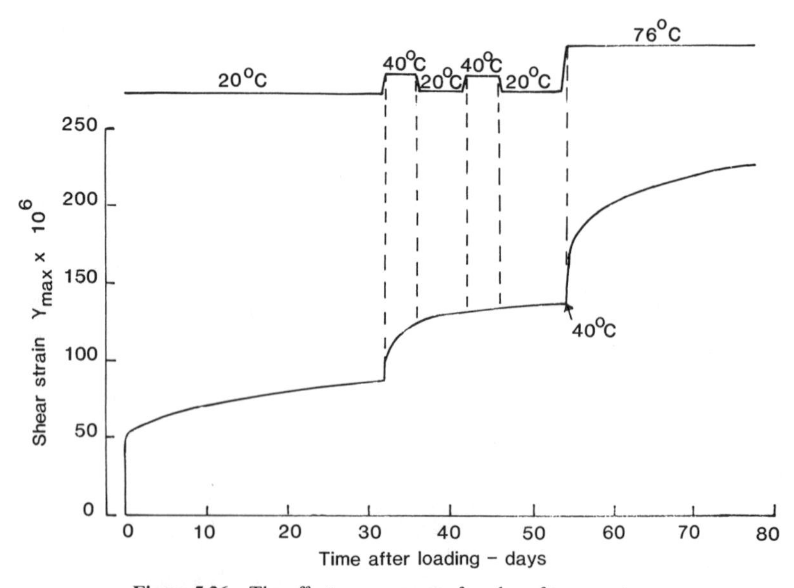

Figure 5.36 The effect upon creep of cycles of temperature

further increase in creep. Such increases due to raising the temperature of concrete under constant load have been labelled 'transitional thermal creep'. The effects are illustrated in Figure 5.36 (Illston and Sanders, 1973).

From this work stemmed the identification of two creep components:

1. constant-temperature creep comprising
   (a) flow – an irrecoverable component whose rate at a given temperature is a function of maturity but not age of loading;
   (b) delayed elastic strain – a recoverable and rapidly-occurring component with a constant limiting value at a given temperature irrespective of maturity; its rate of development declines with increasing maturity;
2. transitional thermal creep as described already; its magnitude is approximately the same at all maturities.

These effects are summarised in Figure 5.37.

Subsequent tests on cement pastes confirmed the model described above but indicated that transitional thermal creep is reduced if the heating *precedes* loading (Parrott, 1979).

The rate of transitional thermal creep seems likely to be age dependent: rapid for young concretes as in Figure 5.36 but much slower for mature

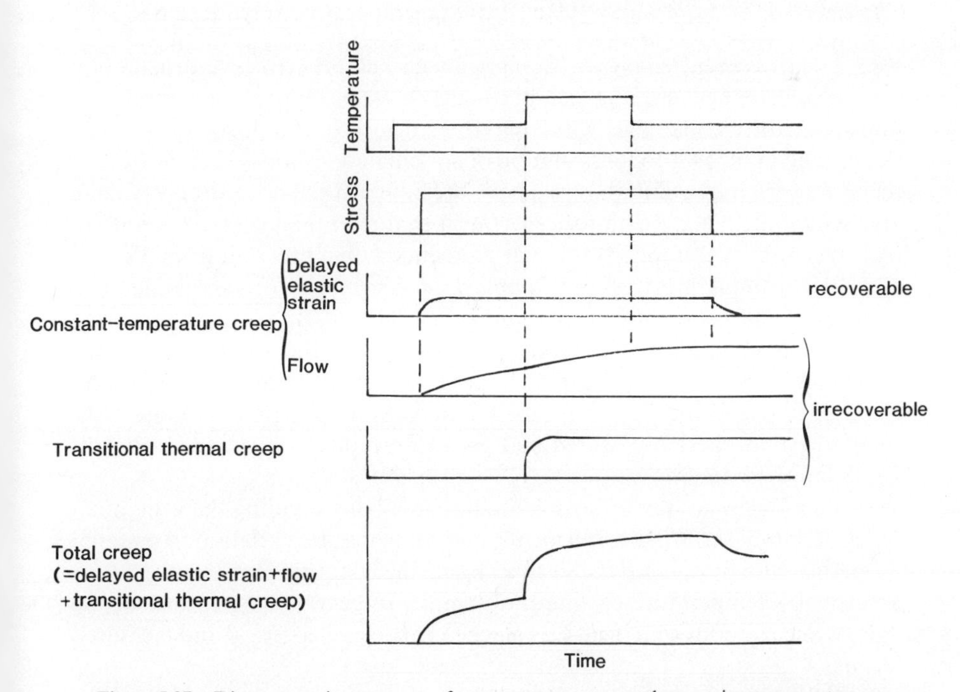

Figure 5.37 Diagrammatic summary of creep components under varying temperature

concretes (Bamforth, 1980). A suggested relationship is

$$c_{tt(t)} = c_{tt(14)}\sigma \cdot 0.85 \log(t+1) \qquad (5.53)$$

where $c_{tt(t)}$ is the transitional thermal creep t days after heating $(0 < t < 14)$; $c_{tt(14)}$ is the value 14 days after heating; σ is the applied stress.

Other work (Parrott and Symmons, 1979) postulates a relationship particularly oriented towards oil storage structures in terms of *specific* transitional thermal creep per degree rise ε_t:

$$\varepsilon_t = \varepsilon_{t\max}[a + b(1 - \exp\{-t/c\})] \qquad (5.54)$$

where a, b and c are constants. As an indicator, for the concrete tested, $a = 0.3$, $b = 0.7$ and $c = 9$ and $\varepsilon_{t\max} = 0.45 \times 10^{-6}$ per N/mm²/°C.

A more extended discussion of the subtleties is to be found elsewhere (Khoury et al, 1985) where the observation is made that 'transitional thermal creep results in relaxation and redistribution of thermal stresses, rendering the elastic stress analysis inappropriate for structures heated for the first time··· Excessive tensile thermal and incompatibility stresses can develop during cooling, when transitional thermal creep is absent, which could lead to the development of cracking and serious weakening of the material.' The evidence also seems to point to its being due to physical/chemical changes in hydrated/hydrating cement as a result of heating, rather than to structural changes following moisture movement.

5.8.2.2 *Stress re-distribution* In oil storage structures stress re-distribution can be severe because the geometry may cause complete restraint and moments can change sign with time due to temperature-dependent creep (Richmond et al, 1980). Confirmation of anticipated response or behaviour is therefore very important and tests have been carried out accordingly (Clarke and Symmons, 1980). Reinforced specimens of similar thickness to the walls of real structures had imposed thermal gradients with a crossfall from 55°C to 10°C. Horizontal prestress was applied of 5 N/mm², corresponding to a differential hydrostatic pressure through the 'wall' of 30 m of water. Restoring moments were applied to maintain straightness of the specimens thereby providing a measure of the bending forces set up by the temperature gradient. Heating and cooling cycles simulated operation of typical storage cells. Internal strains were monitored and parallel creep tests conducted (Parrott and Symmons, 1980).

Using a timewise step by step procedure involving dividing the wall into a series of parallel contiguous elements, and knowing the variation in material properties with time, it is possible to compute the restoring moment history for a particular temperature cycling programme. In general, provided cracking did not occur, good agreement was reached between theoretical and measured results.

It is clear that calculation of peak moments using simple theory and the short-term elastic modulus, results in appreciable over-estimating (by perhaps 20%). Sustained temperature cross-fall moments are significantly (of the order of 30%) reduced by creep. Maximum bending moments occurred about 12 hours after the first application of temperature cross-fall so that there is a gradual reduction with time.

The relationships used to compute behaviour were as follows (the various constants may have to be modified for different concretes: reference should be made to the original sources given in the report):

$$E = \frac{M}{12000 + 27.5\,M} \cdot 10^6 \text{ N/mm}^2 \tag{5.55}$$

where $M = \sum (T + 10)t$ and is the maturity with t the time in days at temperature T°C.

The recoverable creep strain ε_r developing during time t_s after a change of stress and under stress σ is

$$\varepsilon_r = \left[3.45 + \frac{310}{M}(T + 12) \right] \left(0.2 + 0.8 \left[1 - \exp\frac{t}{5.7} \right] \right) 10^{-6}\sigma \tag{5.56}$$

The **irr**ecoverable creep strain developing during a time increment δt is assumed to be a function of the mean temperature during the period and of M:

$$\varepsilon = 4.09 \left(\frac{1000}{M} \right)^{1.25} \exp\left[-\frac{1575}{T + 273} \right] (T + 10)\delta t\sigma \cdot 10^{-6} \tag{5.57}$$

The transitional thermal creep is as given in eq. (5.54) with the quoted values of the constants. It should be noted that transitional thermal creep causes a rapid fall in moment during the first heating cycle.

5.8.2.3 *Effect of cracking* Where prestress is not applied, cracking can be expected to modify behaviour, particularly on moment reversal which occurs during the temperature cycling on removal of the heat source. Fine cracks have been observed on straight, unprestressed specimens and regular cracks on cylinders (Clarke, 1978, 1980). However, at the time of writing details of subsequent tests are not available to the author although it is believed that further information is being published.

Cracking is dependent on elastic behaviour (as distinct from creep) and restraint and will occur when constraint equals or exceeds the flexural strength of the concrete f_t. The flexural moment developed by a temperature crossfall over thickness h is of the form (Fouré et Trinh, 1980):

$$M = \lambda EI \cdot \frac{\alpha \Delta\theta}{h} \tag{5.58}$$

where λ is a restraint coefficient which is zero when the concrete is free to deform and is unity when zero curvature is maintained. For a simply supported beam loaded at two points, in the zone of constant moment $\lambda = 3(a + l)/(2a + 3l)$ where the loading points are distance l apart and a from the ends.

For a rectangular section width b and height h, the cracking moment M_f and the cracking temperature crossfall $\Delta\theta_F$ are related thus:

$$\Delta\theta_F = \frac{hM_F}{\lambda EI\alpha} = \frac{2f_t}{\lambda E\alpha} \tag{5.59}$$

The thermal cracking gradient is therefore

(a) proportional to the flexural strength;
(b) inversely proportional to the modulus of elasticity and the thermal coefficient;
(c) independent of the depth of section so long as the restraint does not depend on it.

Under steady conditions the temperature crossfall is equal to the difference between the temperatures on the opposite faces but in transient conditions an effective crossfall is suggested:

$$\Delta\theta_{\text{eff}} = \frac{12}{h^2}\int_{-h/2}^{h/2} \theta(y)y\,\mathrm{d}y \tag{5.60}$$

where $\theta(y)$ represents the temperature distribution.

5.8.3 *Low-temperature effects*

Somewhat different considerations may prevail at low temperatures. In the previous chapter 'freezing expansion' was referred to, illustrated in Figures 4.47 to 4.49. Where the material is reinforced or prestressed, of course, the steel behaves 'normally' and continues to contract on cooling. Hence the differential thermal strain between steel and saturated concrete may be of the order of 20 μ strain/°C for a water/cement ratio of 0.50 so that for a temperature drop from −20 to −45°C there is an additional thermal strain of 500 μ strain. Since saturated concrete has been presupposed, its elastic modulus and strength are considerably higher than for temperate ambient conditions so that problems with the concrete are unlikely. However, steel does not have similar enhancement of its properties so that the load in the steel will be increased. There is limited experimental evidence to show that this does in fact happen but it is not clear to what extent it is alleviated by creep, for example (Marshall and Drake, 1987). Other calculations indicate that the additional stresses due to such an effect, if occurring at and above the water line in real structures in Arctic conditions, could be comparable with those due to wave and other loads. Consequently they warrant careful examination although much depends on the water/cement ratio and the moisture conditions within the concrete.

As a postscript it may be rewarding to conclude on a more reassuring note. Thermal shock has not so far been alluded to but spillage or limited release of cryogenic liquids is a real possibility to be considered, especially under storm or earthquake conditions. Tests have been conducted on reinforced and prestressed lightweight concrete elements under high-intensity cyclic loading and subjected to cryogenic thermal shock (Berner et al, 1983). They confirm that they can retain their integrity against through cracks and remain structurally sound with only relatively minor reductions in stiffness. Even when heavily reinforced the reduced temperature zone is well confined. Despite some of the uncertainties, therefore, concrete is a good 'thermal' material.

References

Allen, R.T.L. and Forrester, J.A. (1985) The investigation and repair of damaged concrete structures. *Corrosion of Reinforcement in Concrete Construction* (ed. A.P. Crane) Soc. of Chem. Indy., Ellis Horwood, Chichester.

Ammann, W. (1984) Applicability of dynamic plasticity theorems to impulsively loaded reinforced concrete structures. *Proc. Int. Conf. Structural Impact and Crashworthiness, Vol. 2.* (ed. J. Morton) Elsevier, London.

Ammann, W., Mühlematter, M. and Bachmann, H. (1981) Experimental and numerical investigation of reinforced and prestressed concrete beams for shock loading. *Trans. 6th Int. Conf., Structural Mechanics in Reactor Technology*, North-Holland/Commn. Euro. Comm., Brussels.

Anderson, W.F., Watson, A.J. and Armstrong, P.J. (1984) Fibre-reinforced concretes for the protection of structures against high velocity impact. *Proc. Int. Conf. Structural Impact and Crash-worthiness, Vol. 2* (ed. J. Morton) Elsevier, London.

Arthur, P.D., Earl, J.C. and Hodgkiess, T. (1982) Corrosion fatigue in concrete for marine applications. *Fatigue of Concrete Structures, SP-75, American Concrete Inst.* Detroit.

Arup, H. (1985) The mechanisms of the protection of steel by concrete. *Corrosion of Reinforcement in Concrete Construction.* (ed. A.P. Crane) Soc. of Chem. Indy., Ellis Horwood, Chichester.

Bamforth, P. (1980) *The effect of temperature variation on the creep of concrete, Tech. Note UTN 19, CIRIA-UEG.* London.

Banerjee, H.K., Hodgkiess, T. and Arthur, P.D. (1985) Fatigue in prestressed concrete for marine applications. *Proc. Int. Conf. Behaviour of Offshore Structures.* Elsevier, Amsterdam.

Bardal, E. (1985) Effects of marine environment and cathodic protection on fatigue of structural steels. *Fatigue Handbook.* (ed. A. Almar-Naess) Tapir, Trondheim.

Beeby, A.W. (1978) Cracking and Corrosion. *Concrete in the Oceans Tech. Report, No. 1, Cement & Concrete Assocn.* Slough.

Beeby, A.W. (1980) Cover to reinforcement and corrosion protection. *Proc. Int. Symp. Behaviour Offshore Concrete Structures* (*Proc. Coll. Int. Tenue des Ouvrages en Béton en Mer*) CNEXO, Brest.

Bennett, E.W. and Joynes, H.W. (1979) Fatigue strength of cold-worked non-prestressed reinforcement in prestressed concrete beams. *Mag. Concrete Research*, **31 (106)**, March: 13–18.

Berner, D.E., Polivka, M., Gerwick, B.C. and Pirtz, D. (1983) *Behaviour of prestressed lightweight concrete subjected to high-intensity cyclic stress at cryogenic temperatures. Annual Convention, American Concrete Inst.* Los Angeles.

Berriaud, C., Sokolovsky, A., Gueraud, R., Dulac, J. and Labrot, R. (1978) Comportement local des enceintes en béton sous l'impact d'un projectile rigide. *Nuclear Eng. and Design*, **45**: 457–469.

Blundell, R., Dimond, C. and Brown, R.G. (1976) *The properties of concrete subjected to elevated temperatures, Tech. Note 9.* CIRIA-UEG, London.

Booth, E.D., Leeming, M.B., Paterson, W.S. and Hodgkiess, T. (1986) Fatigue of reinforced

concrete in marine conditions. *Proc. Int. Conf. Concrete in Marine Environment, Marine Concrete, Concrete Society*. London.
Boresi, A.P., Sidebottom, O.M., Seely, F.B. and Smith, J.O. (1978) *Advanced Mechanics of Materials (3rd. edn.)* John Wiley, New York.
Braestrup M.W. and Nielsen, M.P. (1983) Plastic methods of analysis & design. *Handbook of Structural Concrete.* (ed. F.K. Kong, R.H. Evans, E. Cohen, F. Roll) Pitman, London.
Brakel, J. and Oostlander, L.J. (1979) Concentrated loading on a thick-walled concrete cylinder. *Proc. 2nd. Int. Conf. Behaviour of Offshore Structures.* BHRA Fluid Engineering. Cranfield.
Brown, I.C. and Perry, S.H. *Transverse impact on beams and slabs, ibid.*
Browne, R.D., Geoghegan, M.P. and Baker, A.F. (1985) Analysis of structural condition from durability results *Corrosion of Reinforcement in Concrete Construction* (ed. A.P. Crane) Soc. Chem. Indy., Ellis Horwood, Chichester.
Buijs, J. (1978) Impact. *Proc. 8th FIP Congress, Cement & Concrete Association.* Slough.
Buijs, J. (1982) Impact effect. *Commission on Concrete Sea Structures, Proc. 9th Congress FIP.* FIP, Slough.
BSI (British Standards Institution) (1972) *Code of Practice for the Structural Use of Concrete, CP 110.* BSI, London.
Burdekin, R.M. and Rothwell, G.P. (1981) *Survey of corrosion and stress corrosion in prestressing components used in concrete structures with particular reference to offshore applications, Cement & Concrete Association.* Slough.
Caldwell, D. and Billington, C.J. (1981) Major ship collision damage to the prestressed concrete towers of offshore gravity structures. *Integrity of Offshore Structures, Applied Science Publ.* London.
CEB-FIP (1978) *Model Code for Concrete Structures, Bulletin d'Information, N. 124/125-E, Comité Euro-International du Béton, FIP.* Paris.
Chiapetta, R.L. and Costelloe, J.F. (1981) Automobile impact forces on concrete walls. *Proc. Int. Conf. Structural Mechanics in Reactor Technology.* North-Holland, Brussels.
Clarke, J.L. (1978) *Thermal stress problems in offshore structures, Oceanology International.* Brighton.
Clarke, J.L. (1980) The behaviour of concrete beams under temperature gradients. *Proc. Coll. Int. Tenue des Ouvrages en Béton en Mer, Actes de colloque No. 11 CNEXO,* Brest.
Clarke, J.L. and Symmons, R.M. (1979) Effects of temperature gradients on walls of oil storage structures. *Concrete in the Oceans Tech. Report No. 3, Cement & Concrete Association.* Slough.
Collard, M.J. and Skillman, J.M. (1985) Deep sea production systems. *Proc. 2nd. Symp. New Technologies for the Exploration and Exploitation of Oil & Gas Resources.* Graham & Trotman (for the Commn. of Euro Comm.), London.
Dallard, P.R.B. and Miles, J.C. (1984) Design tools for impact engineers. *Proc. Int. Conf. Structural Impact & Crashworthiness, Vol. 2.* (ed. J. Morton) Elsevier, London.
Davies, I.Ll. and Mavrides, A. (1981) Assessment of the damage arising from collisions between ships and offshore structures. *Integrity of Offshore Structures.* Applied Science Publ. London.
de Rouvray, A., Arnadeau, F., Dubois, J., Chedmail, J.F. and Haug, E. (1984) Numerical techniques and experimental validations for industrial applications. *Proc. Int. Conf. Structural Impact & Crashworthiness, Vol. 1.* (ed. G.A.O. Davies) Elsevier, London.
Eberwijn, J.J. and de Waal, C.D. (1985) Corrosion aspects of steel in concrete offshore structures. *Proc. Int. Conf. Behaviour of Offshore Structures.* Elsevier, Amsterdam.
Elwell, D. and Pointon, A.J. (1978) *Physics for Engineers & Scientists, 2nd. edn.* Ellis Horwood, Chichester.
Engel, P.A. (1976) *Impact Wear of Materials.* Elsevier, Amsterdam.
Fidjestøl, P. and Nilsen, N. (1980) Field test of reinforcement corrosion in concrete. *Performance of Concrete in Marine Environment, SP-65, American Concrete Inst.* Detroit.
Fjeld, S. (1979) Offshore oil production and drilling platforms: design against accidental loads. *Proc. Int. Conf. Behaviour of Offshore Structures.* BHRA Fluid Engineering, Cranfield.
Fjeld, S. (1983) Design assumptions and influence on design of offshore structures. *Proc. Coll. Ship Collision with Bridges & Offshore Structures, Reports Vol. 41.* IABSE, Zürich.
Ford, H. and Alexander, J.M. (1977) *Advanced Mechanics of Materials 2nd. edn.* Ellis Horwood, Chichester.
Fotinos, G.C. and Hsu, Y-Y. (1984) Durability of concrete in the Arctic environment. *Proc.*

FIP/CPCI Symp., Vol 2, Concrete Sea Structures in Arctic Regions, Canadian Prestressed Concrete Assocn. Ottawa.

Fouré, B. and Trinh, J. (1980) Influence des conditions d'essai dans l'étude experimentale de l'effet des gradients thermiques. *Proc. Coll. Int. Tenue des Ouvrages en Béton en Mer, Actes de Colloque No. 11.* CNEXO, Brest.

Frey, R.P. (1982) Fatigue design concept considering the indefinite state of stress in the reinforcement of RC beams. *IABSE Reports Vol. 37 Coll. Fatigue of Steel and Concrete Structures.* IABSE/ETH-Hönggerberg, Zürich.

Gerwick, B.C. (1986) *Construction of Offshore Structures.* John Wiley, New York.

Gerwick, B.C. and Venuti, W.J. (1980) High and low-cycle fatigue behaviour of prestressed concrete in offshore structures. *J. Energy Resources Tech., Trans. Amer. Soc. Mech. Engrs.* **102(19)**, March: 18–23.

Haldar, A. (1981) Impact loading–damage predicting equations. *Trans. 6th. Int. Conf. Structural Mechanics in Reactor Tech.* North-Holland/Commn. Euro. Comm. Brussels.

Hamilton, W.A. and Sanders, P.F. (1984) Sulphate-reducing bacteria and aerobic corrosion, *Corrosion and Marine Growth on Offshore Structures.* (ed. J.R. Lewis & A.D. Mercer) Ellis Horwood (for Soc. of Chem. Indy.), Chichester.

Haynes, H.H. (1975) Structural considerations and configurations III. *Proc. Conf. Concrete Ships & Floating Structures.* (ed. B.C. Gerwick) University of California, Berkeley.

Haynes, H.H. (1976) Collapse behaviour of pressurised concrete shells. *Proc. Int. Conf. Behaviour of Offshore Structures.* Norges Tekniske Høgskjole, Trondheim.

Haynes, H.H. (1979) *Design for implosion of concrete cylinder structures under hydrostatic loading, Report No. TR-874, Civil Eng. Lab.* US Dept. of the Navy.

Henrych, J. (1979) *The Dynamics of Explosion and its Use.* Elsevier, Amsterdam.

Highberg, R.S. and Haynes, H.H. (1977) Ocean implosion test of concrete (Seacon) cylindrical structure. *Proc. 9th. Annual Offshore Tech. Conf.* Houston.

Highberg, R.S. and Haynes, H.H. (1978) Predicting the maximum ocean depths for submerged concrete structures. *Proc. Euro. Offshore Pet. Conf., Soc. Petroleum Engrs. (UK).* London.

Horii, O and Ueda, S. (1977) Study on fatigue behaviour of offshore concrete structures. *Proc. 9th Annual Offshore Tech. Conf.* Houston.

Hughes, B.P. (1981) Design of prestressed fiber reinforced concrete beams for impact. *J. Amer. Concrete Inst.* July–Aug: 276–281.

Iding, R.H. and Bresler, B. (1978) Effects of normal and extreme environment on reinforced concrete structures. *Douglas McHenry Int. Symp. on Concrete & Concrete Structures, SP-55, American Concrete Inst.* Detroit.

Illston, J.M. and Sanders, P.D. (1973) The effect of temperature change upon the creep of mortar under torsional loading. *Mag. Concrete Rsch.*, **25(84)**, Sept: 136–144.

Incecik, A. and Samoulidis, E. (1982) Analytical and experimental studies on ship-ship and ship-platform collisions. *Design for Dynamic Loading, the Use of Model Analysis.* (ed. G.S.T. Armer and F.K. Garas) Construction Press, London.

Jakobsen, B. (1982) *Effect of cyclic loading on concrete structures after cracking. Nordisk Betong,* **2–4:** 139–144.

Jakobsen, B., Olsen, T.O., Røland, B. and Skare, E. (1983) Ship impact on a shaft of a concrete gravity platform. *Proc. Coll. Ship Collision with Bridges & Offshore Structures, Reports Vol. 41.* IABSE, Zurich.

Kavyrchine, M. and Ashtari, N. (1982) Chocs de bateaux sur les structures offshore en béton: étude de poinçonnement. *Proc. 2nd. Coll. Int. Tenue des Ouvrages en Béton en Mer, Actes Coll. No. 15*, CNEXO, Brest.

Kavyrchine, M. and Ashtari, N. (1983) Punching of concrete shells under collision. *Proc. Coll. Ship Collision with Bridges & Offshore Structures, Reports Vol. 41.* IABSE, Zurich.

Kennedy, R.P. (1976) A review of procedures for the analysis and design of concrete structures to resist missile impact effects. *Nuclear Eng. & Design*, **37**: 183–203.

Khoury, G.A., Grainger, B.N. and Sullivan, P.J.E. (1985) Transient thermal strain of concrete: literature review, conditions within specimen and behaviour of individual constituents. *Mag. Concrete Research*, **37(132)**, Sept: 131–144.

Kufuor, K.G. and Perry, S.H. (1984) Hard impact of shallow reinforced concrete domes. *Proc. Int. Conf. Structural Impact & Crashworthiness Vol. 2* (ed. J. Morton) Elsevier, London.

Lea, F.M. (1970) *The Chemistry of Cement, 3rd edn.* Arnold, London

Leick, R.D. and Bode, J.H. (1978) Implosion strength of concrete shells: a comparison of theoretical and experimental results. *Proc. 10th Annual Offshore Tech. Conf.* Houston.

Lin, C.Y. (1980) Bond deterioration due to corrosion of the reinforcing steel. *Performance of Concrete in Marine Environment, SP-65, American Concrete Inst.* Detroit.

Lovegrove, J.M. and El Din, S. (1982) Deflection & cracking of reinforced concrete under repeated loading and fatigue. *Fatigue of Concrete Structures, SP-75, American Concrete Inst.* Detroit.

Mackay, K.M. and Mackay, R.A. (1986) *Introduction to Modern Inorganic Chemistry, 3rd. edn.* International Textbook Co., London.

Marin, J. (1962) *Mechanical Behaviour of Engineering Materials.* Prentice-Hall, London.

Marshall, A.L. and Drake, S.R. (1987) Thermal and dynamic response of concrete at low temperatures. *Proc. 9th. Int. Conf. Port & Ocean Engineering under Arctic Conditions.* Univ. of Alaska, Fairbanks.

Montague, P. and Goode, C.D. (1979) Some aspects of double-skin composite construction for sub-sea pressure chambers. *Proc. Int. Conf. Behaviour of Offshore Structures.* BHRA Fluid Eng., Cranfield.

Morf, U. (1982) Fatigue strength of weldable high strength reinforcing steel *IABSE Reports, Vol. 37, Coll. Fatigue of Steel & Concrete Structures.* IABSE/ETH-Hönggerberg, Zürich.

Naaman, A.E. (1982) Fatigue in partially prestressed concrete beams. *Fatigue of Concrete Structures, SP-75, American Concrete Inst.* Detroit.

Nagano, H. & Naito, T. (1986) Diagnosing methods for chloride-contaminated concrete structures. *Marine Concrete, Proc. Int. Conf. Concrete in Marine Environement, Concrete Society.* London.

Nilsen, N. & Espeliol, B. (1985) Corrosion behaviour of reinforced concrete under dynamic loading. *Materials Performance, Nat. Assocn. Corrosion Engineers,* July: 44–50.

Nurnberger, U. (1982) Schwingfestigkeitsverhalten von Betonstahlen. *IABSE Reports Vol. 37, Coll. Fatigue of Steel & Concrete Structures.* IABSE/ETH-Hönggerberg, Zürich.

Nurse, R.W., (1968) Cohesion and adhesion in solids. *Proc. Int. Conf. Structure of Concrete, Cement & Concrete Assocn.* Slough.

Okada, K. and Kojima, T. (1982) Fatigue properties of concrete members subjected to torsion. *IABSE Reports, Vol. 37, Coll. Fatigue of Steel & Concrete Structures.* IABSE/ETH-Hönggerberg, Zürich.

Okada, K. and Miyagawa, T. (1980) Corrosion of reinforcing steel in cracked concrete. *Performance of Concrete in Marine Environment, SP-65, American Concrete Inst.* Detroit.

Okamura, H. and Ueda, T. (1982) Fatigue behaviour of reinforced concrete beams under shear force. *IABSE Reports Vol. 37, Coll. Fatigue of Steel & Concrete Structures.* IABSE/ETH-Hönggerberg, Zürich.

Page, C.L. (1985) Corrosion mechanisms. *Proc. First Int. Conf. Deterioration and Repair of Reinforced Concrete in the Arabian Gulf.* Bahrain Soc. Engrs.

Parrott, L.J. (1979) A study of transitional thermal creep in hardened cement paste. *Mag. Concrete Rsch.* **31(107)**, June: 99–103.

Parrott, L.J. & Symmons, R.M. (1979) Deformation characteristics of an oil storage vessel–concrete subjected to fluctuating stresses and temperatures. *Appendix to Concrete in the Oceans Tech. report No. 3, Cement & Concrete Assocn.* Slough.

Partington, J.R. (1954) *General and Inorganic Chemistry.* Macmillan, London.

Paterson, W.S. (1980) Fatigue of reinforced concrete in sea water. *Performance of Concrete in Marine Environment, SP-65, American Concrete Inst.* Detroit.

Paterson, W.S. and Dill, M.J. (1982) Stiffening of reinforced concrete beams under cyclic flexural loading in sea water. *Proc. 2nd. Coll. Int. Tenue des Ouvrages en Béton en Mer, Actes de Colloque No. 15.* CNEXO, Brest.

Paterson, W.S., Dill, M.J. and Newby, R. (1982) Fatigue strength of reinforced concrete in seawater. *Concrete in the Oceans Tech. Report No. 7, Cement & Concrete Assocn.* Slough.

Perry, S.H. (1986) Identification and repair of marine concrete damage by hard impact. *Marine Concrete, Proc. Int. Conf. Concrete in the Marine Environment, Concrete Soc.* London.

Perry, S.H. and Brown, I.C. (1982) Resistance of prestressed concrete slabs to extreme loads. *Proc. Int. Conf. Behaviour of Offshore Structures, Vol. 2.* (ed. C. Chryssostomidis and J.J. Connor) Hemisphere, Washington.

Perry, S.H. and Dinic, G. (1985) Fender layer protection of sub-surface concrete caissons. *Proc. Int. Conf. Behaviour of Offshore Structures.* Elsevier, Amsterdam.

Peyronnet, J-P, Trinh, J. and Seguin, M. (1977) Experimental study on the behaviour of concrete structural elements in natural sea water. *Proc. 9th Annual Offshore Tech. Conf.* Houston.

Rabia, H. (1985) *Oilwell Drilling Engineering.* Graham & Trotman, London.

Rao, S.R. (1972) *Surface Phenomena.* Hutchinson Educational, London.

Richmond, B. (1977) The time-temperature dependence of stresses in offshore concrete structures. *Proc. Conf. Design & Construction of Offshore Structures, Inst. Civil Engineers.* London.

Richmond, B., England, G.L. and Bell, T. (1980) *Designing for temperature effects in concrete offshore oil-containing structures, Report UR-17, CIRIA–UEG.* London.

Robins, P.J. and Standish, I.G. (1982) Effect of lateral pressure on bond of reinforcing bars in concrete. *Proc. Int. Conf. Bond in Concrete.* (ed. P. Bartos) Applied Science Publ. London.

Røland, B. Olsen, T.O. and Skåre, E. (1982) Ship impact on concrete shafts. *Proc. 2nd. Coll. Int. Tenue des Ouvrages en Béton en Mer, Actes Coll. No. 15.* CNEXO, Brest.

Roper, H. (1982) Reinforcement for concrete structures subject to fatigue. *IABSE Reports Vol. 37, Coll. Fatigue of Steel and Concrete Structures.* IABSE/ETH–Hönggerberg, Zürich.

Ross, A.D., England, G.L. and Suan, R.H. (1965) Prestressed concrete beams under a sustained temperature crossfall. *Mag. Concrete Rsch.*, **17(52)**, Sept: 117–126.

Rygol, J. (1983) Structural design: national code specifications for concrete and reinforcement. *Handbook of Structural Concrete.* (ed. F.H. Kong, R.H. Evans, E. Cohen, F. Roll) Pitman, London.

Sakamoto, N. and Iwasaki, N. (1982) Influence of sodium chloride on the concrete/steel and galvanised steel bond. *Proc. Int. Conf. Bond in Concrete* (ed. P. Bartos) Applied Science Publ. London.

Sneddon, R.A. (1985) A comparative review of the performance of different concretes and repaired concrete in the same structure exposed to the Bahrain environment. *Proc. First Int. Conf., Deterioration and Repair of Reinforced Concrete in the Arabian Gulf. Bahrain Soc. of Engineers.*

Søreide, T.H., Moan, T., Amdahl, J. and Taby, J. (1982) Analysis of ship/platform impacts. *Proc. Int. Conf. Behaviour of Offshore Structures.* (ed. C. Chryssostomidis and J.J. Connor) Hemisphere, Washington.

Sørensen, K.A. (1976) Behaviour of reinforced and prestressed concrete tubes under static and impact loading. *Proc. Int. Conf. Behaviour of Offshore Structures.* Norges Tekniske Høgskjole, Trondheim.

Stillwell, J.A. (1983) Exposure tests on concrete for offshore structures. *Concrete in the Oceans Tech. Report No. 8, Cement and Concrete Assocn* Slough.

Stroeven, P. and de Wind, G. (1982) Structural & mechanical aspects of debonding of a steel bar from a cementitious matrix. *Proc. Int. Conf. Bond in Concrete.* (ed. P. Bartos) Applied Science Publ. London.

Taylor, H.P.J. (1983) Structural performance as influenced by detailing. *Handbook of Structural Concrete.* (ed. F.K. Kong, R.H. Evans, E. Cohen, F. Roll) Pitman, London.

Tepfers, R. (1979) Cracking of concrete cover along anchored deformed reinforcing bars. *Mag. Concrete Rsch.* **31(106)**, March: 3–12.

Tilly, G.P. and Moss, D.S. (1982) Long endurance fatigue of steel reinforcement. *IABSE Reports Vol. 37, Coll. Fatigue of Steel & Concrete Structures* IABSE/ETH Hönggerberg, Zürich.

Timoshenko, S. and Goodier, J.N. (1951) *Theory of Elasticity* (*2nd. edn.*). McGraw-Hill, New York.

Trinh, J. (1980) Effets de sollicitations répétitives sur les structures marine en béton–étude experimentale. *Proc. Coll. Int. Tenue des Ouvrages en Béton en Mer, Actes de Coll. No. 11.* CNEXO, Brest.

Trinh, J. and Peyronnet, J.P. (1978) *Etude experimentale du comportement d'éléments en béton en milieu marin, Annales ITBTP, 360.* April.

Tuutti, K. (1980) Service life of structures with regard to corrosion of embedded steel. *Performance of Concrete in Marine Environment, SP-65, American Concrete Inst.* Detroit.

Venuat, M. and Alexandre, J. (1968/9) De la Carbonatation du Béton. *Revue des Matériaux de Construction,* **638–640**, Nov., Dec., Jan.

Waagaard, K. (1977) Fatigue of offshore concrete structures–design and experimental investigations. *Proc. 9th. Ann. Offshore Tech. Conf.* Houston.

Waagaard, K. (1986) Experimental investigation on the fatigue strength of offshore structures. *Offshore Operations Symp., 9th Ann. Energy-Sources Tech. Conf.* ASME, New Orleans.

Waagaard, K. and Kepp, B. (1987) Fatigue of high strength lightweight aggregate concrete. *Proc. Symp. Utilisation of High Strength Concrete.* Tapir, Trondheim.

Warner, R.F. (1982) Fatigue of partially prestressed concrete beams. *IABSE Reports Vol. 37, Coll. Fatigue of Steel & Concrete Structures.* IABSE/ETH–Hönggerberg, Zürich.

Watson, A.J. and Ang, T.H. (1984) Impact response and post-impact residual strength of reinforced concrete structures. *Proc. Int. Conf. Structural Impact & Crashworthiness, Vol. 2.* (ed. J. Morton) Elsevier, London.

Watson, A.J. and Sanderson, A.J. (1984) Cratering and sub-crater fractures produced by explosive shock on concrete. *Proc. 3rd. Conf. Mechanical Properties at High Rates of Strain, Inst. Phys. Conf. Ser. No. 70.* London.

Wilby, C.B. (1983) *Structural Concrete.* Butterworths, Sevenoaks.

Wilkins, N.J.M. and Lawrence, P.F. (1980) Fundamental mechanisms of corrosion of steel reinforcements in concrete immersed in sea-water. *Concrete in the Oceans Tech. Report No. 6, Cement & Concrete Assocn.* Slough.

Wilkins, N.J.M. and Lawrence, P.F. (1985) The corrosion of steel reinforcements in concrete immersed in sea-water. *Corrosion of Reinforcement in Concrete Construction.* (ed. A.P. Crane) Soc. of Chemical Indy., Ellis Horwood, Chichester.

Wilkins, N.J.M. and Stillwell, J.A. (1986) The corrosion of steel reinforcement in cracked concrete immersed in sea-water. *Marine Concrete, Proc. Conf. Concrete in the Marine Environment, Concrete Soc.* London.

Wilkinson, T.G. (1984) Biological mechanisms leading to potential corrosion problems. *Corrosion & Marine Growth of Offshore Structures.* (ed. J.R. Lewis, and A.D. Mercer) Ellis Horwood (for Soc. Chem. Indy.), Chichester.

Withey, M.O. and Washa, G.W. (1954) *Materials of Construction.* John Wiley, New York.

Woodfin, R.L. and Sliter, G.L. (1981) Results of full-scale turbine missile concrete impact experiments. *Trans. 6th. Int. Conf. Structural Mechanics in Reactor Technology.* North–Holland/Commn. Euro. Comm., Brussels.

Zimmermann, Th., Rodriguez, C. and Rebora, B. *Non-linear analysis of a reactor building for airplane impact loadings, ibid.*

Zorn, N.F., Schuëller, G.I. and Riera, J.D. *A probabilistic approach for evaluation of load time history of an aircraft impact, ibid.*

Zorn, N.F. and Reinhardt, H.W. (1984) Concrete structures under high intensity tensile waves. *Proc. Int. Conf. Structural Impact & Crashworthiness, Vol. 2.* (ed. J. Morton) Elsevier, London.

6 Structures

6.1 Introduction

While the behaviour of materials under various types of loading may be intrinsically interesting, its prime significance to the practitioner lies in the use to which they may be put. Behaviour and application are inter-dependent, however, so that just as effective application has to be founded on sound knowledge of behaviour, so study of behaviour must have regard to likely application. Now, therefore, we should turn to the structural use of concrete in the sea, although consideration is necessarily limited due to restrictions on space and because this is not a design manual. Nevertheless, some of the general and particular principles should be discussed as well as the constraints within which the designer has to operate.

Selection is unavoidable and it is a process which may be dictated as much by personal preference as by anything else. However, it is hoped that the examples chosen will be reasonably representative and inherently informative. The term 'structures' covers an enormous range, even when confined to those in the sea: wharves, jetties, sea walls; breakwaters; harbours; pipelines; hydrocarbon storage and production structures; barges and pontoons; structures for energy production from waves, wind, oceanic thermal gradients; power station cooling water intakes and outfalls; tidal barrages; bases for chemical plants; marinas; bridge piers; causeways; floating airports; lighthouses. Hopefully the general approach adopted will demonstrate some of the broad unifying principles which make comparisons a little easier, even though specific details cannot always be given.

Design cannot be isolated from construction and installation, particularly for many marine structures. For instance the connection is obvious when prefabrication is being adopted at some stage—with the argument perhaps being carried to its limit in the case of oil-storage platforms in the North Sea. They could be said to have achieved the ultimate in pre-fabrication with entire structures of an all-up weight of around one million tons built in one location and installed in another. While the sea makes many things difficult, it also makes some possible.

At the outset then it has to be emphasised that the integration between design, construction and installation underlies the whole discussion: it is the fundamental precept. *Constructibility* is implicit although direct references to it will be necessarily brief. The term is a very useful one and embraces concept development, integration of design with construction, selection of construction methods, facilities and stages, procurement and assembly of materials and fabricated components, organisation and supervision of the work, and training of workers. It includes analysis and planning, quality control and assurance, safety engineering, cost estimating, budget control plus weight control in the case of offshore structures (Gerwick, 1986).

Bearing this in mind, as well as the impossibility of doing justice to the totality, the ingenuity, and the variety of structures built and proposed for use in the sea, let us first consider some fairly general aspects prior to seeking an indication of the guidance available in the form of recommendations, codes, rules, and standards. Some illustrative but important examples may then be helpful.

6.2 The design process

'Design' is not entirely easy to define. The most popular conception is perhaps better described (at least in this context) as 'styling' where appearance seems to outweigh utility. A conspicuous example of this (possibly apocryphal) is the plastic nutcrackers which won a design award but fractured in use. At the other extreme is detail design which at times seems to comprise solely (and almost to the complete exclusion of judgement or perception) intimate familiarity and compliance with every clause of a particular code without any real understanding of their justification. In the first of these, and in other cases, the primary requirement of 'fitness for purpose' may seem to be forgotten or ignored unless of course the purpose is simply to adorn or decorate as in some clothes or most jewellery. The latter also ignore another fundamental requirement, that of economy, and probably also the further one of durability (minimising maintenance).

Using such illustrations has the merit of suggesting the distinguishing features of engineering design while avoiding some of the pitfalls of actually defining it. They should not be used, however, to imply that aesthetics are to be ignored—whether or not appearance is important should be established at the concept- or problem-definition/formulation stage. This is what may, in fact, be at the root of the design process, because no solution is possible until the problem is defined, whether it be the entire concept or the detail of joining two components.

6.2.1 *Design sequence*

Design is an iterative process which has been modelled in a variety of ways, one of the best of which is probably the design spiral. This typifies the gradual

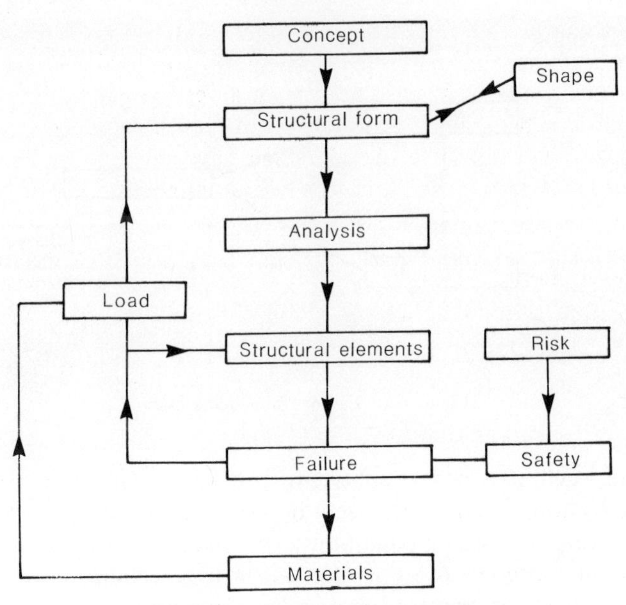

Figure 6.1 Central design sequence

homing-in or centering on the end result. Here, however, it is more convenient to represent it as a sequence of 'steps' or 'blocks' with loops returning to earlier steps, and which has linking subsidiary loops. Thus in Figure 6.1 there is a central or spinal sequence beginning with the *concept*. This may range from the very simple, such as the provision of access to a beach, to the massively complex, such as the production of oil and gas from beneath the sea bed. The concept is instrumental in establishing the *structural form* that is, its configuration, a process in which past experience is normally a guiding feature but occasionally in which there is scope for major innovation. Thus the beach access may be by a flight of stairs or a ramp because that is a frequent or common solution; the extraction of oil, however, may require invention on the scale of the storage/production structures described later—not unlike inventing the first of the great pyramids of Egypt or the Eiffel Tower. Once the *form* and scale of the structure have been determined it can be *analysed* which then enables its *elements* to be designed, using the term 'element' in the wide sense of the previous chapter i.e. a cylindrical vessel is as much an element as is a beam or a slab. Designing the elements may then compel adjustments to, or alteration of, the form of the structure and the process has to be repeated. Here too factors such as safety, risk and mode of failure, and reliability may become of overwhelming importance.

This is all pretty obvious and rudimentary but it becomes more complicated as ancillary loops are added. Consider structural form for example. This is governed by the intended use: for storage of a liquid, some shape of container

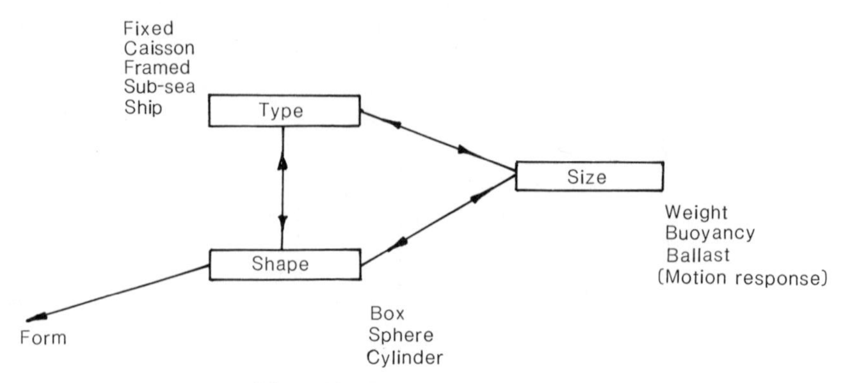

Figure 6.2 Factors influencing form

is required, such as a box, a sphere, or a cylinder. However, the shape is influenced by both function and size—beyond a certain shell size, for instance (and depending on loading conditions), the thickness of plate required in a steel vessel may become too great for satisfactory welding so the shape has to be modified, stiffeners introduced and so on. Furthermore the shape may be affected by the function of the structure: whether it is for transporting the liquid, as in the case of a ship, or processing the liquid or merely for storage or perhaps a combination of any two or even all three. Figure 6.2 shows such a loop.

The analysis which leads to the design of the Elements is not an independent process since it is governed by the nature of the element (linear as in a beam or column, two-dimensional as in a wall or slab or three-dimensional as in a cylinder) and by the materials from which it is made and their response to load. The load may be static, dynamic or from the natural and/or working environment and the response manifested in the form of deflection, deformation, vibration or fatigue for instance. Figure 6.3 illustrates this.

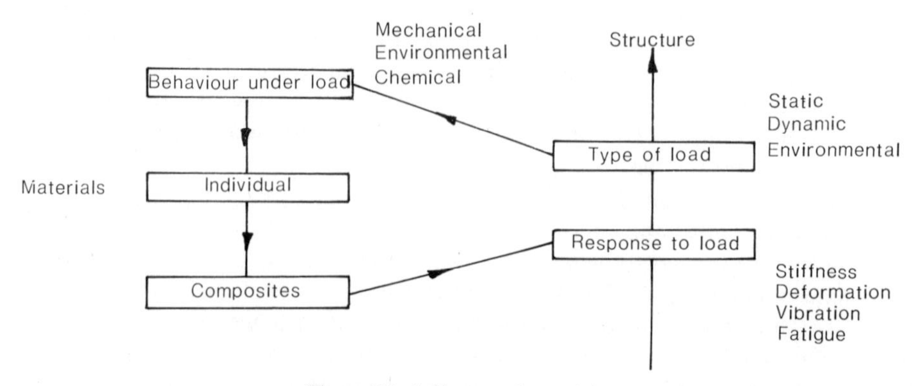

Figure 6.3 Influence of materials

It soon becomes clear that alterations to any block, step, or loop can have repercussions, not just in adjacent parts of the sequence, but in more apparently distant areas. Adding to them the impact on costs, the desirability of minimising the number of joints, the difficulty of effecting repairs and so on, compounds the complexity.

The distinction between construction on land and in the sea can be epitomised by a simple example. If a bolt is omitted from a structure 10 m above the ground then it can be fixed by a man with little more than a spanner and a ladder. On a marine structure 10 m above the seabed might be over 100 m below the water surface and fixing may require divers or the use of a remotely operated vehicle (capital cost $500,000 to $ 1 million) and its mother ship. Fitting one bolt could cost tens of thousands of dollars.

6.2.2 *General design requirements*

There is an important consideration behind the preceding discussion. Although various codes, rules, recommendations, and standards exist, design of structures for the sea is often probably less prescriptive or constraining than most other areas of structural design. While there is an even greater need for economy since the fall—collapse—in the price of oil in early 1986 and its impact on energy prices, there is still a considerable requirement to think in conceptual terms, despite the climate of caution which limits the number of possible coincident innovations. (Proven by experience is generally a guiding principle.) A few years ago 'streams of development' were listed and their realisation was viewed as being essential if concrete structures were to fulfill their potential (Gerwick, 1976). They remain as valid today:

(a) conceptual development for new environments and foundation conditions and new owner requirements;
(b) advanced design procedures, analytical methods and recommended practices (standards);
(c) improved construction methods, both for the structures and their foundations;
(d) material developments, to ensure material properties suitable for the intended use;
(e) improved management organisational and contractual capabilities, adequate for the sophistication, complexity, and size of these future projects.

The main requirements for any industrial structure have been defined as (Lee, 1982):

1. the capability to fulfill its intended purpose;
2. structural adequacy for both operational and environmental loading;
3. practicality of construction;
4. cost-effectiveness.

In the case of offshore platforms, aesthetic and architectural considerations are generally minimised (this is probably much less true of many inshore and coastal structures) and the steps in actual design of an offshore platform are:

(a) conceptual development of the structure based on the proposed method of installation;
(b) selection of layout to satisfy the operational requirements, of size, load carrying capacity, and efficiency of arrangement;
(c) preliminary structural design for the operational and environmental loadings; procedures for the first can follow those for other structures but design for environmental loading is considered more usual because of the complexity of analysis and size of loads involved;
(d) start of 'installation engineering' including a review of the construction procedure which must account for lifting, launching, floating, hydrostatic pressure, etc.
(e) repetition of the design cycle due to probable consequential alterations to the preliminary concept or member sizes.

Primary design objectives have been described as maximising economic attractiveness while minimising risks to personnel, environment, and system (Knecht et al, 1979). The same authors adopted two main guiding principles:

(a) the extension of accepted concepts to minimise the use of radically new technology that would require long development times or high risks;
(b) the integration of accepted concepts into a total working system to avoid duplicating equipment or manpower and to ensure efficient operations.

6.2.3 *Innovation and cost*

If there is an over-riding factor which encapsulates most of the foregoing catalogue, it is, perhaps, that there is a need to innovate but using accepted or acceptable technology. In other words, the innovation is often as likely to be in the way the technology is applied as in the technology itself. Innovative technology is by no means eschewed, of course, but its introduction should be graduated to keep risk manageable: too many concurrent changes incur technological and economic hazard.

Certainly major technological development in the North Sea has resulted in substantial cost over-runs (Moe, 1981). The increases were attributed to underestimation, inflation, government requirements, new operator requirements, and inefficiencies in project execution. For six completed projects up to mid-1979 the cost increases ranged from an increase of 79% on an estimate of $638.2 million to 466% on $352.4 million. Later projects under development indicated a reduction in the size of increase to an average of 29%. It is not

Table 6.1 Statistical analysis of project costs (overspend)

Items estimated	Mean of actual to estimated cost	Number of projects	Standard Deviation
Weapons 1950s	1.89	55	1.36
Weapons 1960s	1.40	25	0.39
Public works			
—Highway	1.26	49	0.63
—Waterprojects	1.39	49	0.70
—Building	1.63	59	0.83
—Ad hoc	2.14	15	1.36
Major construction	2.18	12	1.59
Energy process plants	2.53	10	0.51

possible to generalise on the causes of the increases since their distribution varied among projects but 'it seems to be a general experience that start estimates are biased to the low side for projects involving new technologies'. Values for 'one-of-a-kind' or 'first-of-a-kind' projects are quoted from a study made for the US Department of Energy (Table 6.1).

The large variation between types of project serves to emphasis the uncertainty underlying the whole process of estimation.

Not all projects are as large as major offshore developments, of course, but although the consequences of inaccuracy become less in *absolute* terms with dimishing size, the *relative* effects may be just as severe. There may be some alleviation due to the fact that, the smaller the project, the more likelihood there is of there being more of them. That is, the data base from which projections can be made should be broader, reducing the risk of errors.

The narrowness of the base of past experience for major innovative structures has been reflected in the difficulties for their insurers (Bjelmrot and Maltegård, 1981). Underwriting or coverage is more problematic because of their fewness of numbers, particularly for large concrete structures and especially when it is realised that the insured value for a large platform in the North Sea can exceed the world premium income.

6.3 Safety and risk

The conjunction of financial uncertainties and insurance leads naturally to contemplation of risk, from the analysis of which reliability cannot be separated. This, of course, is closely related to serviceability and safety factors, detailed consideration of which is beyond the scope of this text. However, discussion of safety and risk is an important element of any review of marine concrete and it would be remiss not to include it here, albeit somewhat curtailed.

Unfortunately, some of the terminology is commonplace, which is not unusual, but it can lead to particular preconceptions according to the situation in which it is used. Thus in argument between two people there is risk of causing offence, the gambler risks his money, the mountaineer and racing driver may risk their lives. Subjectivity is difficult to avoid and in any of these examples, acceptability is an inevitable accompaniment. In relation to engineering work (as to many others), however, what is acceptable is not an individual choice but depends on a group of people or even a society as a whole. Since social patterns and mores are constantly changing, it follows that acceptability of risk does not remain constant.

6.3.1 *Categories of risk*

Suggested definitions, suitable for pragmatic use are (Halliwell et al, 1979):

Individual risk—a measurement of the probability of harm to an individual from a hazard.
Group/social risk—a compound measurement of the probability of harm, its severity and its consequences to the group or society.
Hazard—a significant element of danger in a given work situation, community or environment.
Harm is measured in fatalities because it is the most accurate, maintained statistic.

These all relate to the category of *physical* risk. In addition there is *economic* risk. Here, however, the assessment of consequences is less fraught with difficulties, since the criterion is one of financial acceptability which is generally more easily arrived at than is social acceptability. That is not to say that the latter does not include a financial element—one need only refer to the compensation awarded to the victims of accidents involving fatalities to appreciate that.

6.3.2 *Perceptions of safety*

The assessment of safety depends then on three factors (Green and Brown, 1978):

(a) perception of risk,
(b) strength of preference,
(c) monetary valuation of safety.

That perceptions vary has been demonstrated by survey, including some among students, and is illustrated in Tables 6.2 and 6.3, the first concerning the perceived safety of some common activities and situations and the second on the perceived severity of injury.

Table 6.2 Perceptions of safety

Activity or situation	Perceived safety (safest = 100)
Staying at home (fire)	163
living within five miles of major airport	203
swimming in a swimming pool	212
staying in a hotel (fire)	262
travelling by train	272
living within five miles of a nuclear power station	335
crossing the road	361
travelling by coach	368
living within five miles of a major chemical plant	376
driving a car	406
ski-ing	512
travelling by plane	525
riding a motor cycle	808
rock climbing	1056

Table 6.3 Perceived injury severity

Type of injury	Perceived severity (least = 10)
Being unhurt by accident	10
Bruises	16
Sprained ankle	21
Concussion	40
Simple fracture of an arm	44
Broken ribs	79
Compound fracture of an arm	80
Internal injuries	197
Fractured skull	225
Carbon monoxide poisoning	286
Multiple facial lacerations	310
Loss of one eye	349
Loss of right arm	559
Severe burning over 1/3 of the body	683
Loss of one leg	735
Radiation sickness	780
Paralysis from the waist down	1643
Loss of sight of both eyes	1917
Death	5676
Paralysis from the neck down	6213
Brain damage	7243

These tables are not necessarily representative of the population as a whole but they do illustrate first, the variation in perceptions of safety and in acceptability, and second that death is not always the worst outcome. However, as has been indicated already, it is the most easily quantifiable and can be used to fix the possible acceptance limit of a design.

6.3.3 *Risk index*

Figures 6.4(a) and (b) (Halliwell et al, 1979) show the relationship between frequency of occurrence and number of fatalities for various man-made and natural disasters. The 'target' line for nuclear power stations is its acceptance limit i.e. the one below which conceivable failure should fall. 'Risk' is the event frequency so that the risk of 100 fatalities from dam failure is rather less than 0.1 while for meteors it is about 0.00001 or 1 in 10^5, referred to as a fifth-order risk. If hazard is defined as the number of fatalities in one event then a **risk index** can be produced from the product of hazard and risk, as shown in Table 6.4 which is from the same source.

It is well recognised that the hazard to a single group registers more in the public consciousness than the hazard to an equivalent number of individuals. For example, the death of 100 people in an air crash has more impact than that of 100 people in separate automobile accidents. From the viewpoint of the structural designer it is the group hazard which is probably more often of concern but before proceeding further it may be worth quoting

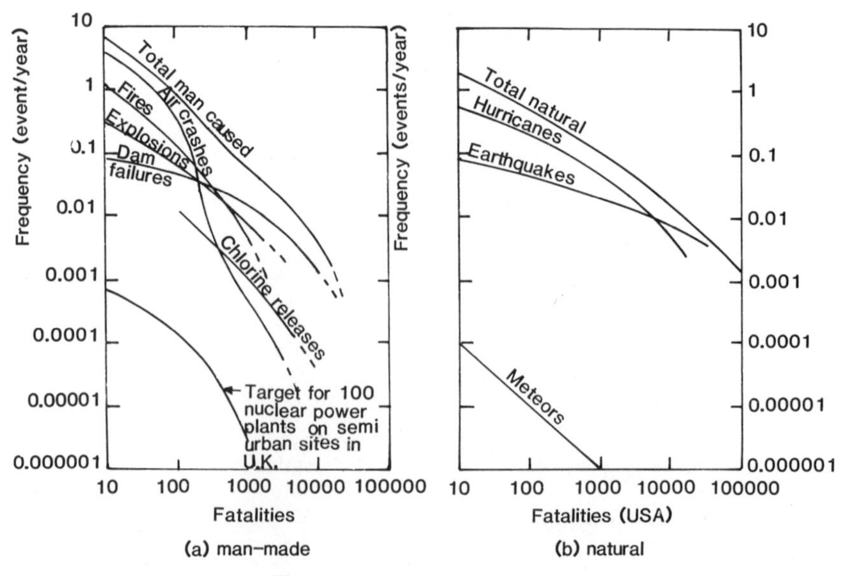

Figure 6.4 Incidence of disasters

Table 6.4 Derivation of risk index from defined hazard and risk

Event	Hazard	Risk	Risk index	Relationship between hazard and index
Air crashes	100	1	100	Index decreases as
	1000	0.001	1	hazard increases
Nuclear	10	0.001	0.01	Index decreases as
power plants	1000	0.000001	0.001	hazard increases
Chlorine	100	0.01	1	Index remains constant
releases	1000	0.001	1	as hazard increases
Dam	10	0.1	1	Index increases as
failures	100	0.03	3	hazard increases
Earthquakes	10	0.1	1	Index increases as
	10000	0.01	100	hazard increases

Table 6.5 Comparative annual risks of death

Activity	Exposure	Annual risk/person
All causes		0.0092
General aviation	200 hrs/yr	0.0054
Pleasure flying	100 hrs/yr	0.0045
Mountaineering (international)	100 hrs/yr	0.0027
Air travel (crew)	1000 hrs/yr	0.0012
Offshore operations (UK) (excluding construction, diving and vessels)	—	0.0011
Boating	143 hrs/yr	0.0007
Mining, quarrying	—	0.0006
Construction	—	0.0006
Agriculture	—	0.0006
Motorcycling	5000 miles/yr	0.0006
Motoring	12000 miles/yr	0.0004
All industries	—	0.00015
Air travel (passenger)	100 hrs/yr	0.00012
Homicide	—	0.00010

some individual statistics (Table 6.5) which are germane to construction and operational aspects (Stahl, 1986).

It is interesting to note that, in the earlier table, travelling by plane was *perceived* to be less safe than by car whereas statistically it is much safer. Despite being well known, this does point towards the potential conflict between the need to design cost-effectively for realistic probabilities whilst having to accommodate public or group perceptions.

In some instances there should be zero risk—for example, at a busy road crossing there should be no risk at all that traffic lights show green in all directions. However, in most cases the question is one of risk *reduction* rather than elimination, and cost-effectiveness becomes a question of balancing the

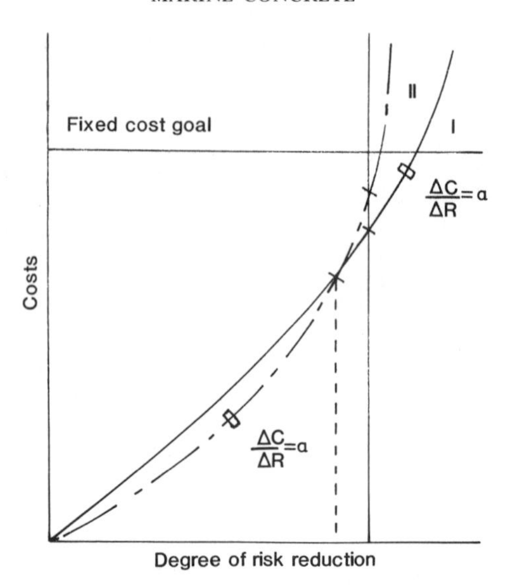

Figure 6.5 Cost-effectiveness of alternative strategies

cost of reduction against the benefit gained, as indicated by Figure 6.5 (Rowe, 1983). Various risk criteria are indicated on the figure. Other approaches are discussed in the reference but this one in particular demonstrates the law of diminishing returns.

6.3.4 *Structural failure*

The point has been made that the 'incredible' incidents in technology are often made up of combinations of quite ordinary, credible events, so that 'incredible' and 'credible' accidents cannot be distinguished logically (Abrahamsen, 1976). Disasters often stem from the combined effect of a sequence of conceivable events. Thus a systematic means is required for studying complex phenomena, as formulated by Figure 6.6. (Dahl, 1981). If the size of consequence of a possible event is W_i and its probability is p_i, then the risk of occurrence for that part of the system is $p_i W_i$ and the total risk to the system will be some function $f_{i=1 \text{ to } n}\,(p_i W_i)$, generated by the procedure outlined in Figure 6.6. Generally structural systems can be categorised as being in *series* or *parallel* (Moses, 1976). A series system fails if any single element fails (the weak-link-in-the-chain principle) whereas the parallel system enables the load to be re-distributed. 'The strength or resistance probability distribution for a parallel system can be calculated by summing the element strengths while for a series system it must be calculated from a product of the probability distributions of the elements.'

The manner of failure is also important: load level is maintained by a ductile element while it may be progressively lost by a brittle or semi-brittle

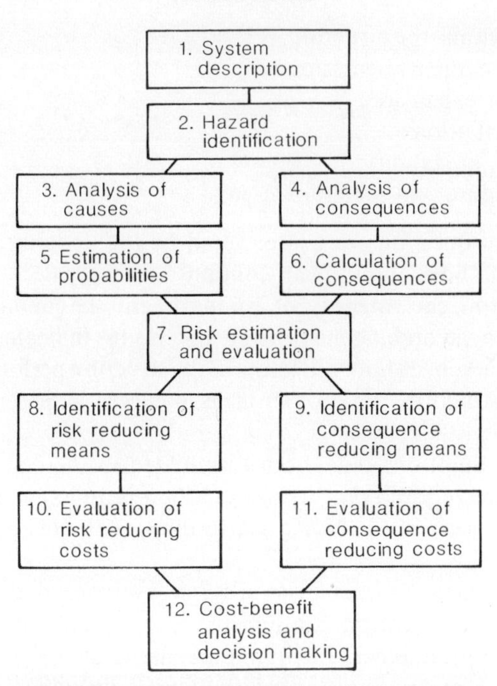

Figure 6.6 Sequence of steps in a risk evaluation

element. Structural continuity is a further factor which has been shown to be beneficial in the case of fire (Forrest, 1986). It is tempting, in consequence, to claim some advantage for concrete marine structures over steel, since they tend to be monolithic in contrast to the multiplicity of members in steel jackets and the like. On the other hand the failure of any one 'element' will be more critical in the concrete structure since they would be fewer in number (as well as greater in size). But again the failure-causing event or accident would, presumably, have to be greater and thereby rarer. Reference to response to fire introduces additional considerations which prompt qualifications to notions of 'safety'.

6.3.4.1 *Fire* *Absolute* safety implies that a structure should remain safe under *any* foreseeable (or even un-foreseeable) circumstance, but safety against *what* and for *whom*? Building occupants will have different expectations from fire fighters: some occupations involve greater risks than others and so an oil production platform would be expected to have totally different criteria were it to be open to free access by the general public. Time is an additional factor, so a structure may be designed to behave in such a way that collapse occurs sufficiently slowly to permit it to be evacuated without human damage. In the case of fire, the following factors need to be taken into account (Forrest, 1986):

continuity within the structure;
retention of compartmentation;
provision for expansion;
load re-distribution;
maintenance of stability;
type of structure and materials used.

Fundamental questions to be answered by the designer include: can the building expand horizontally? can columns and walls flex adequately in the desired direction? can integrity of compartments be maintained? can load be redistributed via arching/suspension systems (as in beams and slabs)? can stability be maintained? how will joints in the structure perform? what services penetrate the structure? what other aspects affect the integrity of the whole structure?

Whilst such questions have been formulated in the context of buildings on land they are very relevant to hydrocarbon production structures, as the Piper Alpha tragedy in the North sea testifies so eloquently.

6.3.4.2 *Safety index* Target levels of safety (average values) for the ultimate collapse of major structural components in a lifetime have been listed (in ascending safety) (Faulkner, 1984) and are shown in Table 6.6.

The numbers are values of the safety index β and the relation between it and the notional probability of failure p_f is given in Figure 6.7 from the same source. The safety (or reliability) index β is defined by Figure 6.8 and the following:

'Demands' (D) on the structure in service are exemplified by loads, and 'capability' (C) by its strength. 'Failure' occurs when $C < D$

$$p_f = p(C < D) = p(Z < 0) = \phi(-\beta) \tag{6.1}$$

Table 6.6 Safety or reliability index for various structures

Structure	Safety Index β
Torpedo boat destroyers C1901	2.0
Royal Navy frigate deck collapse	2.2
Fixed jacket platforms	2.3
North American buildings	3.0
Tubular columns and beam-columns	3.3
UK offshore thinking	3.7
German steel bridges	3.7
Existing semi-submersible	4.4
UK steel bridges	4.8
North Sea tension leg platform	5.3
Merchant ship deck collapse	5.3

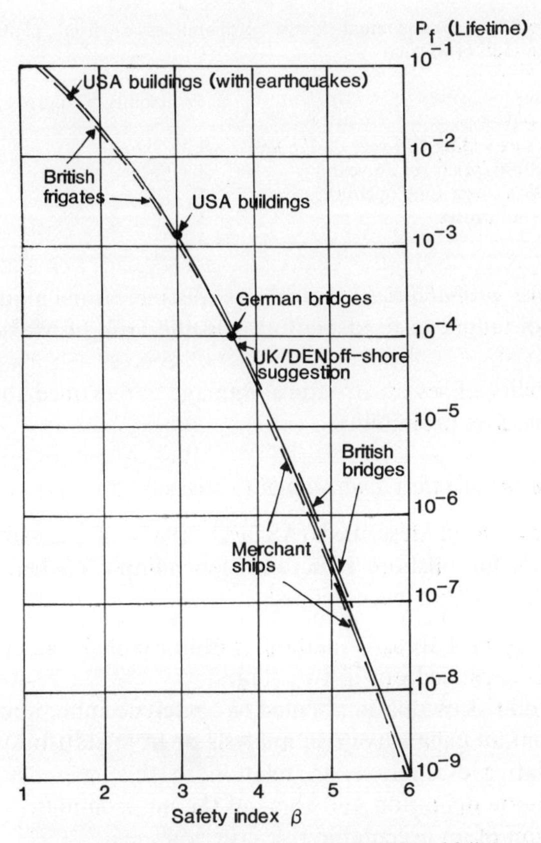

Figure 6.7 Saftey index versus probability of failure

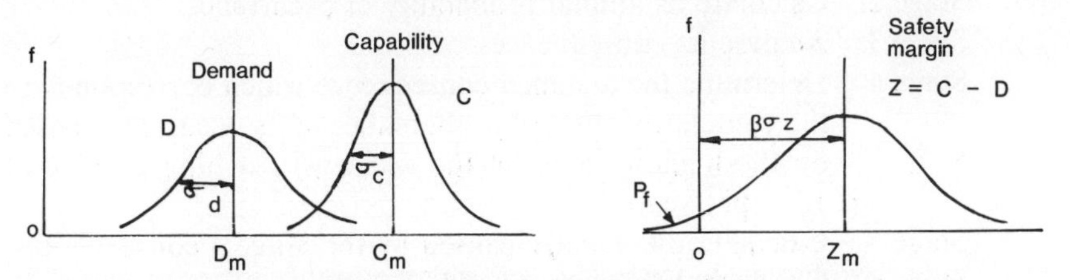

Figure 6.8 Demand, capability and safety margin distributions

where ϕ is the standard normal distribution function

$$\beta = \frac{C_m - D_m}{\sqrt{\sigma_C^2 + \sigma_D^2}} = \frac{\theta - 1}{\sqrt{(\theta V_C)^2 + V_D^2}} = V_z^{-1} \tag{6.2}$$

where $\theta = C_M/D_m$ is the central safety factor and V_C and V_D are the coefficients of variation (or random uncertainties) of the two distributions.

Table 6.7 Actual annual failure probabilities of fixed platforms in the Gulf of Mexico

Cause	Probability of failure $\times 10^{-3}$
waves exceeding 100-year design level	1.3
structural loss (over 16 years)	1.3
collision losses (one operator)	5
loss—all causes	3

6.3.4.3 *Failure probabilities* **Actual** (as distinct from notional) annual probabilities of failure of fixed platforms in the Gulf of Mexico are shown in Table 6.7.

The probability of severe structural damage is reckoned to be about 1.5 to 3.0 times the loss probability.

6.3.5 *Assessment of safety and reliability levels*

Two ways have been described (ASTEO, 1987) of assessing safety and reliability levels for offshore structures depending on whether or not an overall risk criterion has been defined:

1. The use of probabilistic methods without global risk criteria
   Analysis is carried out in two stages—
   (a) a full risk evaluation related to a reference unit; it may be derived from an exhaustive risk analysis or from statistics;
   (b) relative evaluations in relation to this reference to eliminate criteria definition and some of the uncertainties.
2. Definition of an acceptable risk criterion
   Stage 1. Define the accident considered.
   Stage 2. Calculate its annual probability of occurrence.
   Stage 3. Analyse its consequences.
   Stage 4. Determine the assumed consequence which corresponds to the cumulative total of probabilities of its occurrence based on the highest value (of the accident) reaching a value of 10^{-4} per year.
   Stage 5. Calculate the damage caused by the Stage 4 consequences.
   Stage 6. Check that the damage still allows:
   —at least one escape path to remain intact for at least one hour;
   —refuge zones to remain intact until evacuation can be undertaken;
   —the supporting structure to continue to fulfill its function.

Where innovative concepts are involved some events will clearly be difficult to quantify since the relevant statistics obviously cannot exist and so an appropriate testing programme will be required.

6.3.6 *Limit states*

Such considerations lie behind limit-state design although it is often recognised that insufficient data exist for probabilistic methods to be adopted fully and so semi-probabilistic methods are frequently used incorporating partial safety factors. Concrete sea structures, however, have been considered (Eriksson, 1979) to be more amenable to a statistical approach than most structures because:

'1. the dominant loads, wave action and wind action, are generally represented on the basis of statistical data;
2. the structures are large but with a comparatively simple layout, which makes it possible to calculate with reasonable accuracy the resistance against failure, including coefficients of variation related to material properties, geometrical imperfections and deficiencies in the methods of calculation.'

For completeness, and as a reminder, the limit states are re-stated here as given in the Common Unified Rules for Different Types of Construction and Materials (JCSS, 1978).

The *ultimate limit state* (corresponding to the maximum load-carrying capacity) may be reached due to:

(a) loss of equilibrium of a part or the whole of the structure or excessive deformation;
(b) transformation of the structure into a mechanism;
(c) buckling due to elastic or plastic instability;
(e) fatigue.

The *serviceability limit states* (related to the criteria governing normal use and durability) concern the appearance and performance of a structure and include deformation, local damage, vibration, and such other limit states as may be required for special or unusual functions.

Other considerations to be assessed are overall stability and robustness (including accident), fire resistance and durability.

6.3.7 *Safety categories*

The Rules refer to two consequences of failure which have been alluded to already:

(a) risk to life or concern for public reaction (or aversion) to possible failures;
(b) economic consequences due to
    (i) loss of use of structure and all ancillary costs;
    (ii) need for replacement or repair.

For *sea* structures, however, there is an important third category which comes between these two in priority:

(c) environmental damage.

Recognising this, therefore, the FIP Recommendations list three safety categories (FIP, 1985):

Safety Class 1: very little risk of human injuries (unmanned structures); little risk of environmental damage.
Safety Class 2: limited risk of loss of human lives; severe environmental damage (oil spill).
Safety Class 3: great risk of loss of human lives.

The emphasis in all this is on structural integrity, but there are other risks for other types of structure, which do not necessarily affect the strength of the structure itself. For example, a sea wall may be designed to repel or deflect a particular size of wave but that accepts the risk that the wave be exceeded at some time and inundation occur behind it. In designing a harbour or yacht haven it may be expected that shelter is provided for storms up to a certain rating, but again, that has a particular likelihood of being exceeded with consequent potential damage to small craft within. Sub-sea structures on the sea-bed can be the cause of damage to, or loss of, fishing gear.

Structures should not be regarded in isolation. They have some purpose and that helps to define consequence. Where the structure supports some sort of process plant, then fault in or failure of any part of the structure may damage part of the process and that, rather than the structural flaw itself, may be the source of severe hazard. On the other hand, failure in part of the process may result in severe damage to the structure which in turn further damages the process. There is an inter-dependence which should not be ignored. Indeed it has been argued that concrete used as a *passive* safety system (which confines damage as does a firewall, for example) is more valuable than *active* safety systems which are subject to human error: the passive systems have an almost deterministic reliability (Bomhard, 1988).

6.4 Design guidance

There can be few structures of significance anywhere in the world which have not been designed and built in accordance with some prescription, be it code of practice, statutory regulations, or whatever. Where there are statutory requirements, the 'rules' will be unique to a specific country since 'enforcement' is necessarily confined to national boundaries. However, it is likely that they will have been prepared with an eye to international trends and this will be particularly so for *recommended* practice. (Codes of practice of course can be legally binding if adherence to them is a contractual stipulation.)

Certainly offshore structures have been subject to a considerable degree of internationalisation attributable, probably, to a number of factors rather than to any one specific cause. They are primarily (and not in any particular order):

(a) the size and scale of operations of the major oil companies;
(b) the legislative need for certification of the structures, influenced by the extension of the functions of ship classification societies which traditionally operate in all parts of the world;
(c) developments in the North Sea which depend on fabrication and construction industries internationally, particularly in North West Europe and the United States; in its initial stages, experience was extrapolated from that gained in the Gulf of Mexico until the stage now, where much of the world effort is based on North Sea experience.

6.4.1 *Certification systems*

Certification systems have been adopted by governments to safeguard the environment and the people involved but they have been developed through consultation between the authorities and companies (via representative industrial groupings as well as individually) (Furnes, 1979). In general a Certificate of Fitness is required before any hydrocarbon installation can be established or operated in territorial waters, with authority for issuing the certificate being delegated to a Certifying Authority which is responsible to the appropriate department of state. UK and Norwegian waters typify world-wide practice, for example, six UK Certifying Authorities have been appointed after comprehensive and thorough investigations by the UK Department of Energy (Olbjørn and Foss, 1980). They are:

(a) American Bureau of Shipping;
(b) Bureau Veritas (from France);
(c) Det norske Veritas (from Norway);
(d) Germanischer Lloyd (from Germany);
(e) Lloyds Register of Shipping (UK);
(f) Offshore Certification Bureau (UK).

With the exception of the sixth, which is a grouping of consultants, these are all ship classification societies which, of course, have had many years experience in naval architecture and marine engineering, especially in relation to safety at sea. Statutory authority is vested in such as the UK Department of Energy, the Norwegian Petroleum Directorate, the US Geological Survey, and equivalent organisations in other countries which may be provincial or state, independently of or in conjunction with national government departments as in Canada and Australia.

General legal regulations are issued covering, for example, environmental considerations; foundations; primary structures; secondary structures, fittings; materials; construction; equipment. In addition to these, the Department of Energy in the UK, for instance, has issued a set of guidelines, Offshore Installations: Guidance on Design and Construction, which is amended at intervals. For concrete structures, this set gives brief guidance and sources of information as well as recommendations which should be followed. More detail is provided in the British Standard BS6235 Code of Practice for Fixed Offshore Structures, first issued in 1982 but it refers to the much more detailed British Standard Code of Practice CP110, The Structural Use of Concrete of 1972, subsequently overtaken by British Standard BS8110, Structural Use of Concrete, in 1985. The Norwegian Petroleum Directorate similarly has issued Rules for Fixed Structures on the Norwegian Continental Shelf with further reference to relevant Norwegian Standards. This pattern has been followed in other parts of the world.

6.4.2 *DnV Rules*

Detailed guidance is not confined to that issuing from state bodies because the certifying authorities also produce documentation themselves, some of which is quite comprehensive. For example, *Det norske Veritas* in 1977 produced *Rules for the Design, Construction and Inspection of Offshore Structures.* The stated intention is 'to lay down minimum requirements regarding structural strength, serviceability and inspection of offshore structures'. They are of course intended to be used for the society's own operations in supervising design and construction but, where it is functioning as a certifying authority, the Rules supplement any existing national regulations. Apart from general regulations the Rules incorporate environmental conditions; general design requirements; loads; requirements for steel structures and concrete structures; hydrostatic stability, watertight integrity and anchoring; foundations; marine operations; maintenance of Certificate of Approval. A number of appendices corresponding to the principal sections of the Rules complement the main document. On concrete structures the Rules stipulate concrete properties and design parameters while the requirements are detailed for ultimate limit state, fatigue limit state, progressive collapse limit state, and serviceability limit state. Constructional and testing requirements are also given. Appendix D Concrete Structures, then considers cement; admixtures; permeability; tensile strength of concrete; temperature effects; structural instability; shear design; fatigue analysis; testing of concrete in the structure. To illustrate the document's comprehensiveness, the section on structural instability covers stress-strain relationships, linear buckling of circular cylindrical shells, tangent modulus approach, effect of initial imperfections, verification of resistance considering large displacements, and effect of creep.

At the outset design life is defined as 'the period of time from commencement of construction till condemnation of structure' dividing it into five phases:

construction;
transportation;
installation;
operation;
retrieval.

This is a useful reminder that any structure, not just one intended for hydrocarbon production, must be designed to withstand the likely conditions in any and all of the phases which are appropriate, wherever in the world they occur. Thus, as has indeed happened, a structure may be built in Japan, transported to the Arctic and installed and operated in the Beaufort Sea. It must be sound under any conditions prevailing in any one and all of these situations.

The Veritas Rules are not mandatory but other methods and procedures used must result in an equivalent standard of quality and safety (Waagaard, 1982).

6.4.3 *FIP Recommendations*

A great deal of European practice owes much to the CEB-FIP Model Code (CEB-FIP, 1978) and it has been used as the basis for a major international document, *Recommendations for the Design and Construction of Concrete Sea Structures*, the fourth edition of which was published in 1985 on behalf of *FIP* (Fédération Internationale de la Précontrainte—International Federation for Prestressing). Despite their origin, the recommendations are not confined to prestressed concrete and 'refer to structures, or parts of structures, built in reinforced and/or prestressed concrete, intended for use in a marine environment at an offshore location, and which will remain stationary while in service'. Furthermore, it is worth drawing attention to the global composition of the FIP Commission on Concrete Sea Structures, which produced the guide so that it should provide a distillation of current world practice. (The countries represented at the time of publication were Australia, Canada, China, Federal Republic of Germany, France, India, Japan, New Zealand, Norway, Sweden, the Netherlands, UK, USA, and, USSR.) However, it should also be pointed out that the recommendations 'are intended to serve as an application of the general codes to concrete sea structures, but contain only basic aspects for sound design and construction of such structures. Full details with design values, etc. are assumed to be taken from codes prescribed by national authorities, classification societies and so on'.

Sub-division is fairly predictable, dealing in turn with loads; materials and durability; structural design and detailing; foundations; construction and installation; inspection and repair. Limit states are categorised according to the Unified Rules (JCSS, 1978) (see above). The limit state of rupture or excessive deformations should take account of overturning, sliding, buckling, or implosion. Individual sections should have sufficient load-carrying capacity but redistribution can be applied given sufficient ductility. Failure in compression or shear of primary components is to be avoided. Progressive collapse should not result from the rupture of one or more critical sections by sliding or overturning or by failure due to elastic or plastic instability. The structure should be sufficiently ductile to absorb collision and accidental impact but progressive collapse analysis is required for extraordinary loads and exceptional environmental loads, such as seismic or iceberg collision. Overload of members is permitted in such cases provided damage and its operational consequences are not disproportionate to the cause. The structural system remaining after cessation of the load should be able to survive until repair is completed. The guiding figure for probability of occurrence of such loads is 10^{-4} per year.

Checking for fatigue is unnecessary if the total cycles are fewer than 1000 but otherwise **all** load cycling effects should be incorporated although they can be condensed into discrete blocks. If the maximum compressive stress does not exceed 50% of the 28-day characteristic strength for extreme environmental conditions of loading in excess of 1000 cycles, a fatigue check is *not* required. Other exceptions are permitted also. Miner's rule can be applied to random variations summing from one to the number with 50% probability of exceedance.

For the serviceability limit state of durability the most sensitive area is the splash zone, which is defined as 'that between the highest and lowest water levels reached by the wave with a statistical return period of six months, when superimposed on the highest and lowest levels of spring tides.' (It will be noted that this differs again from those quoted previously.) Crack widths in reinforced concrete are limited to 0.3 to 0.4 mm and 0.1 mm in prestressed concrete. No through-cracking is permitted in thin-walled sections under long-duration or frequent tensile loading.

Displacement, motion, and vibration should be limited. There should also be a safety margin against permanent damage, achievable by limiting reinforcement tensile stresses to 80% of yield.

Structural tightness is particularly important for the containment of fluids so that through-cracking has to be avoided. Operational thermal stresses should be catered for as well as the effects of accident spillage or release.

For design loads and properties of materials, partial safety factors and characteristic values are subject to agreement with the owner unless otherwise specified but one statement is worth repeating with some emphasis, particularly in view of some of the comparisons referred to later in this text:

'Code figures for strength, modulus of elasticity, etc., and values for γ_m are often related. Therefore strength values, material properties and material coefficients should be consistently adopted from *one* single source'.
(γ_m is the appropriate partial safety factor for strength.)

General design considerations are catalogued with attention being drawn particularly to the need to consider loads on the complete or partially completed structure at all stages of construction, towing and installation, as well as breakout and retrieval should it be moved from one location to another during its working life. Reference is made to motion response and how it affects structure, personnel and plant and to the effects of ice accumulation, not previously discussed here, but which can be significant in some situations. During installation particular considerations are highlighted:

(a) uplift on horizontal surfaces as they are submerged;
(b) changes in weight over time due to water absorption;
(c) compression of air or other gases or fluids within the structure;
(d) variations in ballast water level;
(e) hydrodynamic effects in shallow water and on landing.

In-service inspection and repair should be considered from the very early stages of design with attention being drawn to the need for provision of clear location reference points for divers.

Slamming on the underside of the deck support structure should be avoided by the provision of adequate freeboard under the worst conditions.

The Recommendations go on to discuss foundations, construction and inspection and repair but it is hoped that sufficient of their flavour has been indicated here to give a proper impression of their content.

6.4.4 *ACI Guide*

The *American Concrete Institute* has prepared a '*Guide for the Design and Construction of Fixed Offshore Concrete Structures*' last reaffirmed in 1988. It acknowledges reference to the Recommendations of FIP and the Rules of Det norske Veritas referred to above and also to another US code used world-wide, mainly for steel structures, the Recommended Practice for Planning, Designing and Constructing Fixed Offshore Platforms, API RP2A of the American Petroleum Institute.

The ACI Guide treats in sequence, materials and durability; loads; design and analysis; foundations; construction, installation and relocation; inspection and repair; there are two appendices covering environmental loads and design for earthquakes. The structure should be strong enough to resist extreme forces without sustaining permanent damage. 'Load multipliers' ('safety coefficients' in the European Model Code) differ slightly from European practice as do the stress-strain relationships for the materials.

Progressive collapse is to be avoided and ductile (non-brittle) failure ensured. Failure conditons to be considered should include loss of overall equilibrium, failure of critical sections, instability resulting from large deformations, and excessive creep or plastic deformation. Serviceability limit states are: excessive cracking, unacceptable deformations, corrosion of reinforcement or deterioration of concrete, undesirable vibrations, excessive leakage. Design for seismic effects (not dealt with in the FIP Recommendations save for reference to the ACI Guide) is discussed but by avoiding specifics, preference being given to a 'check list' approach which leaves the designer to exercise judgement in the light of experience. The point is made that offshore structures differ significantly from onshore in both global and local characteristics with submarine landslides, tsunamis and acoustic water waves among the hazards faced by marine structures, although ground shaking remains the most important. Selection of design criteria, dynamic analysis and stress analysis are included in the topics considered.

6.4.5 *Code comparisons*

Comparison of the documents cited above is tempting but not entirely straightforward, particularly as it ought also to incorporate any produced (whether in outline, draft or finished form) by such as the American Bureau of Shipping, Bureau Veritas, Lloyd's Register, Nippon Kaiji Kyokai (NKK) of Japan, and so on. Selection too is tempting but must be done with considerable caution—any aggregation of clauses apparently most favourable to the selector has traps for the unwary. For example, design safety coefficients should not be isolated from their appropriate material factors. In short, the documents ought to be considered as a whole although clearly parts can be separated within limits. Strictly speaking the only effective way to compare different sets of rules would be to apply each of them to the design of a hypothetical structure and then compare the results for economy, safety and durability. Of course the outcome might well vary with the type of structure and loading criteria selected.

Despite such reservations, comparisons between codes *have* been published and it is worth looking at one or two. Doubtless there are others, including those **un**published and prepared for internal purposes as in the author's own organisation.

6.4.5.1 *Basis for comparison* One comprehensive comparison (Subra and Kavyrchine, 1980) classifies the information necessary for design and construction and which is contained in the ensemble of specifications and recommendations:

- A determination of actions (the generic term which includes loads and accommodates durability, etc.);

B the calculation of response to the actions;
C verification by calculation of the safety or reliability of the structure;
D durability requirements and effects under the conditions of use;
E the quality control necessary to guarantee fitness for purpose in relation to the requirements deduced from these considerations.

In determining the actions, loads are categorised as

(a) permanent actions on the structure such as self weight and loads from fixed equipment; hydrostatic pressure from the sea or stored liquids;
(b) variable actions from use or construction;
(c) environmental actions under which work must be able to continue normally (waves, currents, wind, ice, etc.);
(d) rare environmental actions under which operations may have to be interrupted;
(e) indirect actions due to imposed deformations as a result of temperature differences, creep, shrinkage, prestress;
(f) particular exceptional actions such as earthquakes;
(g) accidental actions such as explosion, fire, collision, for which stability must be maintained even if the structure is out of service.

The first two of these present no great difficulty but, of the others, loads from the sea are the trickiest as well as the most important. A sea state must be defined by a wave spectrum with dynamic response of the structure being dependent on both wave height and period. The return period of extreme wave height to be taken into account varies from 50 to 100 years, according to the set of rules adopted with its value being governed by local conditions. The wave period for a given height can vary of course. Normal service conditions concern waves with a return period of one to three months.

In general, calculation methods for dynamic effects are not quoted in the various rules although classical methods can be used for static loads and service surcharges. Methods for calculating wave loads differ, as do dynamic responses, according to the wave theory adopted and the stiffness of the structure.

Verification of structural soundness is achieved by ensuring that in well-defined circumstances certain values are kept to a safe level. For example, the moment produced by the actions should be kept below the moment of resistance, cracking must be limited, and so on. Load multipliers/safety coefficients vary as has been indicated already but it is considered that the *net* results are close enough. Material characteristics are generally the same (although see below) but control sampling rates vary. Fatigue also is referred to in more detail below.

Because of the uncertainties underlying the behaviour of the sea, probability is an important consideration and so design rules based on semi-probabilistic theories are the more easily adaptable to marine conditions. However, due to deficiencies in our knowledge, there are gaps and divergences in the sets of rules. For example, while Miner's law is acceptable, the quotation of numerical values for fatigue calculations is avoided (but more of that anon).

These comments give the essence of a verbal commentary on an extensive table in the reference, reviewing twenty criteria from eight sets of rules: API, ACI, FIP, DnV, NPD, UK Dept. of Energy, all of which have been discussed above to varying extents, plus Bureau Veritas and proposals from a specific project. It is important to recognise that they were all contemporary with each other so, in invoking particular criteria now or in the future, it would be wise to refer to the most recent edition of any given set of rules in case of modification.

6.4.5.2 *Comparison of concrete specifications* A slightly earlier comparative study (Carneiro, 1979) reviews the semi-probabilistic process and the various limit states, tabulating values of partial safety and security factors as well as characteristic values. Some of the specifications also have been reviewed more recently (Browne, 1986) from which Table 6.8 has been drawn.

Table 6.8 Specifications for offshore and marine concrete

	DnV (1977)	NPD (1977)	ACI (1978)	BS6235 (1982)	Aus. (Draft) (1983)	BS8110 (1985)
MIX DETAILS	Cement content kg/m³ and water/cement ratio					
Atmospheric zone	300 (0.45)	400 (0.45)	— (0.40)	400 (0.40)	400 (0.45)	400 (0.45)
Splash zone	400 (0.45)	400 (0.45)	— (0.40)	400 (0.40)	400 (0.45)	400 (0.45)
Submerged	300 (0.45)	400 (0.45)	— (0.40)	360 (0.40)	330 (0.50)	
COVER TO REINFORCEMENT	mm—rebar and (prestressing steel)					
Atmospheric zone	40 (80)	50 (70)	50 (75)	75 (100)	75 (100)	60 (60)
Splash zone	50 (100)	50 (70)	65 (90)	75 (100)	75 (100)	60 (60)
Submerged	50 (100)	50 (70)	50 (75)	60 (75)	60 (75)	60 (60)
CHLORIDE CONTENT	% Cl⁻ by weight cement					
Reinforcement						
BS 12 cement or similar	0.19	—	0.10	0.35	0.20	0.40
BS 4027 cement or similar	0.19	—	0.06	0.06	0.20	0.20
Prestressing steel						
all cements	0.19	—	0.06	0.06	0.10	0.10
CRACK CONTROL Max. crack width (mm)	by agreement	—	—	0.3 or 0.004 × c	— 0.004 × c	0.3
Max. steel stress (mPa)	160	—	120	—	—	—

c—cover

These source documents have all been mentioned above except for AUS(1983) which is AS 1480, Draft Australian Standard, Unified Concrete Structures Code. DnV—Det norske Veritas, NPD—Norwegian Petroleum Directorate, BS—British Standard. Note also that the definition of the splash zone may vary as has been discussed already. BS 12 is for ordinary and rapid-hardening Portland Cement, BS4027 for sulphate-resisting Portland cement.

The quantities in Table 6.8 relate primarily to serviceability/durability requirements. They do imply, therefore, that the concrete to be used will have a 'minimum' characteristic strength, that is, if a minimum cement content of 400 kg/m^3 and a maximum water/cement ratio of 0.45 are required to produce a material of adequate durability, there is little point in basing design upon a required strength of, say, 20 N/mm^2, although the strength does depend on several factors of course and not just on water/cement ratio and cement content.

6.4.5.3 *Comparison of partial safety coefficients* Partial safety coefficients for a number of codes have been summarised and compared by using them to compute a so-called 'effective strength' (Leivestad et al, 1987). This makes adjustments for specimen type (cube or cylinder), conversion to strength in structure, reduction factors for sections in pure compression and differences in load and material coefficients. The coefficients are given in Table 6.9.

(T) in the country list refers to new codes where a reduction in material coefficients is permitted when geometrical tolerances on section dimensions

Table 6.9 Summary of partial safety coefficients

	Material coefficient		Load coefficient		
	Concrete	Reinforcement	$p=0$	$p=0.5$	$p=1.0$
Onshore Rules					
Norway	1.4	1.25	1.2	1.4	1.6
Norway (T)	1.25	1.15	1.2	1.4	1.6
Sweden	$1.5\gamma_n$	$1.1\gamma_n$	1.1	1.15	1.3
Sweden (T)	$1.5\gamma_n/1.1$	$1.1\gamma_n/1.05$	1.1	1.15	1.3
Denmark	$1.8\gamma_n$	$1.4\gamma_n$	1.1	1.15	1.3
Finland	1.9	1.35	1.2	1.4	1.6
Germany	Not used			1.75–2.1	
UK	1.5	1.15	1.4	1.5	1.6
USA	1/0.9 =	1.11 Bending	1.4	1.55	1.7
	1/0.85 =	1.18 Shear	1.4	1.55	1.7
	1/0.7 =	1.43 Compression	1.4	1.55	1.7
Canada	1/0.6 = 1.67	1/0.85 = 1.18	1.25	1.38	1.5
Model Code	1.5	1.15	1.35	1.43	1.5
Offshore Rules					
Norway	1.4	1.25	1.2	1.15	1.3
Norway (T)	1.25	1.15	1.2	1.15	1.3
USA	1.5	1.15	1.2	1.25	1.3
DnV	1.4	1.15	1.2	1.15	1.3

Table 6.10 The effective strengths for the offshore rules.

Origin	Load combination 1	2	3
Norway	22.0	23.0	20.3
Norway (T)	24.7	25.7	22.8
USA	23.6	22.7	21.8
DnV	25.3	26.4	23.4
Onshore range	14.9–27.8	14.9–26.6	14.9–23.5

are considered directly in the design. The coefficient γ_n is a variability coefficient related to the reliability class of the structure. The three values of load coefficient correspond to the three loading categories, firstly where the actions are well-defined or well-controlled, secondly where there is a reduced probability that all actions will occur simultaneously at their characteristic values and thirdly, where they probably will.

Bearing in mind that in limit state design the actions are to be multiplied by their appropriate coefficients and the material resistance divided by theirs the effective strengths were found to be as in Table 6.10.

Concrete shear strength, reinforcement and reinforced concrete also displayed substantial variations although it is noticeable that the offshore rules were more comparable with each other. However, this is perhaps less surprising when it is considered that only four sets were studied, of which three were from the same country.

The point again is made that comparisons of load and material coefficients should bear in mind differences in:

(a) definition of concrete strength in the structure;
(b) required testing methods and evaluation;
(c) definition of characteristic loads e.g. recurrence interval;
(d) conversion factors included in design formulas.

6.4.5.4 *Fatigue checks* Reference has been made above to fatigue which, in the past, has not been a major consideration in the design of concrete structures. However, the choice of structural concrete for a wider range of applications and the design of more slender members due to better use of structural resistance at the ultimate limit state, may lead to greater need to examine fatigue resistance (Thielen, 1982). Certainly some analyses indicate a potential for damage in particular parts and types of marine structure (Price *et al*, 1982; Tricklebank *et al*, 1982). For offshore gravity platforms with towers, while fatigue is not a dominant criterion, hot spots may develop at stress concentrations (for example, around holes or at sudden changes of cross section) where stresses approach the static strength under serviceability limit state conditions. Moreover, in structures such as articulated columns and pontoon hulls, stresses may be as sensitive to small waves as to larger ones and in these circumstances, fatigue could be a problem. Wave energy type structures such as the spine hull and raft, which are essentially box-type in construction, may also be sensitive to fatigue loading. Cyclic local bending at corners and wave pressure induced shear forces in prestressed membranes are likely to cause problems.

It has been suggested that the design action safety coefficients for the fatigue limit state should be the same as that for the ultimate state while the material coefficients correspond to those for the serviceability limit states (Kakuta et al, 1982). Although fatigue checks may be unnecessary for reinforced normal weight concrete in air where γ_f (the action coefficient) = 1.5 and γ_{mc} (the concrete material coefficient) = 1.5, it is pointed out that lightweight concrete and concrete in water have lower fatigue strengths so the same criterion may not apply.

Stress ranges beyond which fatigue checks become necessary have been suggested for US practice (Hawkins and Shah, 1982) (all stresses in psi, pounds per square inch):

1. concrete in compression, under maximum loading

$$f_{cr} = 0.5f'_c - \frac{2}{3}f_{\min} \tag{6.3}$$

2. deformed reinforcement in tension or a combination of tension and compression

$$f_{rr} = 20 \times 10^3 \tag{6.4}$$

3. f_{rr} is to be reduced by 50% in the region of bends or tack welding of auxiliary to main reinforcement
4. Prestressing tendons in tension:
   4.1 where the nominal tensile stress in the precompressed tensile zone $\not> 6\sqrt{f'_c}$ and the member is uncracked

$$f_{tr} = 0.10 f_{pu} \tag{6.5}$$

   4.2 where that stress $> 6\sqrt{f'_c}$ or the member is cracked

$$f_{tr} = 0.04 f_{pu} \tag{6.6}$$

f_{cr} = stress range in the concrete; f_{rr} = stress range in deformed bars; f_{tr} = stress range in prestressing steel; $f_{\min}$ = minimum compressive stress; f'_c = compressive strength.

The Norwegian Veritas Rules also give limiting conditions beyond which fatigue checks are not necessary:

1. the design loading effect does not exceed half the design resistance for any load combination and no tension exists for any local combination;
2. the number of load cycles $\not> 10\,000$;

3. the design stress range at 10 000 cycles < fatigue strength range at 2×10^6 cycles.

Formulae are given to calculate the concrete fatigue strength range, the number of cycles to failure in combined axial and flexural compression, in shear and in bond, as well as to failure of straight and bent bars. Limiting conditions for prestressed concrete are:

1. sufficient prestress to ensure no cracking;
2. tendon anchorages remote from critical sections;
3. limited stress range in tendons.

6.5 Marine structures

Thus far the discussion has been concerned with fairly general considerations such as

conceptual design;
financial implications of innovation;
safety and risk;
statutory requirements;
design factors;
recommended practice.

Direct relation of these to concrete marine structures in general is not entirely straightforward since their relative significance will vary according to the type of structure, although design factors and recommendations feature prominently for all, of course. Their potential scope might instead be more effectively demonstrated by taking a sample representative of the imagination and inventiveness which have been applied to the solution of a wide variety of problems. In some respects too, progress has been stimulated as much by unfulfilled ideas as by those reaching fruition but it is not even necessary always to be revolutionary and this is certainly not intended to be a vade mecum of innovation.

Problem definition is an important first step so it might be helpful to take a few structures and relate them to some aspects of the behaviour of the sea. A fairly crude attempt to do so is made in Table 6.11 which is largely self-explanatory although the last three columns warrant some comment. It must be emphasised, however, that it is the relationship between the *structures* and the sea which is of concern here rather than what the structure supports. Moreover, fluctuations in water level might be as important in some situations as force in others.

'Energy' concerns wave force primarily (although currents and tides must obviously have a role) and the characteristic sought from the structure might be transparency in the case of an oil production platform or absorbency

Table 6.11 Some aspects of the behaviour of the sea in relation to the working environment & typical structures

Structure	Function	Worst Exposure	Maximum Water depth	Waves	Tides	Storm surge	Energy	Motion response	Station keeping
Wharves and jetties	Provision of ship berth and level base for cargo transfer	Sheltered	Shallow	Not important	Important	Possibly important	Not applicable	Fixed structure	Fixed structure
Sea walls, breakwaters, harbours	Provision of sheltered water and potential anchorages and moorings	Severe	Shallow	Important	Important	Important	Absorbent	Fixed structure	Fixed structure
Hydrocarbon production and storage	Stable platform for control of oil and gas flow to shore &/or ship	Severe	Deep	Important	Important	Important	Transparent	Possibly important	Possibly important
Barges and pontoons	Cargo transport, plant base	Moderate	Deep	Important	Unimportant	Unimportant	Part transparent Part absorbent	Important	Possibly important
Wave energy devices	Conversion of wave energy	Severe	Moderate	Important	Important	Possibly important	Absorbent	Important unless of fixed type	Important unless of fixed type
Tidal barrages	Maximise pressure head variation from tidal range and regulate flow	Moderate	Shallow	Depends on location	Important	Probably not important	Absorbent	Fixed structure	Fixed structure

in the case of a break-water. That is, the configuration of the transparent structure aims to minimise wave force while the absorbent structure should reflect, deflect or otherwise absorb it. Wave energy devices as converters (from wave energy to some other transmissible form) are clearly absorbers. 'Motion response' signifies that operation is affected by movement of the structure: a floating production plant might have to suspend operations in sea states beyond a certain severity. An elaborate tethering system might also be required. 'Station keeping' implies that the structure **must** remain on station in some cases (if it has oil conductors connected to the sea bed, for example) and be relatively more free to move in others, as in an ocean thermal energy conversion plant, for instance, or completely free in yet other situations—as in barges, perhaps.

Establishing the function of the structure i.e. what it is intended to provide, is part of the conceptual process during which also its size or capacity will be determined. Thereafter its configuration will become fixed by criteria such as water depth, wave climate, structural feasibility, operational features and so on. It is possible retrospectively to see what the logic of the evolutionary process might have been in some instances, it is less easy to see what prompted the innovative step at any particular stage. In many cases the primary objective is to create a controlled space, whether for oil storage, for trains or vehicles to pass through (in a tunnel) or for some process to take place within. The space has to be located at a certain place which will determine the circumstances from which the space has to be protected or isolated such as the limits of those listed in Table 6.11. The 'rules' and 'recommendations' are all intended to enable this to be done successfully and safely. Let us now look at what has been done or intended to be done.

Concrete marine structures of one type or another have been built in most parts of the world. Some truly merit the label 'spectacular' and inevitably there is a substantial body of literature describing them. This is much too large to review in its entirety but a small group of reviews known to the author (and there are bound to be others) is particularly valuable in indicating the imaginative scope and range of potential applications for the material.

Not confined to concrete, a survey of offshore storage systems (for crude oil primarily but including other commodities) was published in 1972 (Burton and Smith). They ranged from submerged to floating systems of a variety of shapes and sizes: cylinders, boxes, cones, rings or doughnuts, cellular constructions, articulated towers, are among those illustrated. In the same year a milestone FIP Symposium (Maxwell-Cook, 1973) was held in the USSR. Proposals described included a floating airport, LNG carriers, and floating immersed tunnels. Among projects implemented were immersed tunnels, bridge caissons, lighthouses, oil storage, a tidal power station, and floating dry docks. Three years later a 'presentation of the expanding use of prestressed concrete for ocean structures and ships' was produced in the United States (Gerwick, 1975). Although incorporating useful brief comments on various technical aspects of the subject, a noteworthy feature of the booklet is the series of (black and white) photographs of structures in different stages of construction. Finally of this small group of reviews, a comprehensive survey was published early in the 1980s encompassing such diverse schemes as offshore windmills, concrete LNG carriers, hydrocarbon platforms, and floating airports' (UEG, 1982). Some of its conclusions are worth reiterating here (the numbers are those of the original document):

'2. Design and construction of most of the proposed concrete structures is within existing capabilities.
3. In general, economic rather than engineering problems are the major obstacle to the development of concrete offshore structures.
4. Reinforced or prestressed concrete construction becomes more attractive as offshore structures become larger.
5. In many cases the structural concrete part of an offshore development involves relatively low technology and serves only as a support for the mechanical and electrical equipment. In this context the prime requirement is to develop design and construction procedures which will minimise the cost of the structural units.
6. Many of the alternative-energy proposals are based on an array of similar concrete structures and the economic and engineering advantages of multiple construction need to be studied carefully.
7. Offshore 'work platforms', for many uses and in many forms, appear to be an area with major potential for concrete in the longer term. The main application is in the oil and gas industry but there are other opportunities in other industries, such as fisheries, mineral extraction, sub-sea mining etc.'

Here it is proposed to have a closer look at three groups of structures:

1. Oil (and gas) platforms;
2. Arctic structures;
3. 'Miscellaneous' structures.

6.5.1 *Oil platforms*

6.5.1.1 *Design types* There are two principal types of design of concrete platform which have been employed in the North Sea (the predominant area for their use—similar platforms have been installed in shallower waters offshore Brazil, for example). **Tower** platforms have a cluster of cells at the base with two, three or four surface-penetrating towers rising from them to support the deck—Figure 6.9 (Larsen and Fosker, 1979) shows the Shell Brent 'B' platform, typical of this arrangement. **Manifold** platforms (which are fewer in number) comprise a group of concentric cylindrical or lobate shells as in Figure 6.10 (Parat, 1980). Artist's impressions of the Condeep

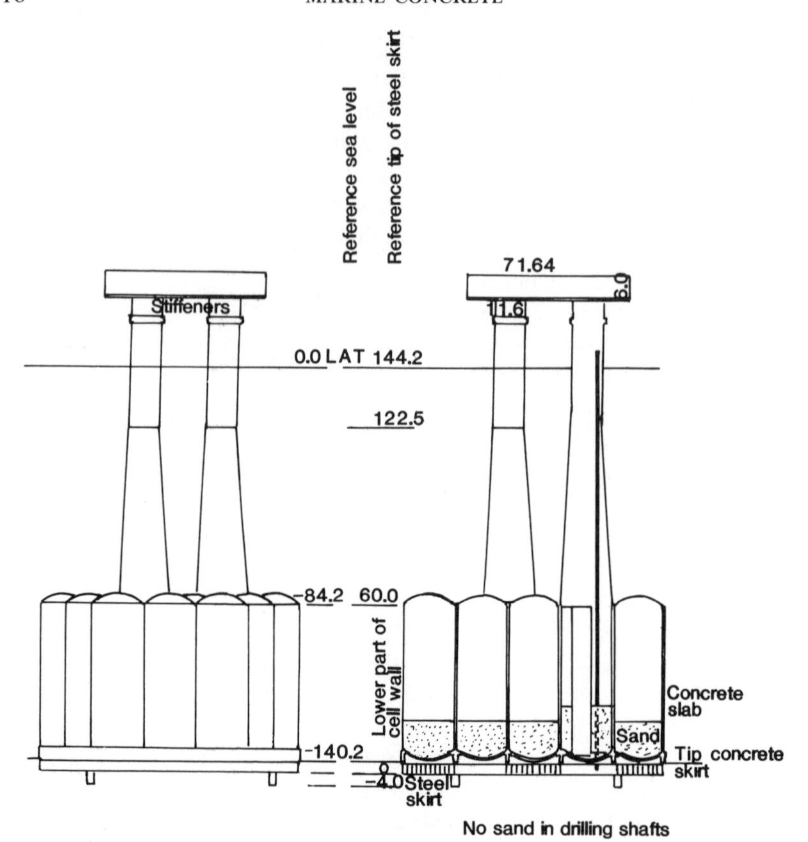

Figure 6.9 Condeep platform (Shell Brent 'B')

tower design and the Ninian Central platforms illustrate the differences more effectively in Figures 6.11(a) and (b) (Ridley, 1978). The tower platforms are further sub-divided into two main categories where the Condeep has a hexagonal array of base cells while the Andoc and Sea Tank designs (not shown) have a square array of a (greater) number of smaller, mainly square cells. This second group has conductors external to, i.e. between, the towers or shafts whereas the Condeep has them contained within one or more of the shafts. These arrangements have different thermal consequences, bearing in mind that the oil is hot, indicated by Figure 6.12 (Richmond et al, 1980). Heat will be transmitted through the conductor casing walls into the base caisson of (a) and into the tower and tower caisson in (b).

Some idea of the size of the structures can be obtained from the very basic statistics in Table 6.12. A few random facts also help to set the context: the design wave load for the Ninian Central platform is 102 900 tons from an extreme wave height of 31.2 metres (C.G. Doris, 1977); the Gullfaks C platform will be the heaviest man-made structure ever moved at around 1.1 million tons; Cormorant A was built on the west coast of Scotland, towed 1630 km

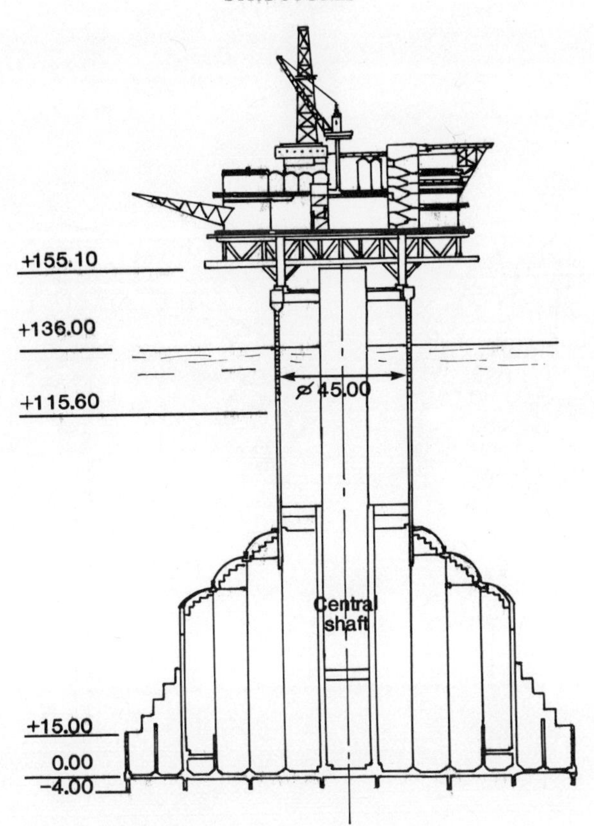

Figure 6.10 Ninian central platform

to Norway to have the deck and superstructure added prior to installation, the deck having been built on the Mediterranean coast of France then towed to Norway (FIP, 1978).

These platforms can carry substantial topsides loads, 40,000 tons for Statfjord B for instance so that they have greater flexibility in accommodating under-estimated or unexpected equipment loads. Their basic features are listed as (Eriksen, 1982): corrosion resistance: little or no maintenance; unanticipated toploads acceptable; offshore piling eliminated; inshore completion of the entire platform; large deck area/top loads possible; oil storage capacity.

The reference to inshore completion points to one of the key advantages of such platforms over conventional steel construction: installation of the concrete structure with its full topsides equipment already fitted takes a matter of days rather than the months of its steel rival where the deck and modules have to be fixed on location (Sjoerdsma, 1975). On the other hand the concrete platform needs a deep water anchorage for completion whereas steel jackets can be built on any level site adjacent to shallow water.

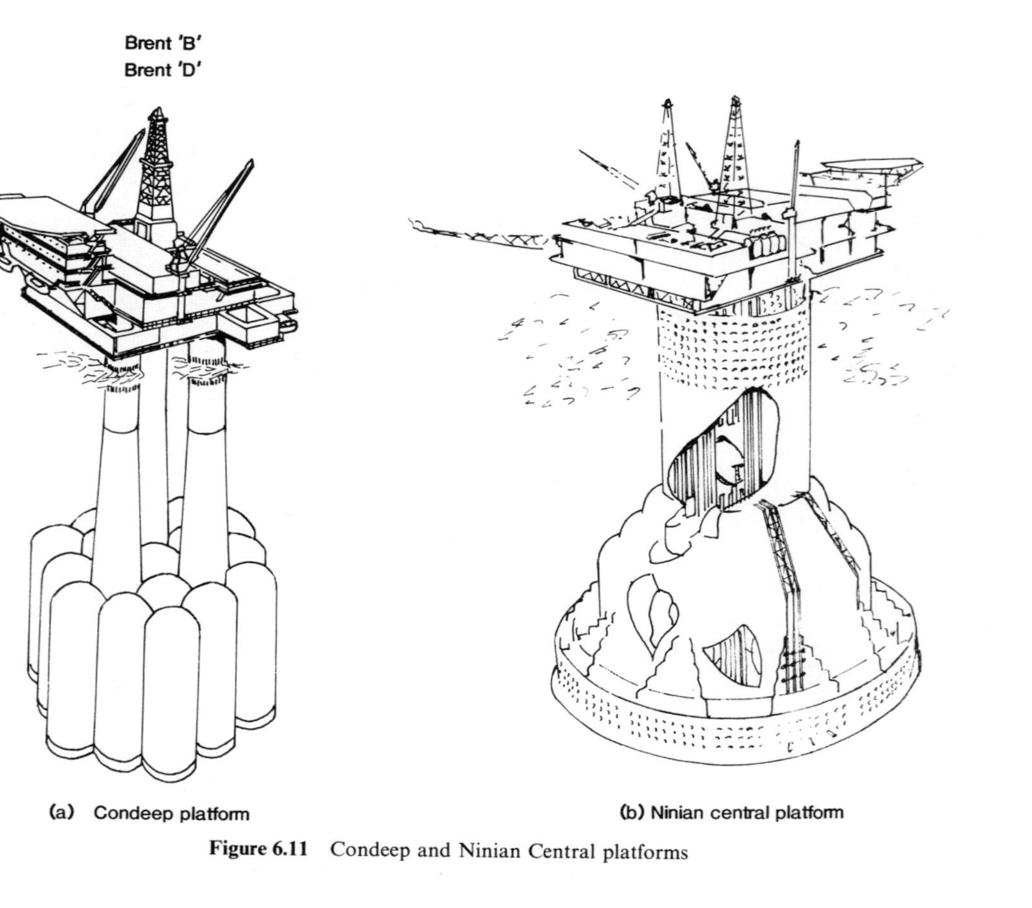

Figure 6.11 Condeep and Ninian Central platforms

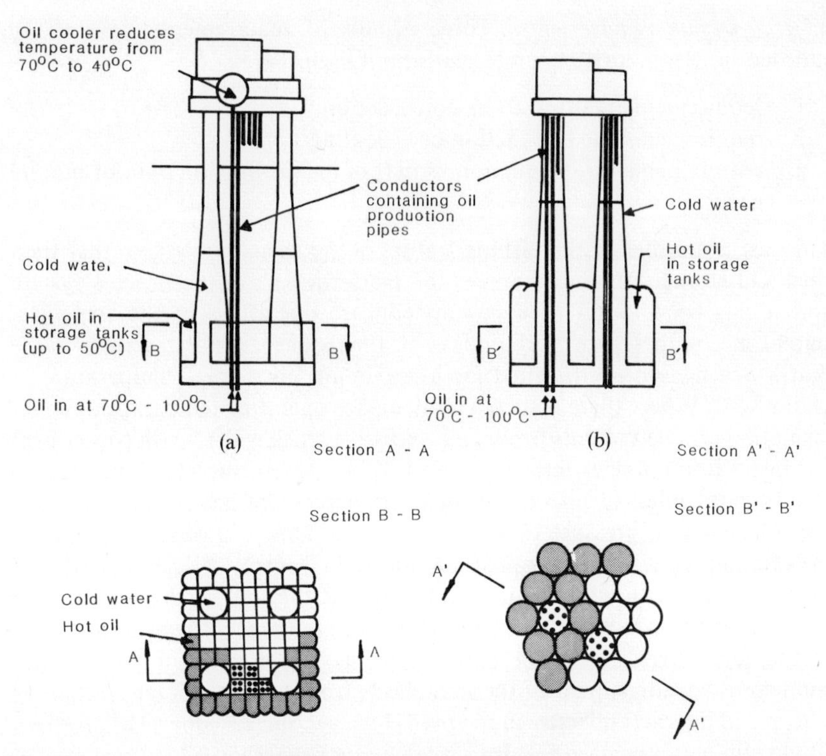

Figure 6.12 Storage arrangements

Table 6.12 Basic structural data for a number of concrete oil and gas platforms in the North Sea

Platform	Design	Towers	Cells	Water depth m	Concrete volume m^3
Beryl A	Condeep	3	19	118	52 × 10^3
Brent B	Condeep	3	19	140	64
C	Sea Tank	4	64	142	105
D	Condeep	3	19	140	68
Dunlin	Andoc	4	81	151	90
Ninian Central	Howard Doris	—	—	143	140
Cormorant	Sea Tank	4	64	150	120
Ekofisk	CG Doris	—	—	70	80
Frigg CDP1	Howard Doris	—	—	104	60
TP1	Sea Tank	2	25	104	49
TCP2	Condeep	3	19	104	50
MP2	CG Doris	—	—	94	60
Statfjord A	Condeep	3	19	145	87
B	Condeep	4	24	145	140
C	Condeep	4	24	145	130
Gullfaks A	Condeep	4	24	135	
B	Condeep	3	19	141	
C	Condeep	4	24	216	
Oseberg A		4	30	109	

6.5.1.2 *Design requirements* Three groups of requirements have to be satisfied in designing a concrete platform (Olsen, 1975):

1. requirements relating to its construction;
2. requirements for its function as a floating body;
3. requirements for its function as part of the production unit for oil and gas.

This last is another distinguishing feature of the use of concrete rather than steel. Oil storage capability makes the platform part of the process system and it has been found necessary to educate operatives in the structural significance of their actions (Faulds et al, 1983). Shell Expro structures in the North Sea have been designed for a maximum oil storage temperature of about 37°C. Where there is such a limit, higher operating temperatures could result in thermal gradients producing excessive crack widths with consequent corrosion in the longer term if allowed to persist. Furthermore the storage facility is intended to operate at an internal pressure head some 40 m less than the external pressure of normal sea level. This is in order to maintain the cell and leg walls in compression, but tight control of such 'drawdown' causes problems in equipment maintenance at the foot of the legs. It also means that the structures will be sensitive to any incident causing flooding of the legs. Stress variation due to changes in drawdown will require compensatory adjustment of thermal effects from hot oil storage so there is a degree of interaction between the two. However, increased level of prestress would eliminate the need for drawdown and a new crude oil storage system has been proposed which means that ballast water can be opened to sea pressure—made possible because oil contamination of ballast water has proved to be so low that treatment is unnecessary (NC, 1987) thereby simplifying the whole process.

6.5.1.3 *Loading conditions* Six load conditions have to be considered (Furnes, 1978), with some of the critical conditions arising during construction:

1. construction in drydock;
2. launching (dependent on ballasting);
3. immersion for deck installation;
4. towing;
5. installation;
6. operation.

It has already been pointed out that design and construction are inter-dependent and the list should emphasise that. However, for it to be appreciated properly, some indication of the construction sequence is required. It may be regarded as a seven-stage process:

1. construction of base raft and lower caisson walls in dry dock;
2. flooding of dry dock and float out;
3. completion of caisson and slip forming of towers (structure afloat);
4. tow to deep water deck-mating site;
5. ballasting of structure until deck can be floated in and mated to towers;
6. de-ballasting and tow-out to field location;
7. installation.

6.5.1.4 *Base and caissons* Specific criteria associated with certain of these stages govern design. Thus, while the base of the structure must be capable of transferring the operational, environmental and self-weight loads to the foundation, it also must carry bending stresses due to non-uniform weight distribution when being floated out. The caisson cells are domed on the bottom (see Figure 6.9) but the base also has steel and/or concrete skirts which penetrate the sea bed on installation, confining any soft top layers and securing the bearing capacity (Larsen and Fosker, 1979). Skirt layout varies but a 'typical' arrangement is shown in Figure 6.13 (Kjekstad and Stub, 1978). During floatout from dry dock compressed air is injected under the base, the air cushion reducing the draught required, so that the skirts enable the cushion to be controlled, any tilt being corrected as required, for example. The peripheral skirt also helps to protect the structure against scour. The dowel shown in Figure 6.13 (there are 3 in all for this particular structure) is to facilitate touch down during installation. While the primary function of the skirts is to carry the horizontal load down through soft layers to stronger and supporting soils, they also act as formwork during grouting beneath the platform. Moreover, control of pressure within the skirt controls rate of settlement and penetration load during installation and will facilitate breakout for eventual platform removal (Olsen, 1978). (Of the platforms listed earlier, the Frigg CDP 1 and MP 2 do not have skirts while the Ekofisk tank has 0.4m concrete ribs rather than skirts.) Skirt lengths are generally of the order of 3–4 metres.

Several factors influence the overall shape of the caisson. Cellular construction derives from operational and stress limitation criteria while the need to limit foundation stresses demands a large bearing area (foundation areas range from, for example, 5,600 m^2 for Frigg CDP-1, through 7,800 m^2 for Statfjord A and 9,700 m^2 for Cormorant A, to 15,400 m^2 for Ninian Central). However, flotation stability during tow-out for installation requires the caisson to be high rather than low (Olsen, 1978). High top loads mean that substantial ballast loads are needed near the bottom of the structure when floating, requiring a large volume which attracts large wave forces and consequent large foundation area. The cylindrical shape of cells and towers arises naturally from the efficiency of using concrete in membrane

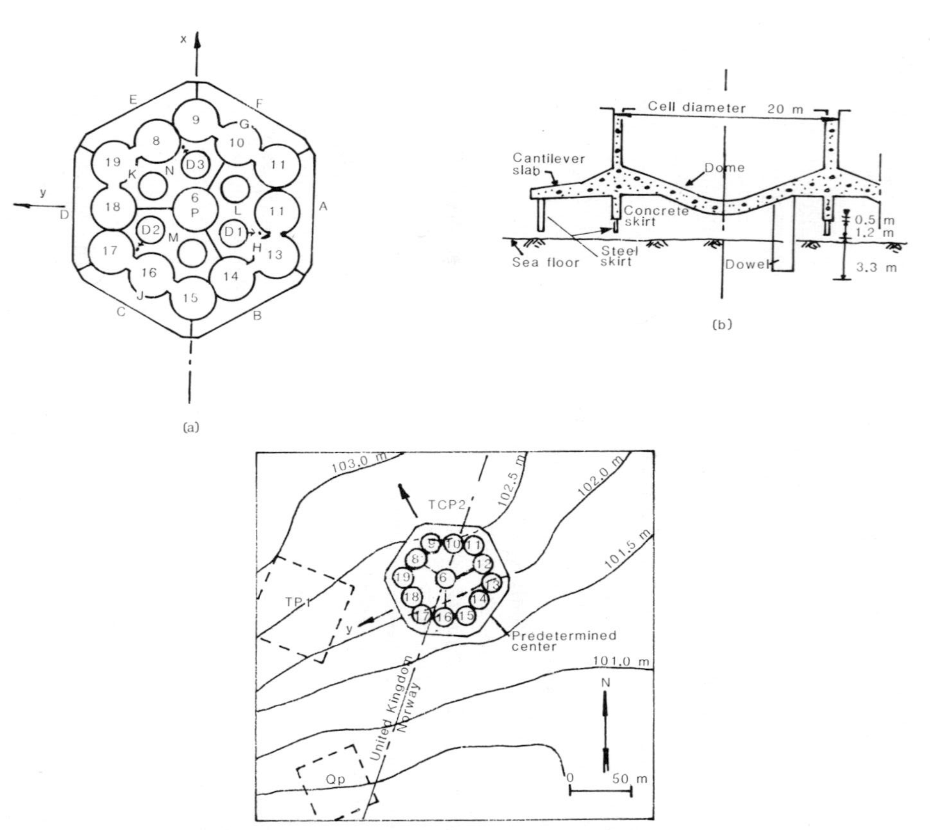

Figure 6.13 Platform base structure details (a) Arrangement of skirt compartments: 1–19, cell number; A–P, skirt compartment; D1–D3, dowel no. 1–3. (b) Detail of bottom section.

compression as well as convenience in slip-forming. Domes are used top and bottom to close the cells.

6.5.1.5 *Deck mating* The operational hydrostatic head is limited to 40–50 m through drawdown as indicated earlier: maintaining an external pressure differential and consequent compression reduces any risk of oil leakage and subsequent pollution. Higher prestress has a similar benefit of course. Deck mating produces the highest hydrostatic load with immersion of the structure to a freeboard of around 6 m and some parts empty. Sometimes air pressure is increased inside the structure to limit this load—4 atmospheres gauge in the case of Brent B, for example (Larsen and Fosker, 1979).

In fact 'mating loads' caused in-plane cracking in certain parts of the Statfjord A platform (Naesje et al, 1978). Reference to Figures 6.12 and 6.14 will show that there are small triangular cells 'between' the main cylindrical cells where full hydrostatic pressure develops. The solid common wall of adjacent cylinders (shaded in Figure 6.14) develops transverse tensile stresses if the cylinders are empty, sufficiently high towards the centre of the caisson to produce virtual in-plane cracking and necessitating repairs in the case in question. Leakage of sea-water inwards occurred because the cracks in fact were inclined to the vertical which meant that they appeared to be horizontal on the wall faces whereas they represented a certain amount of delamination. Similar stress patterns occur in similar locations elsewhere in the structure but are insufficient to cause cracking.

Figure 6.14 (Uherkovich and Lindgren, 1983) shows tendons in the peripheral walls of the structure which of course are the only ones unsupported by adjacent cylinders, hence hoop prestressing is not necessarily required in cells in general. However, it may be required in the domes, in and near the rims particularly, because of the moments and shears developed

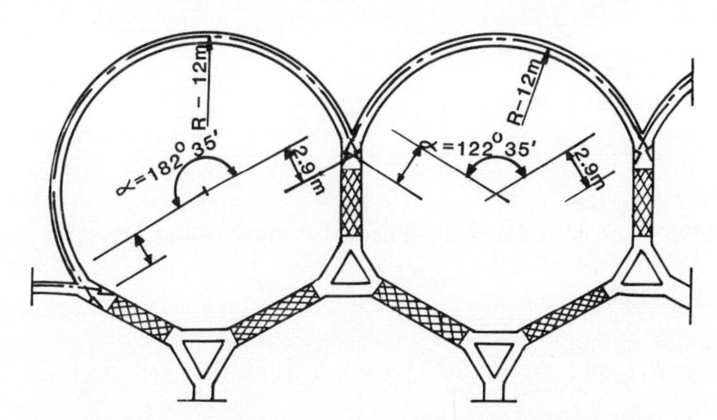

Figure 6.14 Tendons in the cell wall of the Condeep Structure

at their junction with the cylinders. The lower part of the cell walls is shear reinforced and a haunch employed at the top (Olsen, 1978).

6.5.1.6 *Shaft base* The concrete dimensions are governed largely by loads during deck mating but there are exceptions. One is at the base of the shafts where wave forces are the most important. Vertical prestressing is therefore required to ensure no cracking, even from the 100 year wave (apart from possible corrosion problems in the longer term, opening of cracks means penetration of water at full hydrostatic pressure so that the assumptions regarding loading may be affected).

5.1.7 *Elastic instability (buckling)* An important feature of cylinders under compressive loading is their potential for elastic instability (buckling) although because of their inter-support it is considered only to be a problem around the caisson periphery (Olsen, 1978). Such instability leads, or may lead, to implosion, of course. Design techniques and formulae are provided in the DnV Rules with a more extended discussion provided separately (Furnes, 1978). What follows is based primarily on these two sources.

Buckling in concrete structures is a complex phenomenon and second order effects due to construction imperfections can be significant. However, simplified methods can be used provided they are verified with experimental data. The basic formula is

$$\sigma_{cr} = \phi E \tag{6.7}$$

where σ_{cr} is the critical stress, E is the elastic modulus and ϕ is the buckling number which clearly is the critical strain ε_{cr}. In some notations $\phi = 1/\beta$ where β is a so-called buckling factor. For a column length L, moment of inertia I and cross sectional area A, loaded without eccentricity:

$$\phi = \frac{\pi^2 I}{AL^2} \tag{6.8}$$

For an arch, angle $2\theta_0$, mean radius R

$$\phi = \frac{(\pi^2/\theta_0^2 - 1)}{\lambda_R^2} \tag{6.9}$$

where $\lambda_R = R/\sqrt{I/A}$

For a complete cylindrical shell subjected to hydrostatic pressure

$$\phi = \frac{r^4 + s(r^2 + m^2)^2(r^2 + m^2 - 1)^2}{(r^2 + m^2)^2(m^2 - 1 + r^2/2)} \tag{6.10}$$

where $r = \pi R/L$ and $s = h^2/12R^2(1 - \nu^2)$, h being the shell thickness and L its length.

In the so-called tangent-modulus approach this value of ϕ can then be

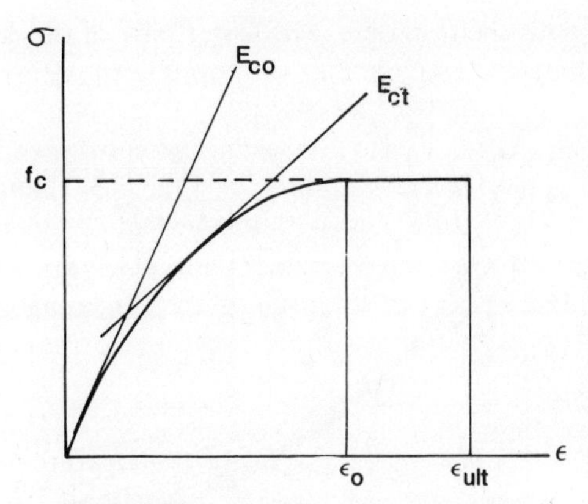

Figure 6.15 Illustrative stress-strain curve

used in the equation

$$\frac{\sigma_{cr}}{f_c} = \alpha = \frac{2}{1 + \sqrt{1 + (\varepsilon_0/\phi)^2}} \tag{6.11}$$

where f_c is the concrete strength ($=0.85 f_{ck}/\gamma_m$ in the DnV rules, f_{ck} being the characteristic cylinder strength and γ_m the material factor), and ε_0 is as defined in Figure 6.15, normally taken as 0.002 for short-term loading or modified by the creep factor ψ for long-term loading so that $\varepsilon_0' = \varepsilon_0 (1 + \psi)$ where $0.5 \leqslant \psi \leqslant 1.0$.

Reinforcement may be allowed for by changing α to α':

$$\alpha' = \alpha\left(1 + \frac{f_s A_s}{f_c A_c}\right) \tag{6.12}$$

where f_s is the steel strength and A_s and A_c the steel and concrete areas.

An approximate prediction of second-order effects may then be obtained by using a magnification factor η:

$$\eta = \frac{1}{1 - \sigma/\sigma_{cr}} = \frac{1}{1 - p/p_{cr}} \tag{6.13}$$

where σ, p are characteristic values of stress and hydrostatic pressure ($\gamma_f = 1.0$) and σ_{cr} and p_{cr} are characteristic values of critical stress and pressure from eqs (6.11) to (6.13) using $\gamma_m = 1.0$.

The total moment is obtained from

$$M_f = \eta M_0 = \eta N_f W_0 \tag{6.14}$$

where M_0 is the direct moment due to the buckling force N_f, eccentricity W_0.

Other, potentially more accurate methods are discussed in the references which should be consulted for more detail: some of the fundamental aspects are also discussed in Høland, 1976.

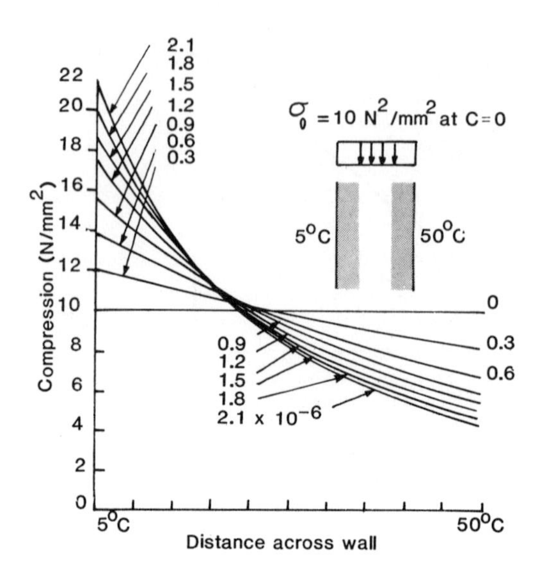

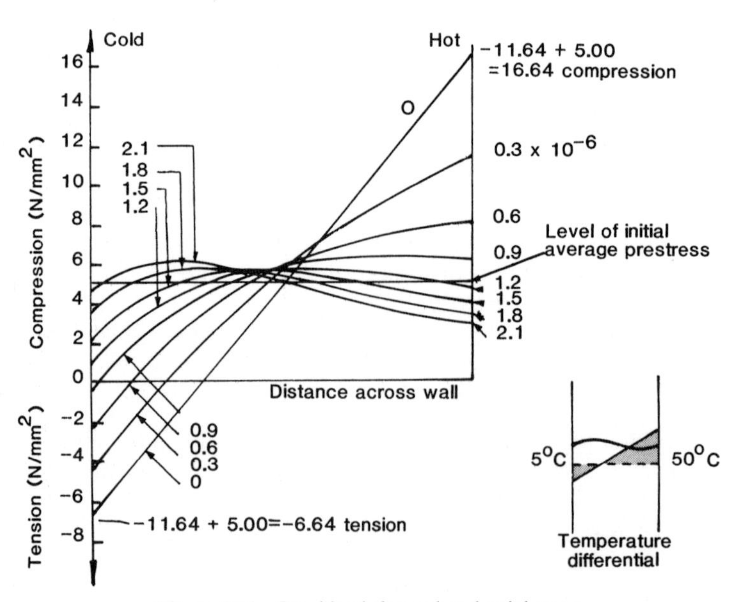

Figure 6.16 Combined thermal and axial stress

Table 6.13 Specific creep and Normalised creep at various times of loading

Times (days)	1	10	100	1000	10000
Specific creep strain 40°C × 10^{-6}	18	28	42.2	59.1	80.3
Normalised creep (specific creep × 1/40)	0.45	0.72	1.06	1.48	2.01

6.5.1.8 *Thermal stress* Thermal stresses have already been mentioned as being significant but analysis and design for them cannot be condensed to two or three paragraphs. Some indication of the complexity of the problem was given in the preceding chapter and it merits a much more extensive discussion. A typical example is of a prestressed tower leg or caisson wall with a vertical prestress of 5 N/mm^2 and a temperature crossfall from the sea water outside at 5° to hot oil inside at 50°C (Richmond et al, 1980). Figure 6.16 taken from the reference, shows the stress distribution at different pseudo times (normalised specific creep). From creep curves the specific creep and hence normalised creep at different times of loading for a temperature of 40°C are as given in Table 6.13.

Hence after 28 days, interpolation gives a normalised creep (pseudo time) of 0.825×10^{-6}. The maximum compressive stress for the (hot) inside is 7 N/mm^2 and the maximum tensile stress for the (cold) outside is 1 N/mm^2 from Figure 6.16.

Operational conditions mean that in storage cells there will be an oil/water interface as oil replaces water or vice versa when filling or withdrawing. Since the liquids are at different temperatures there will be related bending and shear stresses in the cylinder wall(s) due to the consequent vertical temperature gradient. Methods of solution of such problems are given in the reference which also provides tables enabling stresses to be calculated in this particular case, adequate at least for preliminary purposes.

6.5.1.9 *Fatigue* Deterministic calculation of fatigue damage has been illustrated in references quoted earlier (Tricklebank et al, 1982; Price et al, 1982). The procedure is outlined in Figures 6.17(i) to (v), applied to the base of a tower. Wave exceedances are given in (i); (iii) gives the maximum (upper curve) and minimum (lower curve) stresses for each wave which, combined with (i), gives the stress/frequency diagram in (ii); the S–n curve for actual occurrences can now be plotted—the heavy black line of (iv) where the C_1 and C_2 (S–N) lines have been drawn according to the relationship

$$\log N = C\left[1 - \frac{S_{max} - S_{min}}{S_0 - S_{min}}\right] \tag{6.15}$$

C_1 is based on Dutch (TNO) rules with $C = 12.6$ and an endurance limit at $N = 2 \times 10^6$ of $(S_{max} - S_{min})/(S_0 - S_{min}) = 0.5$. C_2 is based on Norwegian (DnV) rules with $C = 10$ and no endurance limit. S_0 is based on a characteristic

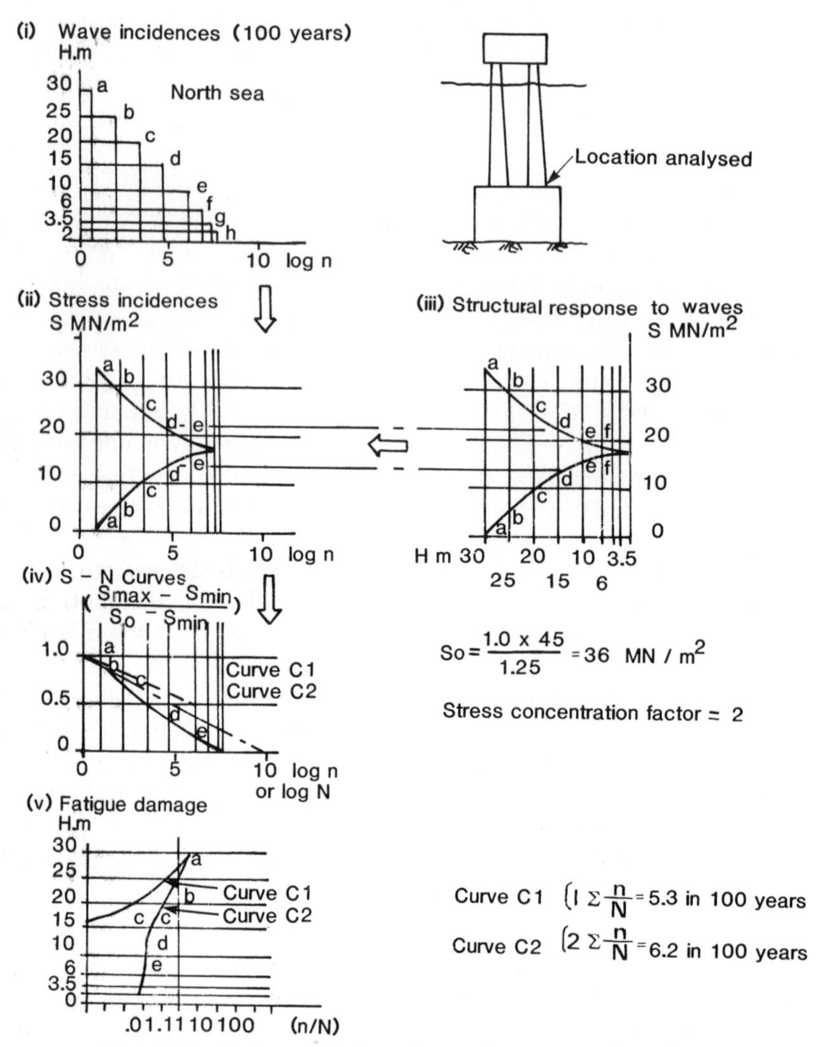

Figure 6.17 PSC tower platform: hotspot in concrete at base of tower

concrete strength of 45 N/mm², a stress gradient factor of 1.0 and a material factor of 1.25.

From the curves of (iv) those of (v) can be prepared showing n/N for various wave heights. The area between this line and the y-axis (if plotted to a linear scale) would give $\sum n/N$. (If there were a hole at the base of the tower with a stress concentration factor of 2, the fatigue damage would be considerably increased). For the particular example quoted the 100-year fatigue damage values in the concrete for both reinforced and prestressed concrete was 0.00 (for C_1) and 0.01 (for C_2) but with the hot spot (stress concentration) the respective values increased to 5.3 and 6.2 for a prestressed tower. The limiting value of course is 1.0 while a recommended value is 0.5 so clearly there are potential problems should provision for any penetrations be made at the tower base.

6.5.1.10 *Collision and impact* Collision and impact have been discussed previously and design guidance has been provided by the Norwegian Veritas (DnV 101, 1982 and DnV 202, 1982). 'The calculations of the impact load should be based on energy distribution between boat, structure and possible fendering system. The energy absorption through the boat should be based on force-indentation characteristics for the assumed point of impact for the maximum authorised vessel.' However, alternative force—indentation curves are given. Operational ship impact load is defined as that from the maximum authorised vessel with a velocity of 0.5 m/s occurring between 6 m below lowest astronomical tide (LAT) and 8 m above the highest (HAT). For an accidental impact the velocity should be 0.5 H_s m/s (H_s is the significant wave height) but not less than 2.0 m/s for North Sea conditions and the level limits are increased to 10 m below LAT and 13 m above HAT. Ship geometries are provided to enable contact areas to be calculated.

Design against accidental loads is categorised as event control, indirect design and direct design. The first of these means that risks should be identified and the design checked against them while avoidable incidents should be forestalled. Indirect design means incorporating adequate ductility by such measures as:

(a) connections of primary members to develop a strength greater than that of the member;
(b) provision of structural redundancy in order to develop alternative load distributions;
(c) avoidance of dependence on energy absorption in slender struts with a non-ductile postbuckling behaviour;
(d) avoidance of pronounced weak sections and abrupt change in strength or stiffness;
(e) avoidance of dependence on energy absorption in members acting in bending;
(f) use of non-brittle materials.

For concrete structures it means keeping reinforcement quantities above code minima but low enough for failure through yielding. Excessive use of prestress should be avoided.

Punching failure and inelastic behaviour are both important considerations and much depends on whether collision from a ship is head-on (bow or stern) or side-on, in which case the ship may bridge any hole and load the undamaged part of the structure. However, the ship also absorbs energy which limits its effect on the structure. Figure 6.18 (DnV 101, 1982) shows

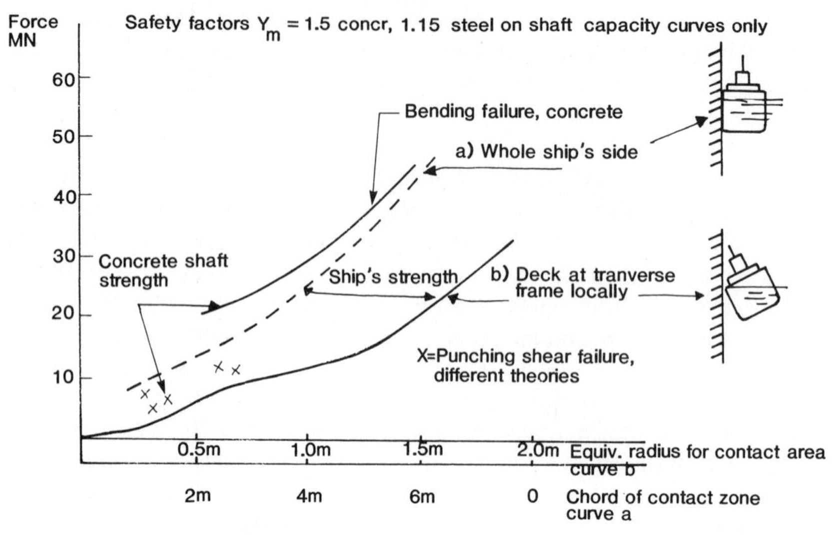

Figure 6.18 Force-contact zone curves

the estimated effects for a 2500 tonnes supply boat colliding with the shaft of a typical platform.

6.5.1.11 *Manifold platforms* Much of the foregoing discussion has centered on tower platforms although the comment in general is relevant to most, if not all types and, indeed, many other categories of structure. At this point, however, it may be useful to have a look at the other principal design so far adopted. This has five main structural elements (Doris, 1979) (see also Figure 6.10):

1. a circular raft foundation incorporating radial and circumferential skirts;
2. a flotation enclosure comprising a circular or lobate vertical wall with radial stiffening diaphragms resting on the raft and surmounted by a torus or several tori;
3. a second enclosure placed either above or around the flotation enclosure; the upper part is perforated (the so-called Jarlan wall or break-water) to dissipate wave energy;
4. a cylindrical (hollow) central shaft supporting the radial beams of the deck;
5. a substructure of one or more decks with large surface areas (up to 10 000 m^2).

There may also be a low perforated peripheral wall around the raft to reduce or eliminate possible scouring from currents.

The primary structure of this type is the Ninian Central Platform, an important aspect of the construction of which was the amount of precasting used. Units for the stiffening diaphragms, the tori and the Jarlan walls were all pre-cast while the vertical walls were slip-formed (Parat, 1980). With a base slab 140 metres diameter, there is a progressive stepped reduction to the 45 m diameter breakwater wall, which rises from 75 m to a level of 149 m above the sea bed (Parat and Maher, 1978). This takes the greater part of the load from the deck, reducing the degree of cantilever required, the remainder of the load being carried by the 14 m diameter central shaft which provides storage space for fresh water and diesel as well as accommodating services and ballast pipework. The lobate annulus making up the flotation unit is segmented making possible differential ballasting. Walls and diaphragms are proportioned to have similar stress levels during the towing and immersion phases while the base was analysed as 'a beam-slab system loaded by distributed foundation reactions and bearing on rigid supports formed by the walls and diaphragms.' Prestressing is used to counteract tensile membrane stresses and shear forces although general design is for reinforced members in bending.

An interesting subsidiary feature of the construction of this platform (and others of its type) was the dry-dock closure. A common practice is to have a sheet-piled embankment or bund which then has to be excavated for float-out. However, at the Kishorn yard in North West Scotland two cellular prestressed concrete gates were employed each 82 m long, 13 m wide and 16 m deep: it seems somehow fitting that such concrete structures should play an important part in the construction of their larger kin.

6.5.1.12 *Loads from foundation* Adequacy of support from the foundation is self-evident and, although it cannot be dealt with properly here, foundation design is clearly an important feature. Conforming with the passage of a wave a 'rocking' load will be applied with intermediate stages of uniform vertical load under the crest and the trough. This means of course that there is a corresponding reaction on the base of the structure which has to be designed accordingly—more important perhaps on structures with a bottom raft. There are additional factors to be considered during installation, however, necessitating the provision of appropriate drainage and grouting facilities.

'The main concern during the skirt penetration phase is to avoid piping below the skirts and to keep the platform vertical' (Eide et al, 1979). Slow penetration rates (0.1 to 0.2 m/hr) can help to achieve this and eccentric ballasting may be necessary to keep the structure level. Water beneath the base and trapped by the skirts has to be evacuated but liquefaction also has to be avoided so the drainage system is also designated as the anti-liquefaction system (Eide et al, 1982).

Structures with bottom domes have contact pressures which can vary considerably from dome to dome: Figure 6.19 (Eide et al, 1979) illustrates

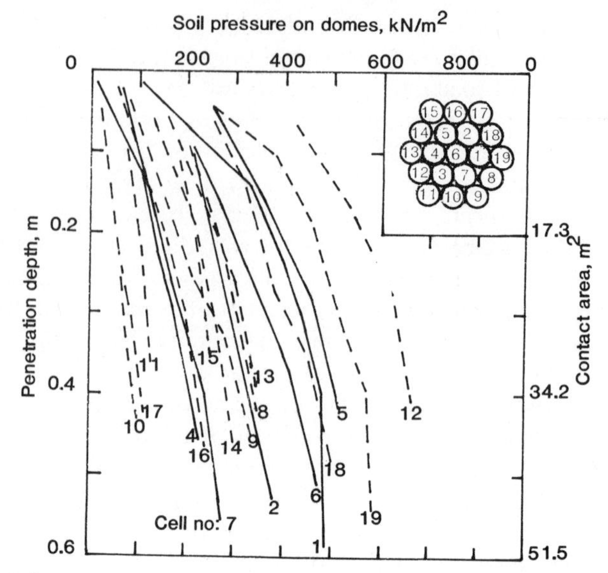

Figure 6.19 Soil pressure on domes as a function of penetration depth, Statfjord A Condeep structure

this graphically for the Statfjord A platform (the domes may penetrate up to 0.6 m). Once installed the under-space generally must then be grouted, especially where the sea floor is uneven or sloping:

1. to avoid further penetration and keep the platform vertical;
2. to secure uniform soil stresses on the slab and avoid over-stressing of any structural elements;
3. to avoid piping from water pockets below the base during environmental loading.

Recommended properties for a suitable grout are:

(a) strength and compression characteristics $\nless$ for supporting soils; (500–3000 kN/m² at 28 days);
(b) bleeding $\nless$ 2%;
(c) low heat emission;
(d) setting time $\nless$ 6–8 hours;
(e) low viscosity;
(f) high sweep efficiency to displace sea water with minimal mixing with it.

Frig TCP-2 required 13 700 m³ over 13 days whereas Brent C needed 'only' 4500 m³ over 13 days.

The drainage system is not intended for operation over the whole lifetime of the structure (Eide et al, 1982). It is intended also to speed up consolidation settlement but it is particularly important during the first storm periods before full consolidation has taken place (1–2 years on very stiff clays). Drainage wells (in earlier platforms) and horizontal and vertical filters have been used, connected to standpipes equipped with valves and operated at a level corresponding to 11–12 m below mean sea level or 10–20 m below lowest astronomical tide under the base slab.

6.5.1.13 *Weight control* Much emphasis is given throughout this text on 'indeterminacy' since clearly many actions and types of behaviour are difficult to estimate. However, one might well suppose that topsides loads are much more deterministic and less problematic. Unfortunately this is not necessarily so: 'it is not unusual... for a "well-defined" module to escalate in weight by a factor of about 1.8 between conceptual design and hook-up [whilst] the weight of a "poorly-defined" module may escalate by a factor of 3 or more over the same period' (UEG, 1984). Weight control and weight management is, self-evidently, an important aspect of design which deserves more attention than can be given it here. Suffice it to remark that there must be careful monitoring at all stages of superimposed loads to ensure that they are within the designed capacity of the structure.

The cautionary tone of that remark does provide the lead-in for a note of optimism, however, especially in relation to weight reduction. It is clear that development of high-strength lightweight concretes is providing scope for significant technological advance in concept generation (Jackobsen et al, 1987). 'The engineer thus has a firm and well-founded basis for further development and optimisation. If human fantasy and innovativeness are in keeping with this basis, new, tailor-made and highly competitive concrete platforms will emerge in the years to come.'

6.5.2 *Arctic structures*

So long as oil and gas are major energy sources, so long will Arctic and sub-Arctic regions be likely for development as oil provinces. Although the pace slowed following the oil price fall of 1986, work has continued, particularly in the Beaufort Sea and in preparation, offshore Newfoundland. Since concrete has certain merits as a construction material for such regions, there is some logic in following consideration of oil platforms with a look at perhaps more esoteric applications but which nevertheless extrapolate from North Sea experience.

At the present stage of development it is convenient to sub-divide or characterise discussion according to the two areas mentioned; the Beaufort where water depths are relatively shallow but with substantial ice loading and low temperatures and Newfoundland where water depths and waves approach North Sea conditions but where seasonal ice may occur and, in particular, where icebergs are a hazard. A slightly more detailed view categorises development areas as indicated in Table 6.14 (Fjeld, 1983).

Table 6.14 Ice, and water depth categorisation of development areas in the Beaufort and Newfoundland

Sheet ice infested areas	
Cook Inlet	20–40 m depths
Sakhalin Area	30 m
Po Hai Bay	20 m
Iceberg infested areas	
Labrador Coast	Large depths
Newfoundland	Planned production < 100 m depth
High Arctic areas	
Beaufort Sea	0–60 m
Barents Sea	

The high Arctic is exposed of course to ice conditions of varying severity. Apart from ice loading the Arctic environment is said to be characterised in addition by

—temperatures down to −50°C resulting in steel material problems;
—dense and frequent fog;
—icing;
—large tides (>4 m) due to shallow waters;
—seismic exposure;
—breaking waves;
—relic permafrost hampering pile driving and also jeopardising gravity structures with wells heating the soil;
—soil conditions, to a large extent consisting of fine silt, unfeasible as material for gravel islands and foundation for gravity structures.

6.5.2.1 *Sheet ice areas* Construction work clearly is difficult under such conditions, increasing the need to prefabricate as much as possible elsewhere. Unfortunately, while water depths in the Beaufort are relatively shallow, structures are even more limited in draught since the maximum depth available for tow-in when ice clears from Point Barrow is of the order of 10 m. This has particular implications to be discussed a little later.

Despite some of the difficulties for gravity structures hinted at above, concrete is claimed to have advantages over steel because of its lower cost, local rigidity, resistance to abrasion and corrosion, favourable behaviour at low temperatures, ability to use the shear lag effect which inhibits progressive collapse and its ductility when suitably reinforced (Engel and Pajouhi, 1985). However, cost-effectiveness drives the designer towards maximising the use of prestress and lightweight concrete. Prestress is particularly useful when applied to shear and tension-sensitive elements. Supplementary reinforcing will still be needed for shrinkage, cracking control and ultimate strength and

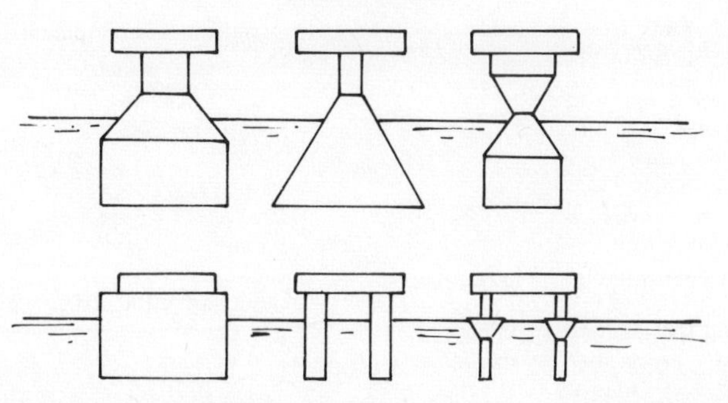

Figure 6.20 Proposed principal configurations for ice resisting structures

of course some areas may need to be heavily reinforced to resist the ice (Mast et al, 1985).

Since the ice sheet fails (i.e. is broken up) more easily in flexure than in crushing this has led to various design configurations based on the cone. Unfortunately as the ice rides up it may also tend to pile up and will be forced higher up the structure. A cone *inverted* from the surface upwards would force the ice sheet down rather than up. Variation in level may also have to be catered for so circumstances have led to proposals for modified hour-glass shapes, albeit somewhat flattened. Some stylised configurations are shown in Figure 6.20 (Fjeld, 1983). More will be said later about caisson shape but several designs are for hybrid structures, part being built from concrete and part from steel, the intention being presumably to benefit from the best characteristics of both materials. Doubtful foundations point towards gravity structures and resistance to ice suggests concrete, at least around surface level. Lighter steel structures require less draught.

6.5.2.2 *Artificial islands* Although monocones have been proposed, the most successful structures used so far in the Beaufort Sea appear to be island systems with two in particular making use of concrete. Water depth is (relatively) so small that artificial islands are entirely feasible and are made more mobile by incorporating prefabricated structural components (mobility in this context means the ability to move/transfer the structure rather than the ability of the structure to move—there is a distinction between the two cases). To give an idea of what might be involved some criteria adopted for one non-island design (Engel and Pajouhi, 1985) are shown at Table 6.15.

Artificial islands have been used in gradually increasing water depths in the Beaufort Sea, having been built initially in shallow near-shore locations. With the almost exponential increase in costs in deeper water due to the greatly increased fill volumes, operators have turned towards retained fill by using caissons (Jefferies et al, 1985). Typical island profiles are shown in Figure 6.21 (Wang and Peters, 1985), the primary purposes of the caisson retained island being

Table 6.15 Criteria adopted for a non-island structure design

Maximum water depth	100 ft. (30.48 m)
temperature range	−48°C to 25°C
minimum water temperature	−30°C
maximum wave height	37 ft. (11.28 m)
predominant wave period	10–14 s
maximum surface current	3 knots
maximum bottom current	1.5 knots
tidal range	−0.2 to 0.9 ft. (−0.06 to 0.27 m)
maximum wind speed (1 min sustained)	125 knots
maximum first year unrafted ice thickness	7 ft. (2.12 m)
maximum first year rafted ice thickness	20+ ft. (6.10 m)
multi-year floes	15 ft. (4.57 m)
first year ice ridge thickness	typical 25 ft. (7.62 m) extreme 100 ft. (30.48 m)
design ice pressure	200 psi (1.38 N/mm^2)
design ice force: horizontal	800–1000 kips/ft. (11.68–14.60 MN/m)
vertical	250–350 kips/ft. (3.65–5.11 MN/m)
seismic conditions: maximum ground acceleration	0.05 to 0.10 g
total weight of structure and facilities	191,100 short tons (173,366 tonnes)

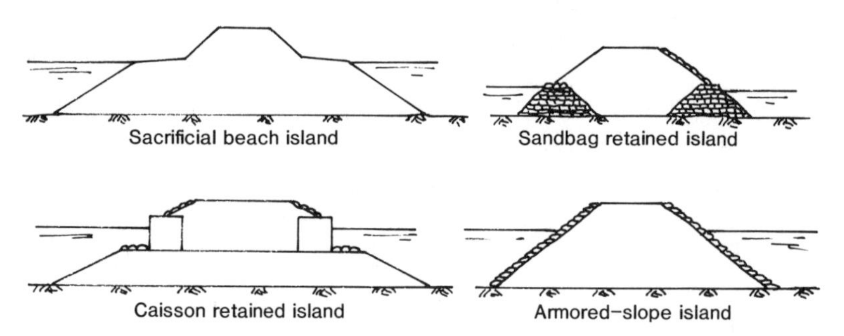

Figure 6.21 Various types of island

(a) to reduce fill quantity (and hence on-site construction time);
(b) to provide adequate shore protection against severe wave action;
(c) in the case of an exploration island, to reduce costs by re-use of caissons elsewhere.

6.5.2.3 *Caisson systems* The Molikpaq (Inuit for 'great wave') is a mobile Arctic caisson installed at Tarsiut P-45 (to the north of Mackenzie Bay) in the Beaufort Sea in 1981 (Jefferies et al, 1985). The cross-section in Figure 6.22 gives the general configuration with the bounding concrete caissons enclosing a sand fill core. Base friction of the caissons transfers 10–25% of lateral ice load into the berm with the remainder being transferred by the sand core. The caisson layout is shown in Figure 6.23(a). (Fitzpatrick, 1984) and a little more detail in Figure 6.23(b). Height, length and width are 11.5,

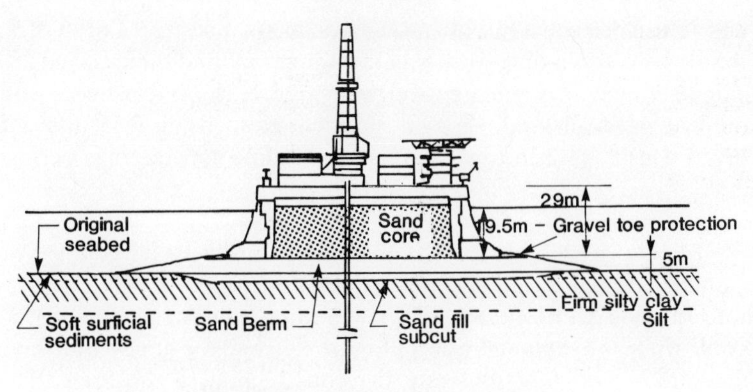

Figure 6.22 Molikpaq at Tarsiut P-45

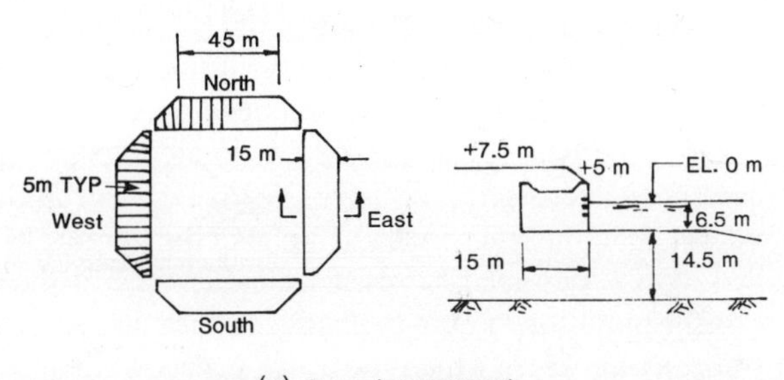

(a) General arrangement

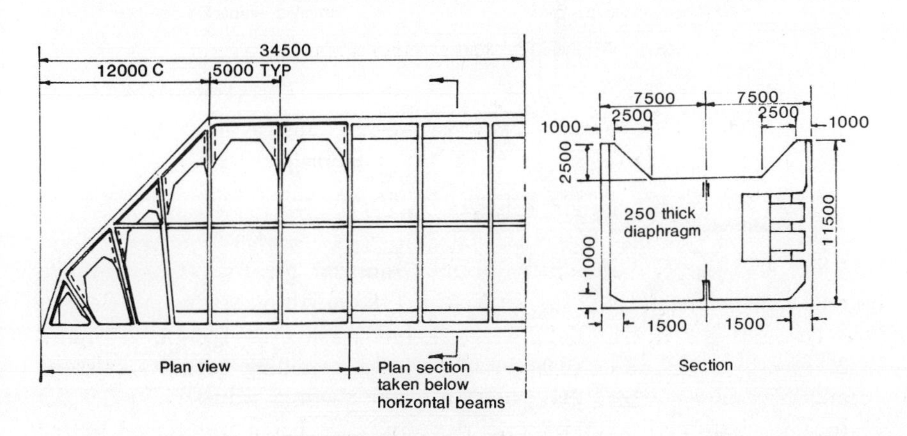

(b) Caisson layout

Figure 6.23 Tarsiut concrete caissons

69.0 and 15.0 m respectively with a weight of 5000 tonnes and a draft of 5.5 m. Concrete density (excluding steel) was 1.95 tonnes/m^3 and the specified 28-day strength 40 N/mm^2. Fly ash, lightweight aggregate, super plasticiser and air entrainment were all used, the water/cement ratio being 0.35 and a final reinforced density of 2.26 tonnes/m^3 attained. The concrete has performed well.

6.5.2.4 *Integrated structures* The Glomar Beaufort Sea I is the world's first Concrete Island Drilling System (CIDS), classed as a MODU (Mobile Offshore Drilling Unit). Figure 6.24 (Wetmore and Borchardt, 1984) gives the profile while the exploded view of Figure 6.25 and the honeycomb layouts of Figures 6.26 and 6.27 illustrate the construction of the brick. This is 71.3 m × 71.3 m × 13.4 m high of 45 N/mm^2 light weight concrete except for the interior wall and shear walls which were of 55 N/mm^2 normal weight concrete. The octagonal shape was intended to be as near the shape of a cylinder as was practical while retaining the advantages of flat-face construction. The honeycomb system, which is encased by the perimeter shear-wall, comprises interconnected cylindrical 'silos' 102 mm thick with 127 mm thick interconnecting walls (Yee et al, 1984). Structural integration with the top and bottom slabs and with the walls is intended to develop the required flexural, shear and torsional resistance and is analogous to the form of construction used in the aerospace industry (Yee, 1975). Variation in height can be achieved by 'stacking' the bricks as indicated in Figure 6.25.

Although the CIDS is designed to resist ice loads, part of the equipment installed is a rubble generation system to produce an ice defence berm around the structure by spraying water which forms ice pellets while in the air. The

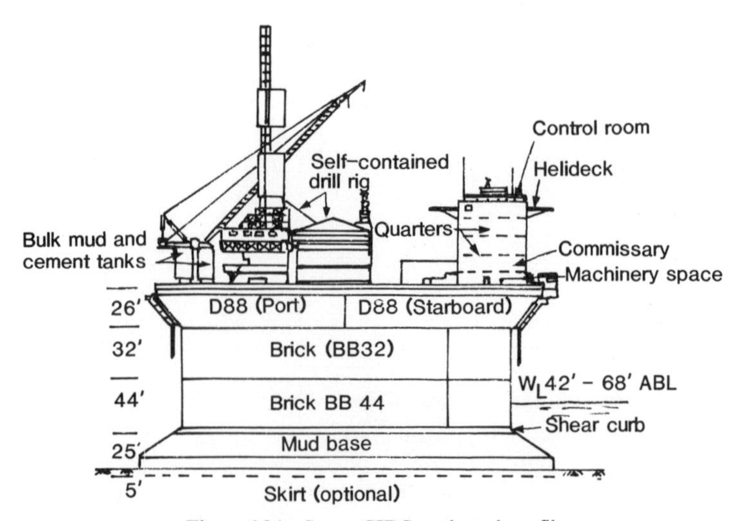

Figure 6.24 Super CIDS outboard profile

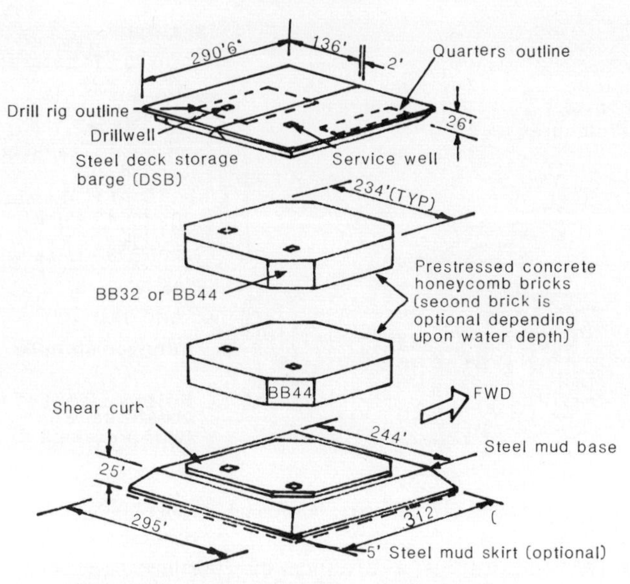

Figure 6.25 Super CIDS-major construction elements

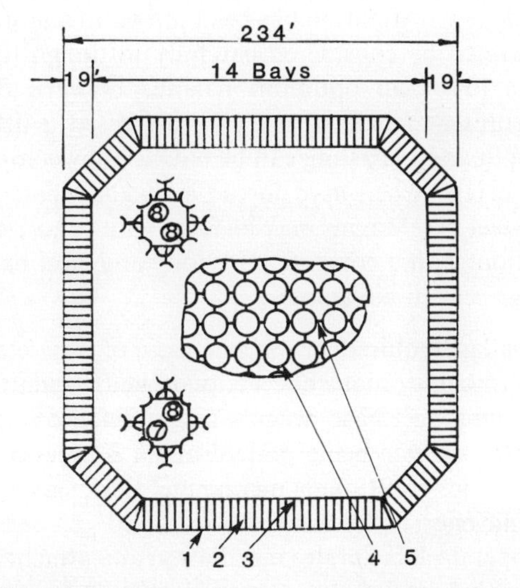

Figure 6.26 Structural arrangement of concrete brick

system is being evaluated since it is capable of substantially reducing ice loads and thereby economising in and/or providing additional protection for the structural design. 'By starting in the early freeze-up period, when the air is cold enough (below 0°F) and the ice is 18 to 24 inches thick, a 300 foot wide grounded ice berm, approximately 40 feet high, and partially

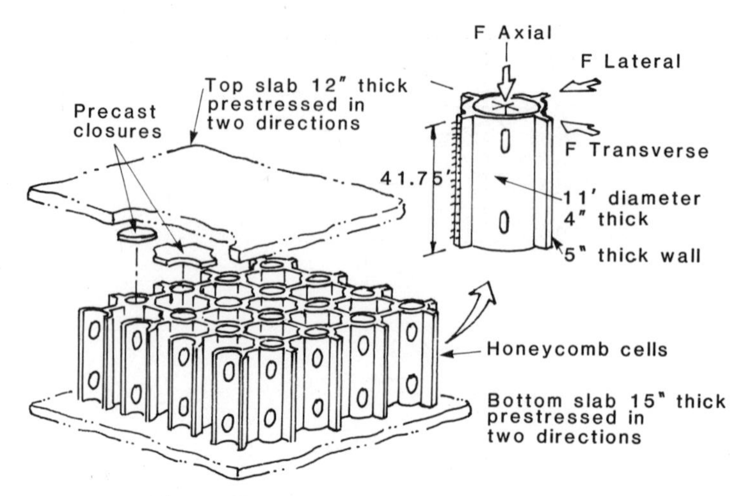

Figure 6.27 Concrete honeycomb design concept

encircling the platform at a distance of 200 feet can be built up in approximately 15 days.' (Wetmore and Borchardt, 1984).

6.5.2.5 *Design factors* Attention has been drawn to a number of particular factors which have to be considered carefully in design (Boyd and Bruce, 1984). There has to be an optimum balance between floating draught, foundation pressures and ballast volumes (discussed a little more below). Minimising draught and ballasting can be balanced by composite use of steel and concrete especially when used in the ice walls where inner and outer steel plates with steel diaphragms may be filled with concrete. The 'sandwich' type of construction, with a concrete mid-body and steel base and top deck, is claimed to have certain advantages:

(a) no solid ballast required; optimising ratio of concrete to steel weights gives low tow-in draught but adequate net foundation pressure;
(b) relatively light steel base permits use of small draught construction facility with concrete being poured afloat if necessary;
(c) concrete in ice wall and mezzanine deck gives structural and thermal efficiency;
(d) steel in upper deck facilitates installation and attachment of operating equipment and services.

(The sandwich arrangement has been proposed for the so-called SHADS design—Sonat Hybrid Arctic Drilling Structure.)

Although the stability and overall dimensions of ice resisting structures are usually governed by global ice forces, it is high local ice loads which determine the strength of the skin. For each element of the ice face the loaded area for maximum stress is found and the contact pressure is then found

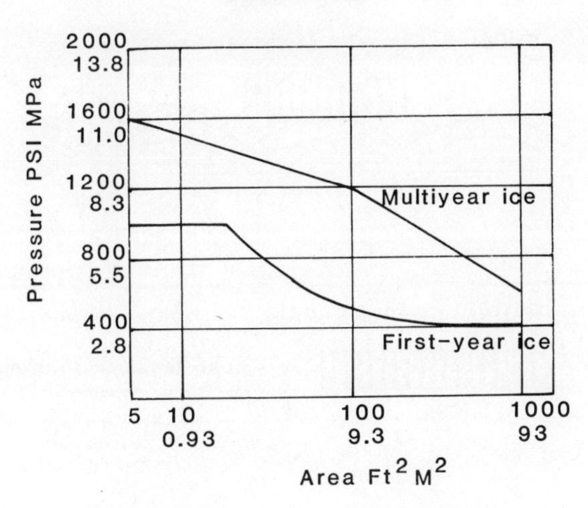

Figure 6.28 Pressure/area curve for ice loading

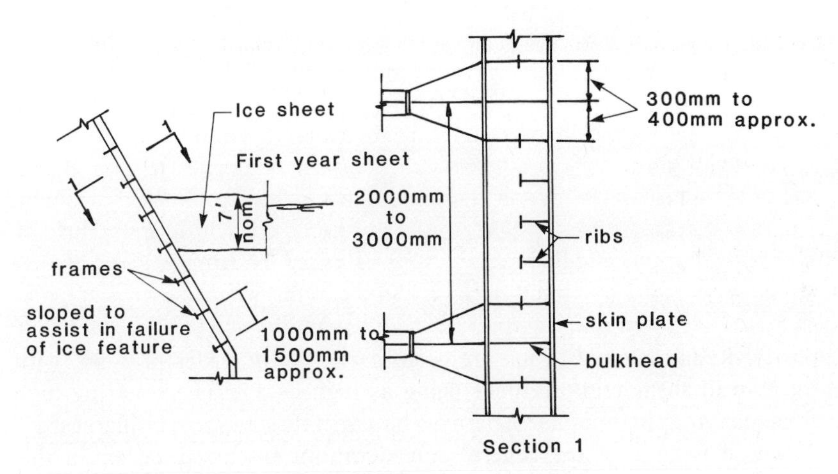

Figure 6.29 Typical steel ice wall

from a pressure/area curve such as that of Figure 6.28. Steel construction would require rib-stiffened plate spanning between frames, in turn spanning between bulkheads as in Figure 6.29. Concrete on the other hand, might have a shell or haunched slab element spanning between bulkheads spaced at 5–10 m centres as in Figure 6.30. Such considerations result in the table of pressures as detailed in Table 6.16.

The reduced ice load for concrete is claimed to give it an inherent advantage over steel.

Shear stresses are considered to be exceptionally high, with the thickness of wall being governed by punching shear, but in-plane compression through arch action reduces the bending and shear loads as well as increasing shear

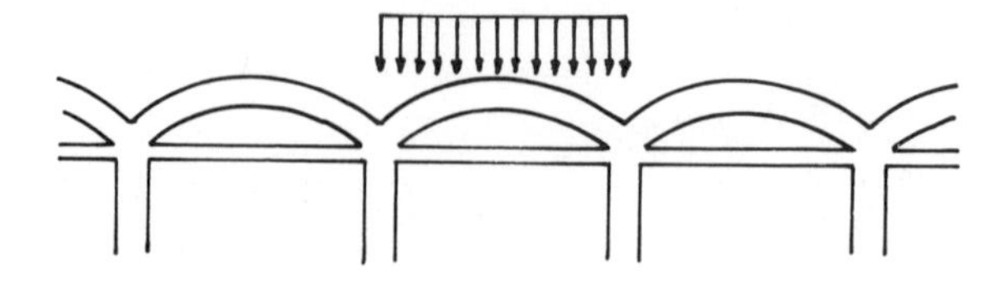

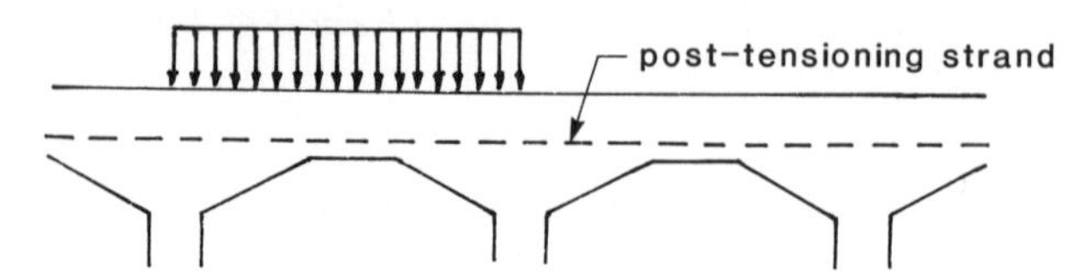

Figure 6.30 Typical concrete ice walls

Table 6.16 Ice pressure design figures for various forms of construction

Material	Technique	Loaded Area (m^2)	Local design ice pressure 1st Year Ice	Multi-Year Ice
Steel	Ribs	0.44	6.9	11.0
	Frames	3.13	5.4	9.3
Concrete	Shell	26.0	2.8	6.4
	Slab	35.0	2.8	5.9

capacity. Reinforcement ratios are customarily of the order of 5–6% with through-wall shear reinforcement being as high as 2%. The resulting high confinement may in time enable permissible concrete stresses to be increased.

Many, if not all, of the general considerations discussed earlier in this chapter apply also to Arctic structures but it has been pointed out (Watt, 1984) that the extrapolations from code approaches may not be strictly applicable to Arctic offshore problems because of their origins. 'Typical examples of this are provisions for conducting serviceability and fatigue checks'. Limit state design with probabilistic derivation of ice forces using simulation techniques is the approach favoured.

Before concluding this section with some comments on ballasting it ought to be remarked that in a design such as the one discussed above the load from the bulkheads has to be transferred somewhere within the structure. In this case load dissipation is achieved through horizontal diaphragms at top and bottom of the concrete mid-body (Steddum et al, 1985). 'The outer concrete shell provides excellent load distribution circumferentially around the structure when combined with a horizontal distribution member located

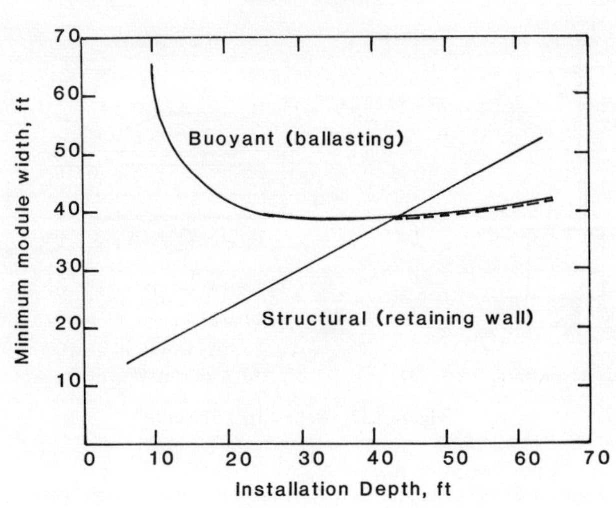

Figure 6.31 Minimum structural width and minimum 'buoyant' width for modules and caissons

at the upper surface of the midbody.' This simplifies the design and construction of the core area of the structure since it is then free of significant horizontal loads.

Returning to the caisson-retained island design; it has been observed (Buslov, 1985) that the normal width to height ratio of gravity walls of about 2:3 is significantly altered by the need for flotation during installation, with the minimum width required being at least doubled.. In fact the lower the draught needed the greater the width has to be. Simple analysis of the Tarsiut caissons produced the graphs of Figure 6.31 showing the minimum widths required for structural reasons and for buoyancy reasons. This indicates that it is only for depths beyond about 42 feet (12.8 m) that buoyancy widths need no longer be considered (the actual numbers/sizes of course apply to that particular design). The Tarsiut caissons were large and were floated in, but the argument presented in this case is that for shallow draughts, less concrete will be required if the walls are designed to structural rather than buoyancy limits. Hence it might pay to make them smaller still and modularise construction, installing the concrete modules with a floating crane.

Taking a contrary line, the same considerations could lead to the argument that the caissons should be widened until they meet from both sides. That is the mid-body should be a complete unit as in the SHADS design. Ballast would then be required and preference has been expressed for water (Boyd and Bruce, 1984) since it is more easily removed than solid ballast when the structure is moved to another location. Figure 6.32 demonstrates very simply the relative volume of ballast required merely to touch-down for a box-like structure compared with a tower platform. The hatched area indicates the additional volume of water required to land which is equal to the additional

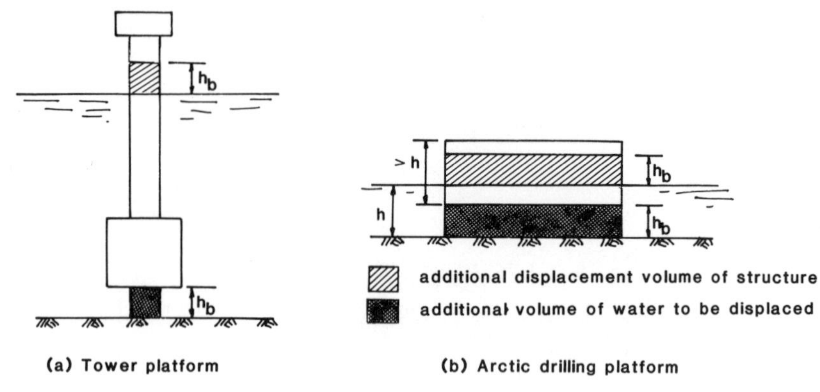

Figure 6.32 Ballast requirements

Table 6.17 Estimated environmental data used for conceptual design work in the Hibernia Field

	Authority	Operator
Water temperature	1°C	1.1–11.6°C
Air temperature	−30 to +30°C	−26 to +31°C
Wave height	25 m	28–35 m
Current	2 m/s	1–2 m/s
Sea ice—rafted	4 m	2–3 m
strength	1–12 N/mm^2	1.6–10.3 N/mm^2
Icebergs—mass	5–15 × 10^6 tonne	5–12 × 10^6 tonne
scour depth	4–8 m	9 m
velocity	1 m/s	0.4–1 m/s
no. crossing 47°N	1000–1500	1200

volume to be displaced (shown cross-hatched). Assuming similar freeboards for the two structures it is clear that a much greater relative capacity is needed for bedding down and stabilising in the case of the box-type structure. Obviously a host of other factors influence the problem and clearly much depends on the original weight of the structure (which equals the weight of water displaced) but this crude examination illustrates one of the major constraints on the designer.

6.5.2.6 *Iceberg infested waters* In the area offshore Newfoundland different conditions prevail, in that water depths are greater, and icebergs the major hazard, since the Hibernia field is on the north-east margin of the Grand Bank. The water depth is about 80 m. Table 6.17 details the environmental criteria for conceptual design estimated by the statutory authority and by the operator (McIntyre, 1981).

Three principal approaches to absorbing inertia energy of the drifting iceberg are shown in Figure 6.33 (Lundrigan and Lindgren, 1981). They are

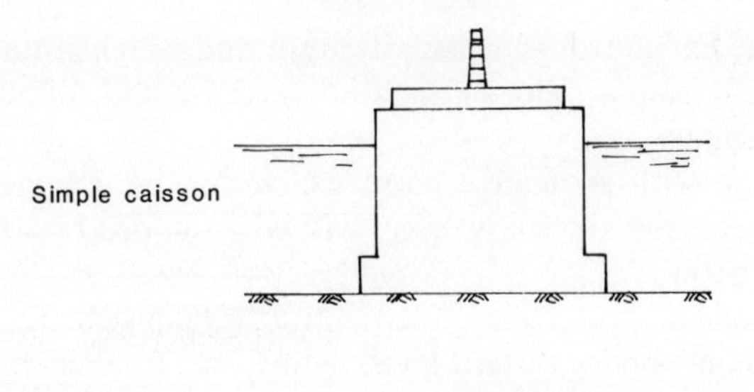

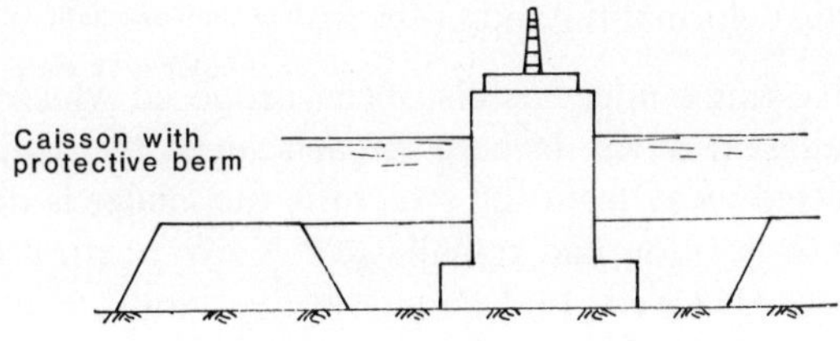

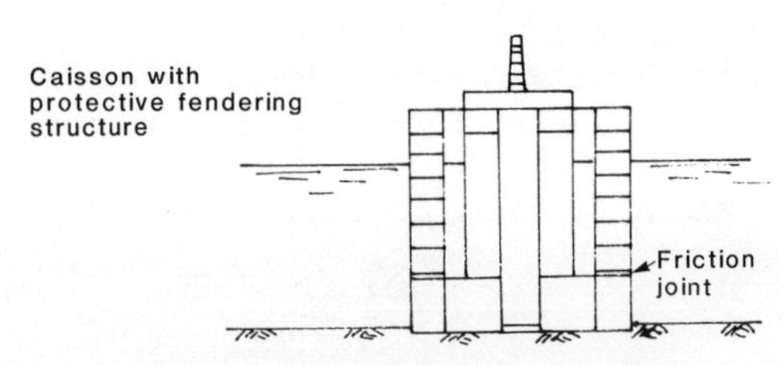

Figure 6.33 Caisson protection methods

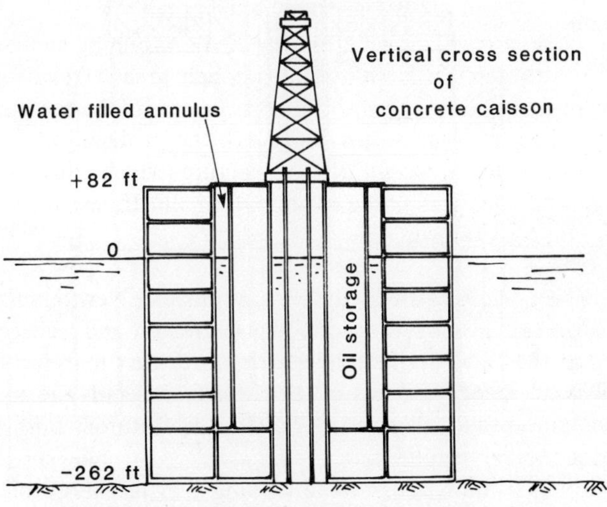

Figure 6.34 Vertical cross section of a concrete caisson

(a) a simple caisson of sufficient strength and weight to absorb iceberg energy by elastic deformations of structure and foundation and by ice crushing;
(b) a caisson with protective berm which can be designed for smaller iceberg impacts since large ones will be grounded on or slowed down by the berm;
(c) a caisson with protective fendering resting on a yielding joint activated by loads beyond a certain level; below that level energy is absorbed by elastic deformation and crushing.

An alternative ring fender has also been proposed which is on a sliding joint and is ballasted down. Iceberg impact causes the fender to slide, the iceberg is deflected away from the structure, the fender is de-ballasted and floated back into position and re-ballasted. A hypothetical cross-section is shown in Figure 6.34 (Weir, 1981); a possible variant is to have the exterior wall of the fender toothed like a cogwheel to facilitate failure of the ice. The 'manifold' design has also been proposed surrounded by a mound analogous to the berm mentioned above: Figure 6.35 (Jarlan, 1981).

Floating platforms have also been suggested which would necessitate disconnection from the oil/gas reservoir and movement off-location should

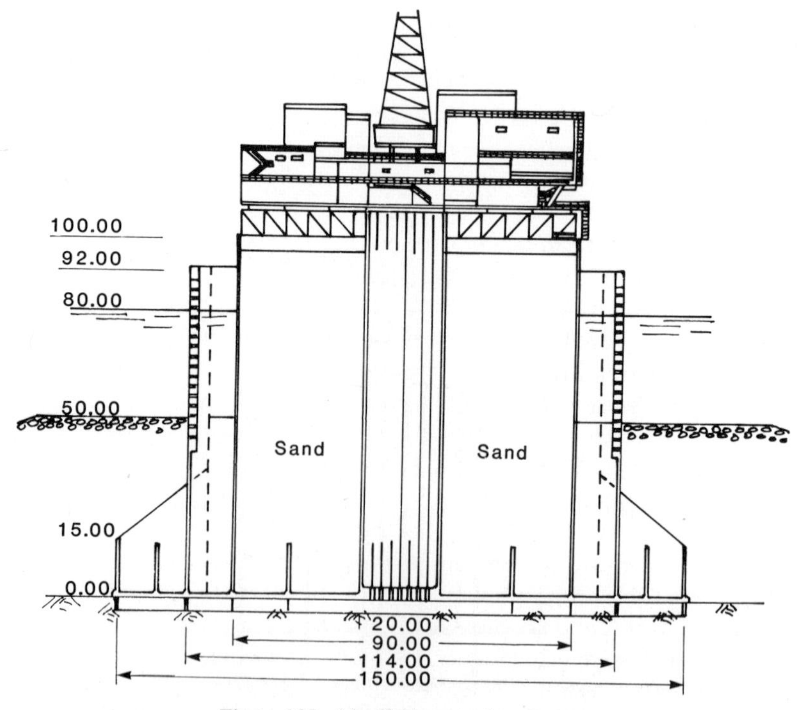

Figure 6.35 Manifold protection system

a major threat arise. Iceberg management is also a possibility in any of the protection programmes although anticipation and diversion or towing out of the way may be not entirely easy in performance.

Clearly Arctic and sub-Arctic operations offer considerable scope for ingenuity.

6.5.3 *Miscellaneous structures*

To use such a term is not to be dismissive of any, but is rather a convenience to keep the discussion within reasonable bounds. It might have been preferable to have a further grouping of 'coastal structures' but they cover such a range in themselves that the question remains arguable. A suggested classification even for them is (Slagter, 1978):

(a) constructions in ports such as piers, jetties, docks, locks, etc;
(b) constructions along the coast in the form of harbours and harbour protection works or communications e.g. breakwaters, harbour lights, bridges, tunnels;
(c) constructions needed to re-shape coastal areas which may be for reasons of safety or tidal power generation.

A parallel is drawn between the first category and riverine construction since both are in (relatively) sheltered waters and problems of construction are analogous. They will not be considered further here. The second category is discussed below but subdivided into sea-walls and breakwaters which are incorporated in this text as coastal structures while the others are lumped together as truly miscellaneous. First though some remarks on the third category as represented by an outstanding example.

6.5.3.1 *Oosterschelde storm surge barrier* In 1953 coastal inundation due to storm surge in the North Sea resulted in damage restoration and improvement works in the UK to a value of £25 million (at 1953 prices). That may seem severe but, in addition to greater damage costs, in the Netherlands nearly 2000 people were drowned and much livestock lost. Consequently the so-called Delta Project was implemented with its final stage the Oosterschelde (Eastern Scheldt) Storm Surge Barrier, recently completed. In essence this involved the construction of three dams 1520, 720 and 960 m long in a water depth of about 40 m with a tidal range of up to 4 m (Slagter, 1978 and 1982).

The dams comprise 66 concrete piers, 45 m apart with steel sliding gates (normally raised i.e. open) spanning between them. In addition there are concrete sill beams and concrete bridge beams for a roadway along the top of the barrier. Figure 6.36 (Slagter, 1978) shows the general arrangement of a pier with a cross-section through the special installation floating crane with its legs resting on the sea-bed: the pier is up to 18 000 t DWT, the sill or

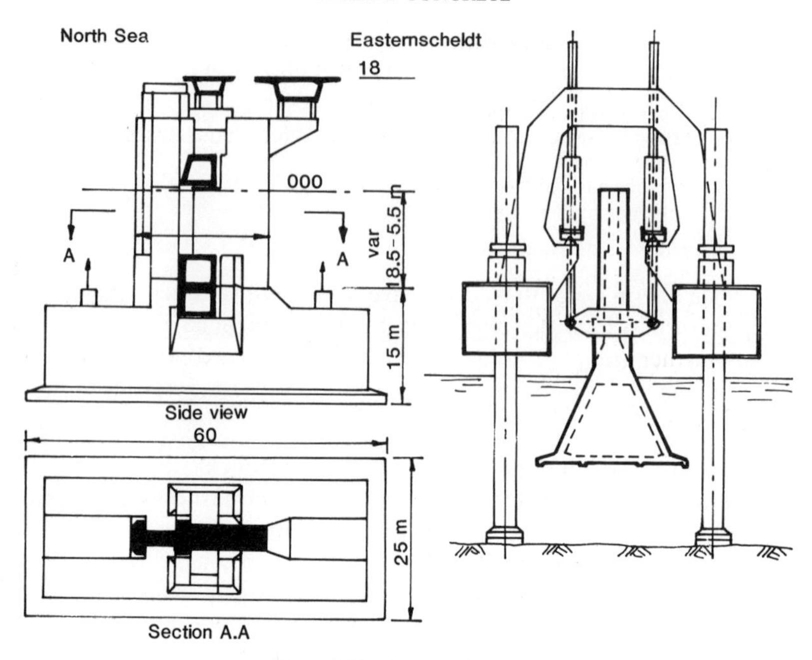

Figure 6.36 Oosterschelde piers

threshold beam 2500 t, the upper beam 1200 t and the bridge beam 1250 t. All these components were precast, the piers being built in dry dock then floated in on to a prefabricated mattress as foundation, the base being grouted after installation. Most of the concrete is prestressed, the piers both horizontally and vertically (de Graaf et al, 1978). The wave zone dimensions are minimised and the sluice gates placed as far forward as possible so that the piers project as little as possible. Rearward projection also is kept short to reduce the transverse stresses should a gate fail to operate or waves break over it. The sills (gaps between piers) were built up from various mattresses including articulated concrete blocks to form an embankment profile topped by the sill beams.

This is a very brief account of what is by any standard a major engineering achievement. Tidal flow regulation and stability of foundations were major factors in the design and therefore largely outside our present remit. A great deal of inspiration and ingenuity went into the construction with special lifting and mat-laying vessels having to be designed, described in summary elsewhere (Gerwick, 1986). It is a singular example of man's response to natural adversity coupled with concern for the environment and which cannot be done justice here.

6.5.3.2 *Coast protection works* Coastal protection works are seldom as spectacular as the Oosterschelde Storm Barrier but nevertheless in total may

Table 6.18 Cost breakdown: coastal protection works (UK)

Cliff protection, grading and drainage, with sea walls, groynes and beach replenishment	£5000/m
Sea walls, offshore barriers or sea embankments	£2000–£4000/m
Beach replenishment	£350–£950/m
Simple revetments and breastwork	£200–£400/m

represent considerable expenditure. To an island nation especially is this true: in the UK it was estimated in 1985 that it was the order of £50 m/annum (say $US 100 m) (ICE/MEG, 1985a) with a gross replacement value of at least £4 billion ($US 8 billion, say).

Cost of course has a significance far beyond that merely of the structures because, as the Dutch project demonstrates, the cost of *not* doing something may far exceed the initial outlay in human and financial loss.

Here we are concerned primarily with concrete but the same design parameters apply as to other structures. They include (Barrett, 1983) design life, tolerance to overtopping (related to return period of extreme tide levels), wave loading and dissipation, abrasion resistance, foundation conditions, ground and tidal water movement, cliff stability (where appropriate), and access for construction. Additional factors to be considered may include public access and safety, amenity use of the beach and appearance—structures may have to be environmentally and aesthetically sound but their engineering reliability cannot always be guaranteed: the worldwide cost of damage to existing breakwaters has been estimated at £250 m ($US 500 m) (ICE/MEG, 1985a).

6.5.3.3 *Sea walls* The term is used here rather broadly and collectively to include associated flood walls, revetments (or aprons) and so on. Figure 6.37 (Barrett, 1983) shows an arrangement for cliff protection using a sea wall with stepped apron and a flood-wall and decking behind in case of overtopping. The filled area accommodates possible cliff falls and the decked area provides access for plant. The apron is of reinforced construction founded on

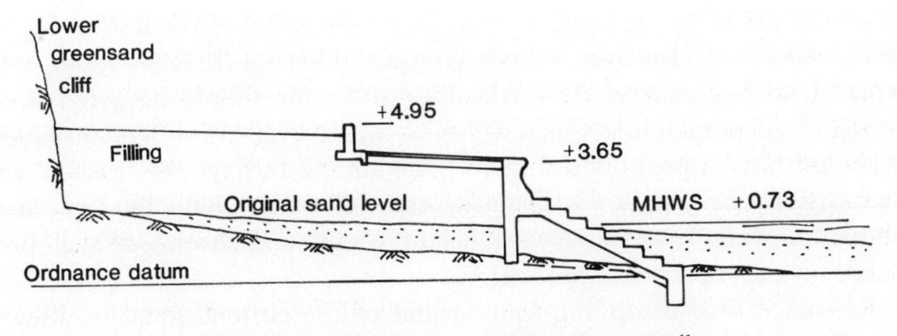

Figure 6.37 Sandown Bay—Littlestairs sea wall

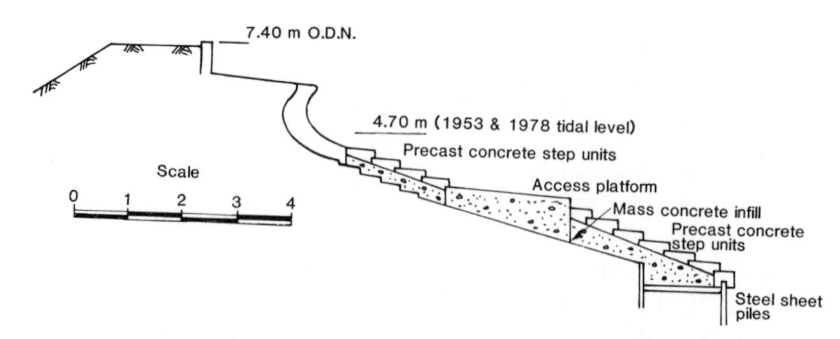

Figure 6.38 Mablethorpe to Huttoft typical stepwork scheme

sandstone. Another arrangement elsewhere is shown in Figure 6.38 (Taylor and Marsden, 1983).

It is scouring by heavy seas at the base or toe which is one of the most common causes of failure of vertical or near-vertical sea walls (ICE/MEG, 1985b) and so some form of protection is needed. Revetments or aprons provide this and are usually set at a slope <1 in 3, being smooth, rough, or stepped. They can be rigid or flexible slabs which are subject to wave impact, high velocities of uprush and backwash and often to abrasion from shingle picked up by the waves. It appears that where precast, interlocking blocks are used, the method of connection or interlock may be significant: the stability of tongue and groove blocks is much greater than that of shiplap blocks (CERC, 1984). The same source also quoted Dutch figures showing that roughness and porosity of the protective layer are important with runup being reduced by 50% if armour units with a 50% void ratio are used instead of a smooth impermeable surface. Even the use of precast blocks gives a small reduction over the smooth slabs.

(It is worth emphasising at this point that maximising the use of precast units helps to increase efficiency of construction by reducing the effects of interruption through inundation from tides and waves.)

There is a great deal of scope for varying block shape and hence apron roughness and porosity. The apron can be stepped as indeed may be the sea-wall which is then, perhaps, more a combination of wall and apron. Less flexibility exists for varying the profile (cross-section) of the wall which may be curved or flat. However, it has been stated (Owen, 1983) that design details which tend to throw water vertically into the air should be avoided since probable accompanying winds will blow water over the wall. Consequently it should have a pronounced throwback aiming to have the 'ejected' water as nearly horizontal as possible, with the wall being about the same height above still water level as the wave height. Figure 6.39 shows one such profile based on experience and model tests.

Reference to overtopping leads neatly to the current trend in design to avoid working on the assumption of a design storm which will give a height above which a wall will not be overtopped (Owen, 1983). There is always

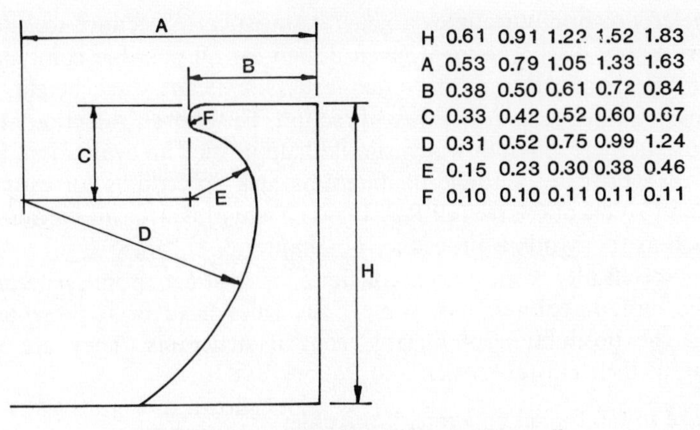

Figure 6.39 Typical form of wave return wall

some statistical chance that it **will** happen and so walls should be designed on the basis of the expected overtopping discharge. Values which should not be exceeded if damage is to be avoided are given in Tables 6.19 and 6.20 which are intended only as general guidelines:

Table 6.19 Threshold overtopping discharge for damage to seawalls

Type of seawall and construction	Threshold discharge $m^3/s/m$ run
Embankment type	
Crown and back slope, unprotected (e.g. clay, compacted soil, grassed)	2×10^{-3}
Crown protected, back slope unprotected	2×10^{-2}
Crown and back slope protected	5×10^{-2}
Revetment type	
Apron (promenade) unpaved	5×10^{-2}
Apron (promenade) paved	2×10^{-1}

Table 6.20 Threshold overtopping discharge for damage or injury to buildings, vehicles or personnel (may be increased by a factor of 10 at 10 m behind the sea wall).

Category	Threshold discharge $m^3/s/m$
For a person to walk immediately behind	
with little discomfort	4×10^{-6}
with little danger	3×10^{-5}
For a vehicle to pass immediately behind	
at high speed	1×10^{-6}
at low speed	2×10^{-5}
For a house located immediately behind	
to suffer no damage	1×10^{-6}
to suffer no structural damage although experiencing partial damage to windows and doors	3×10^{-5}

The reference briefly outlines a general statistical procedure to calculate the overtopping discharge for a given design for all possible combinations of the most significant factors: water level, significant wave height, mean zero-crossing wave period, and wave direction. The Shore Protection Manual gives a useful logic diagram which should help in such an evaluation: Figure 6.40. It also provides a range of diagrams and procedures for estimating run-up and so on of course (CERC, 1984). Wave pressure can be calculated from such as the Minikin procedure of Chapter 3.

More specifically concerning concrete, as distinct from general and hydraulic design requirements, some basic rules have been presented for ensuring the production of durable coastal structures. They are worth repeating in their entirety (Allen and Palmer, 1983):

(a) use unreinforced concrete, if possible;
(b) design to minimise the number of paths by which sea water can reach

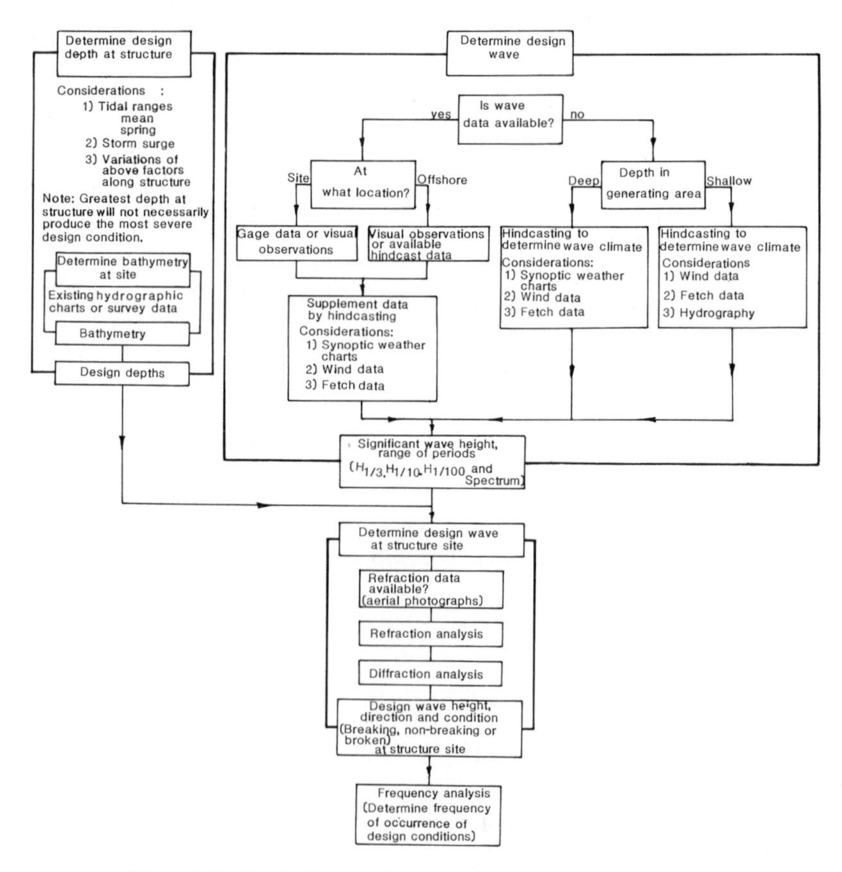

Figure 6.40 Logic diagram for evaluation of marine environment

the reinforcement e.g. use 'wall type' in preference to 'beam-type' structures;

(c) use a cementitious content in the concrete of at least 350 kg/m^3;
(d) reduce the water/cement ratio to 0.45 or less;
(e) provide adequate workability to assist full compaction; this often will need the use of workability aids or plasticisers;
(f) use air entrainment if a number of freeze/thaw cycles is expected during the life of the structure;
(g) provide site supervision which will ensure first class workmanship in mixing, placing, compaction and curing of the concrete;
(h) where abrasion is expected, use a coarse aggregate at least as hard as the abrasive material.

At a detail level, attention is also drawn to the need to minimise the number of, or to avoid if possible, movement joints and the use of surface sealants. Care with arrises is commended (as in much of construction) with the use of chamfers to 'blunt' them.

6.5.3.4 *Breakwaters* While seawalls are generally built on-shore, breakwaters are built into the water being either shore connected or offshore, although the latter term is a little misleading in a text which has included offshore structures of the type we have been considering. *In*-shore is perhaps a better term. They have been defined as 'wave energy barriers designed to protect any landform or water area behind them from the direct assault of the waves' (CERC, 1984). There are several types: rubble mound, composite, concrete caisson, sheet-piling cell, crib, or mobile. Here we need be concerned only with the first three and the last (although only fleetingly).

Figure 6.41 (CERC, 1984) shows a section through a Californian rubble mound breakwater, strictly speaking a composite type, since it has a concrete cap. However, on the seaward-side it has two layers of tetrapods—concrete blocks of a particular configuration—and it is with such as these and 'caissons' that we shall deal primarily. That is, the *armour* units in the case of rubble mound breakwaters and the *concrete barrier* type: breakwater design *per se* is not really within our scope. First then let us turn to the armour units.

Three categories have been suggested (Burcharth, 1984) illustrated in Figure 6.42:

(a) bulky types randomly placed (cubes, Antifer blocks);
(b) complex slender types randomly placed (Tetrapods, Dolosse—plural of Dolos);
(c) hollow blocks placed in patterns (Cob, Shed, Diode).

High permeability is desirable hydraulically. This is helped by random placing but that leads to large variations in contact forces and impact from

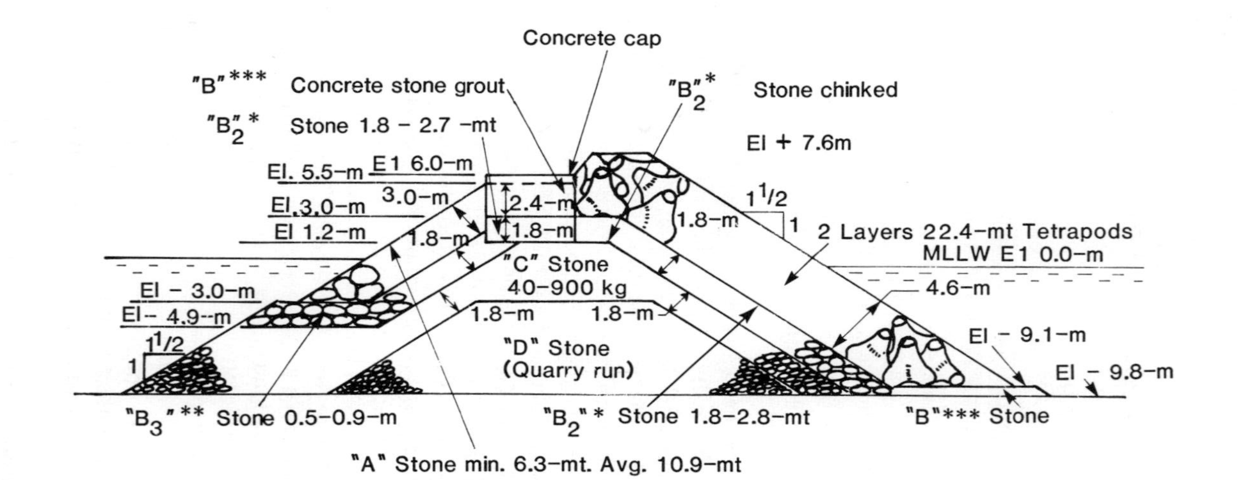

Figure 6.41 Tetrapod and rubble-mound breakwater, *"B_2"—0.9 mt variation to 6.3 mt max; **"B_3"—0.5 to 0.9 mt min, 6.3 mt max as available; ***"B"—0.9 to 6.3 mt or to start depth conditions of seaward toe.

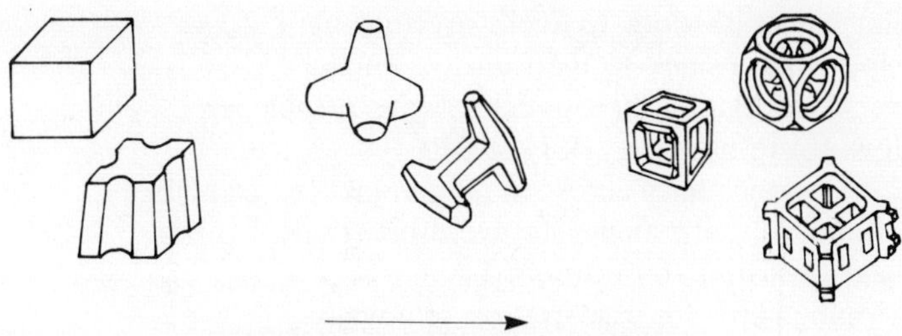

Figure 6.42 Categories of concrete armour units

movement. Hollow cubes and the like give high permeability but require even bedding and accurate placing in the dry which limits their usefulness for breakwaters. Slender, complex units have shown a proneness to damage but may have a resurgence in popularity if they can be designed adequately for deeper water loading conditions. Other factors too can affect construction: thermal stress from heat of hydration may cause problems, for example, since the units can be quite large—50 tons is by no means unusual with twice that weight not unknown. Reinforcement helps to reduce damage, of course, although increasing cost and requiring precaution against corrosion; fibre reinforcement has been employed also. A number of designs are proprietary.

The characteristic local conditions for any armour unit have been listed as (Ligteringen et al, 1986):

(a) casting and transport to the stockyard;
(b) transport to the breakwater and placing;
(c) in-situ static loads due to overlying units;
(d) hydraulic loads due to breaking waves, uprush and downrush;
(e) dynamic loads due to collision between rocking or displaced units.

(a) and (b) are well defined but (c) is random and is increased by settlement. Nevertheless it should be possible to estimate a worst condition according to the design of the unit (such as the corner of another resting on it midspan, with such additional dead loads as seem appropriate). (d) is better assessed from hydrodynamic considerations by taking appropriate combinations of wave conditions and unit orientation. In fact it may be transmitted wave load from adjacent units in contact which could be most critical. (e) might well be the worst loading condition but unfortunately is the most difficult to assess: the criteria for impact are relevant here but a balance of probability against cost is needed—careful selection of unit type might reduce uncertainty since the nature of loading is easier to define for some than others.

A general specification for the 'ideal' armour block has been suggested (Harlow, 1986):

(a) single layer capability–to avoid rocking and rolling;

(b) flat base with slots–to avoid overturning;
(c) slots on the sides–to reduce wave impacts;
(d) very irregular surface–to cause wave turbulence;
(e) low centre of gravity–for stability;
(f) high permeability across the base plane–to trap air;
(g) minimum sharp angles–to avoid breakage;
(h) ease in casting–for economy;
(j) sloping sides–for ease in form removal;
(k) a shape encouraging movement by sliding to promote settlement by 'working' into the slope.

Despite considerable recent criticism and doubts about its validity, the selection of weight of armour unit seems still to be based on the equation derived by Hudson (CERC, 1984):

$$W = \frac{w_r H^3}{K_D(S_r - 1)^3 \cot\theta} \tag{6.16}$$

W is the weight (newtons) of an individual unit in the primary cover layer; w_r is its unit weight (N/m^3) (saturated surface dry); H is the design wave height (m); S_r is the specific gravity of the unit (relative to *sea* water); θ is the slope of the structure (degrees); K_D is the stability coefficient which varies with a number of factors: the quoted range is from about 7 to 32 for the

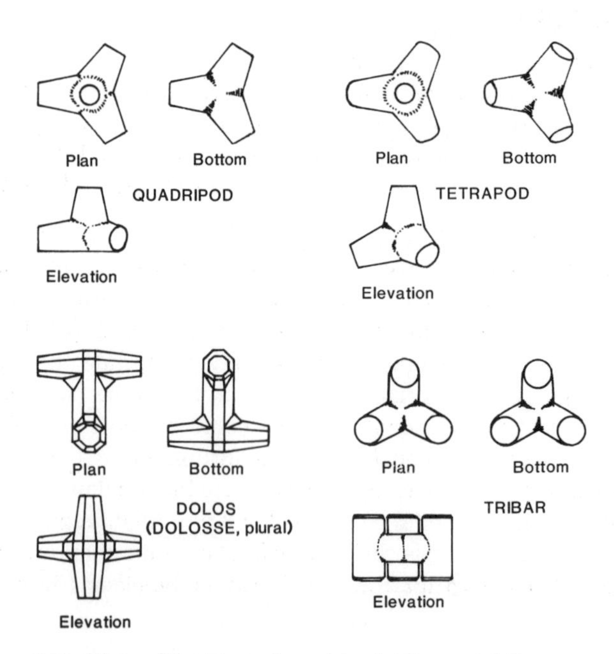

Figure 6.43 Views of the tetrapod, quadripod, tribar and dolos armor units

structure trunk and 4 to 16 for the head. The type/shape of unit affects the K_D value markedly.

It has been suggested (Clifford, 1984) that a low value of K_D might be used: half the advertised value perhaps. Alternatively the design wave height could be increased—the significant wave height H_s is too low and the maximum H_{max} too high: $H_{1/10}$ might be suitable. The ramifications and subtleties are too extensive to be included here and the various sources quoted should be consulted as a starting point.

Some other shapes of unit are shown in Figure 6.43 (CERC, 1984) but at least thirty types are known (HRS Wallchart).

6.5.3.5 *Caisson breakwaters* Caisson breakwaters lend themselves to precast construction although requiring a properly prepared bed and a reasonable weather window for installation. A typical cross-section is given in Figure 6.44 (Eddie, 1982) taken from a general review which provides several useful references. It is pointed out that PIANC (Permanent International Association of Navigation Congresses) 'recommends the rubble mound as generally the preferable form of breakwater because of its inherent ductility.' The consequences of overload or failure of a caisson structure are potentially more serious because of its nature and, presumably, because of the difficulties of repair. Rubble mound breakwaters absorb energy while the vertical caisson reflects it, creating conditions which can be difficult if in a navigation channel. As for other types of gravity structure, the caisson must be designed against overturning, sliding, and foundation failure with precautions against scouring being needed.

Since there is reflection from a vertical face wall the worst condition may not be from the supposed maximum probable design wave but from superposition of, or interference between, smaller incoming waves and their reflections. To alleviate such problems various modified designs have been proposed—the perforated breakwater, for example, of the Jarlan type. This absorbs energy in similar fashion to its application to oil production

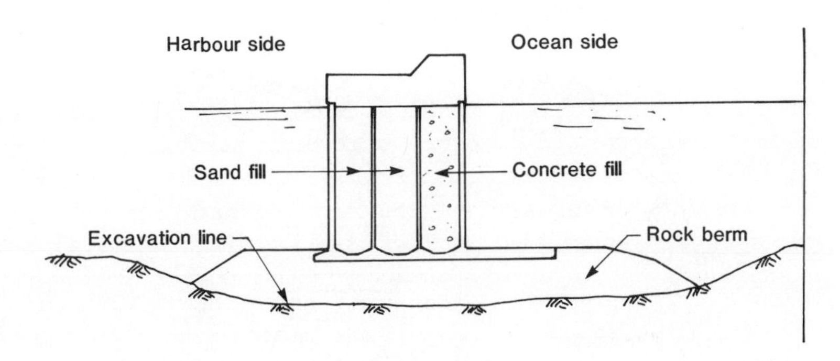

Figure 6.44 Typical caisson breakwater—Taranto Italy

platforms. Another type uses arc-shaped dissipating members as in Figure 6.45 (Kiyomiya and Yakota, 1985) with wave forces distributed as in Figure 6.46. Hydraulic model tests give the following approximate relationship for calculating the wave forces P_w

$$P_w = P_A \cos^2 \theta + P_B \sin^2 \theta \tag{6.17}$$

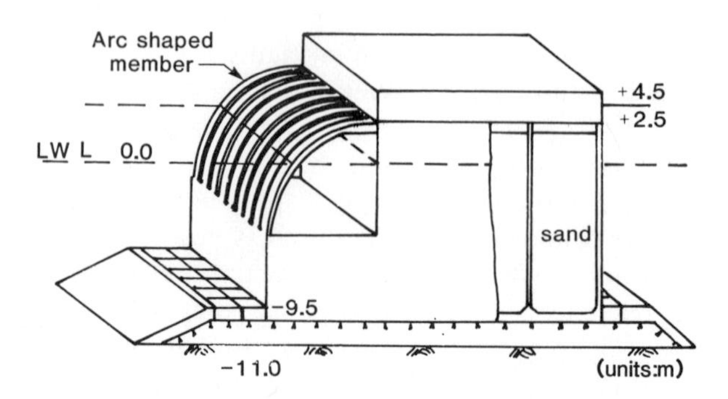

Figure 6.45 A breakwater with arc-shaped wave dissipating members

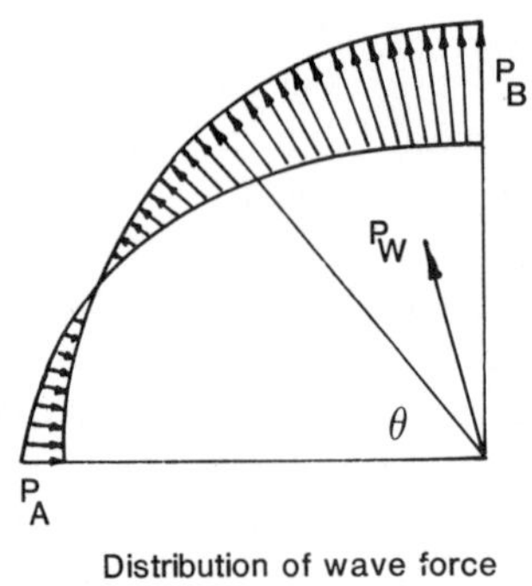

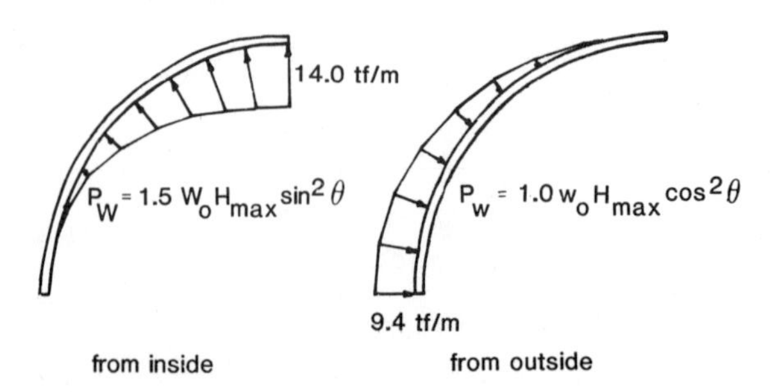

Figure 6.46 Design wave forces for the prototype arc-shaped member

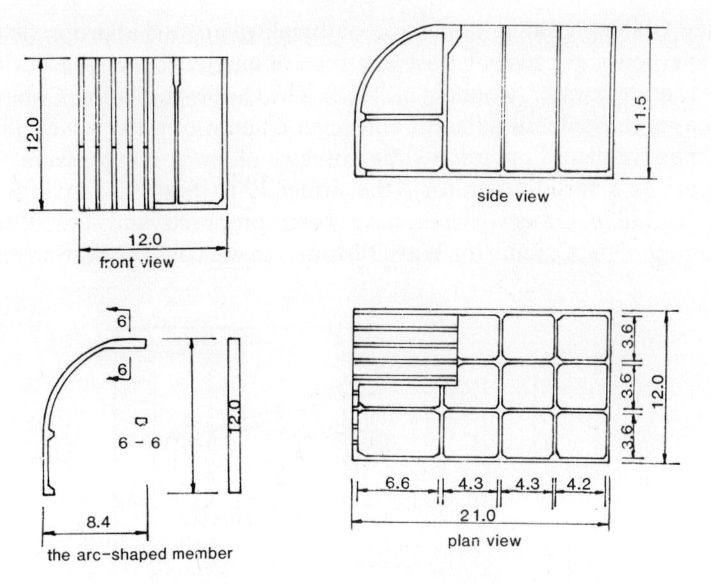

Figure 6.47 Slit caisson breakwater

with the two design conditions being

$$\begin{aligned}&\text{(i)}\quad P_A = 1.0wH, \quad P_B = 0 \quad \text{from the outside}\\&\text{(ii)}\quad P_A = 0, \quad P_B = -1.5wH \quad \text{from the inside}\end{aligned} \tag{6.18}$$

Stresses act on the arc-shaped members alternately from the outside and from the inside and it is therefore preferable to construct them from prestressed concrete, although the caisson may be reinforced of course. The layout of an experimental prototype is shown in Figure 6.47 (Ohtsuki, 1982) from which it can be seen that the slits are one-third of the width of the arc members. This type of design has been considered suitable for water depths of 60 m or so.

Whether or not caissons should be placed on rockfill berms (rubble mounds) seems to be a moot point. The apparent saving in concrete may be offset by the need for greater structural resistance for *breaking* waves since the mound may tend to induce them. Greater uncertainty of estimation may also be introduced thereby.

Sloping front breakwaters have also been adopted with the sloping face starting at about mean sea-level. Compared with a vertical faced type the load can be reduced by one half to two thirds.

6.5.4 *Other structures*

It is perhaps fitting that this chapter should conclude with one or two interesting examples of the variety of structures indicated at its beginning.

In view of the preceding discussion on breakwaters it is appropriate to turn to wave power because at least one type of energy conversion device is in essence a breakwater (Count et al, 1983). The Oscillating Water Column uses the waves to excite an adjacent entrained column of water which then acts as a massive piston pumping large volumes of air at low pressure. The air then drives a turbo-generator—the principle is illustrated in Figure 6.48 (Clarke, 1982). Other devices have been proposed and tested but the advantage of the Column (or Wave Piston) is that it could be bottom mounted

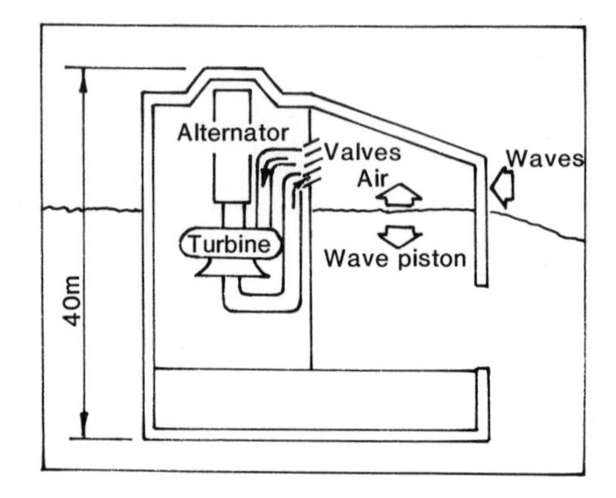

Figure 6.48 Wave piston

Figure 6.49 Oscillating water column/wave piston generator

in (relatively) shallow water and is fundamentally very robust. A variant of this principle is being tested in Norway with a small plant set into a rock face as in Figure 6.49 (Bønke et al, 1986). The oscillating chamber is 10 m square and the total height 20 m producing around 500 kW—judicious hydraulic design of the entrance 'harbour' increases the potential generation capacity.

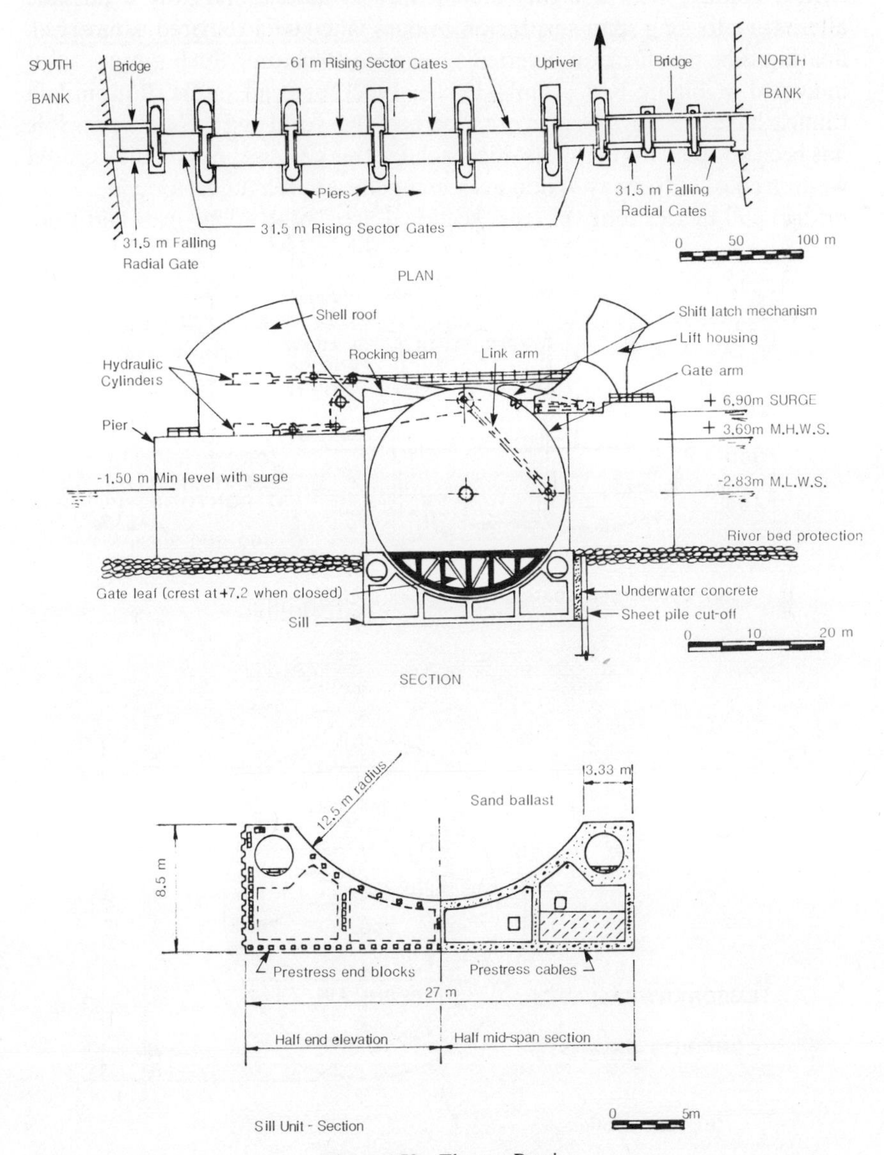

Figure 6.50 Thames Barrier

A totally different type of structure is the Thames Flood Barrier in the UK. As for the Oosterschelde Barrier, the objective is to guard against the effects of storm surge and to provide protection for Greater London. Figure 6.50 (Mitchell and Wainwright, 1983) shows the general arrangement with the piers and abutments being of mass concrete with local reinforcement. The sill units were precast of prestressed and reinforced concrete.

In a country with a highly indented coast such as Norway a possible alternative to long-span suspension bridges is to use a tethered submerged, floating tube (in effect, an inverted suspended structure). Such a proposal is indicated in Figure 6.51 (Tambs-Lyche, 1982) for road traffic. Bottom-laid tunnels have been proposed in other parts of the world and the same principle has been adopted for the shore approach for two gas pipelines on the exposed western coast of Norway. A flooded rectangular-section tunnel (or underwater bridge) 590 metres long, of cross sectional area 30 to 45 m^2, was laid from

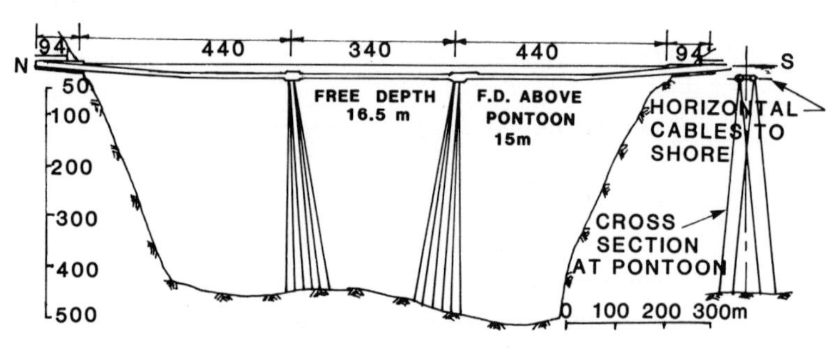

(a) Submerged floating tube across Eidfjord

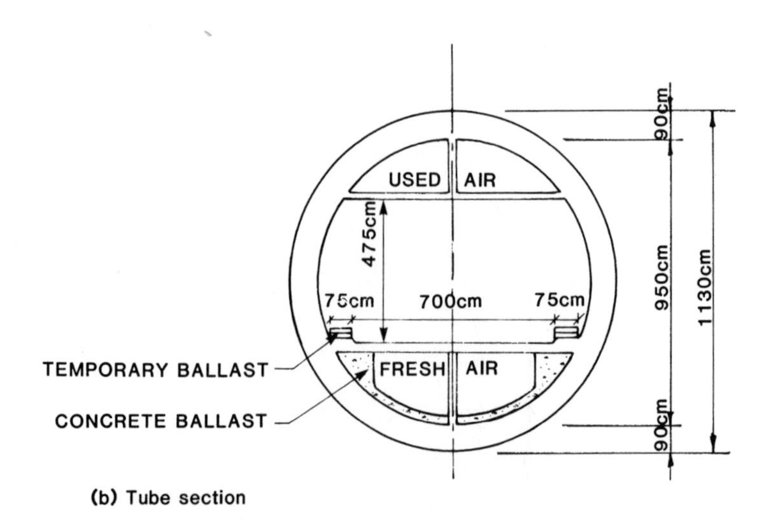

(b) Tube section

Figure 6.51 Floating tube bridge

a depth of 30 m to beach water level (Einstabland, 1986). The tunnel comprised 5 precast, prestressed elements with displacements up to 7000 tons.

There are numerous other structures which could be mentioned but since bridges have been referred to perhaps it would be appropriate to conclude with one in an area which, although not subject to severe sea states, does have other problems which have been alluded to, such as corrosion and high temperatures.

The Saudi Arabia—Bahrain Causeway is 25 km long including five bridges totalling 12.5 km, individual lengths being between 950 m and 5150 m (van Tongeren, 1986). Each bridge has two separate prestressed box girder superstructures giving two traffic and one emergency lane in each direction. They rest on piers as in Figure 6.52 founded on precast hollow piles 3.5 m diameter and up to 40 m long placed generally in pre-drilled boreholes but in some cases on pre-installed caissons. Some (higher) bridges have a cross-beam just below the water surface as shown while all have cross-beams at the pier heads. Because of potential durability problems Portland blast furnace cement was used and clearly prefabrication was a major requirement. The heaviest elements weighed up to 1350 tons and in all some 350000 t of concrete was used with nearly 60000 t of reinforcing and prestressing steel.

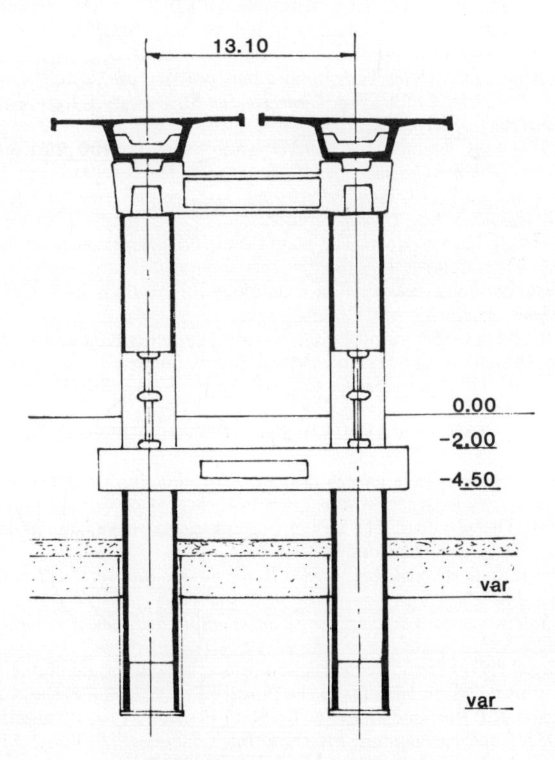

Figure 6.52 Bahrain Causeway piers

Hopefully this demonstrates that the term 'marine structures' has a wide definition and may be extended to those not at first associated immediately with the sea. It also clearly demonstrates that, while the sea provided the obstacle, it also provided the means by which the obstacle could be overcome.

References

Abrahamsen, E. (1976) Safety requirements for offshore engineering. *Proc. Int. Conf. Behaviour of Offshore Structures.* Norges Tekniske Høgskole, Trondheim.

Allen, R.T.L. and Palmer D. (1983) Concrete in coast-protection works. *Proc. Conf. Shoreline Protection.* Thos. Telford, London.

ASTEO (1987) *Risk Analysis for Offshore Structures and Equipment.* Graham and Trotman (for the Comm. of Euro Comms.), London.

Baird, W.F. and Hall, K.R. (1984) The design of armour systems for the protection of rubble mound breakwaters. *Breakwaters, Design and Construction, Proc. ICE Conf.* Thos. Telford, London.

Barrett, M.G. (1983) Coast-protection design: the overall scheme. *Proc. Conf. Shoreline Protection.* Thos. Telford, London.

Bjelmrot, S. and Maltegård, G. (1981) Offshore risk and insurance. *Proc. Conf. Offshore Göteborg* **81**.

Bomhard, H. (1988) *Concrete structures as a preventive answer to environmental catastrophes, FIP Notes 3*: 15–19

Bønke, K., Einstabland, T., Olsen, T.O., Berg, Nic (1982) Wave power project in Norway: Kvaerner Brug's multiresonant oscillating water column—MOWC. *Nordisk Betong*, **1–2**: 43–45.

Boyd, A.D. and Bruce, J.C. (1984) The promise and practice of concrete construction in ice infested waters. *Proc. FIP/CPCI Symp. Concrete Sea Structures in Arctic Regions, Canadian Prestressed Concrete Inst.* Ottawa.

Browne, R.D. (1986) Low maintenance concrete-specification versus practice? *Proc. 2nd Int. Conf. Maintenance of Maritime and Offshore Structures.* Thos. Telford, London.

Brunton, J.D. and Smith, J.R. (1972) *Offshore storage systems—a review for UK industry, Report UR3, CIRIA Underwater Eng. Group.* London.

Burcharth, H.F. (1984) Theme paper: The way ahead. *Breakwaters, Design and Construction. Proc. ICE Conf.* Thos. Telford, London.

Buslov, V.M. (1985) Modular construction technology for Arctic islands. *Civil Engineering in the Arctic Offshore, Arctic '85.* ASCE, New York.

Carneiro, F.L.L.B. (1979) Codes for offshore structures, design criteria and safety requirements. *Proc. Int. Conf. Offshore Structures Engineering, Brazil.* Pentech Press, London.

CEB-FIP (1978) *CEB-FIP Model Code for Concrete Structures, CEB-FIP Int. Recommendations 3rd edition*, CEB, Paris.

CERC (Coastal Eng. Rsch. Center) (1984) *Shore Protection Manual*, Dept. of the Army, Washington.

Clarke, J.L. (1982) Concrete wave energy devices. *Proc. 9th. Int. Congress, FIP, Vol. 1*, FIP, Slough.

Clifford, J.E. (1984) Theme paper: The design process. *Breakwaters, Design and Construction, Proc. ICE Conf.* Thos. Telford, London.

Count, B.M., Fry, R. and Haskell, J.H. (1983) *Wave power: the story so far, CEGB Research No. 15.* Nov: 13–14.

Dahl, H. (1981) *Safety standard settings based on scientific and human criteria, Proc. Offshore Göteborg.*

Doris, C.G. do Brasil (1979) Design, construction principles and setting of one type of concrete gravity platform installed on oil fields in the North Sea. *Offshore Structures Engineering* (ed. F.L.L.B. Carneiro, A.J. Ferrante and C.A. Brebbia) Pentech Press, Plymouth.

Eddie, A.G.J. (1982) Concrete caissons. *Proc. 9th. Int. Congress FIP, Vol. 2.* FIP, Slough.

Eide, O., Andresen, A., Jonsrud, R. and Andenaes, E. (1982) Reduction of pore water pressure beneath concrete gravity platforms. *Proc. Int. Conf. Behaviour of Offshore Structures, Vol. 2.* (ed. C. Chryssostomidis and J.J. Connor) Hemisphere, Washington.

Eide, O., Kjekstad, O. and Brylawski, E. (1979) Installation of concrete gravity structures in the North Sea. *Marine Geotechnology* **3**(4): 315–368.

Einstabland, T. (1986) High Strength concrete: Underwater concrete bridge for the shore approach at Karmøy, Norway. *Nordisk Betong,* **1–2**: 55–58.

Engel, D.J. and Pajouhi, K.K. (1985) Arctic drilling and production platform concept. *Civil Engineering in the Arctic Offshore, Arctic '85, ASCE.* New York.

Eriksen, K. (1982) Design and construction of offshore platforms in Norway. *Nordisk Betong,* **2–4**: 82–86.

Eriksson, K. (1979) FIP Commission on Concrete Sea Structures: Risk Analysis. *FIP Notes* **80**, May/June: 18–23.

Faulds, E.C., Philip, M.G. and Neffgen, J.M. (1983) Operating experience and repair case histories of five offshore concrete structures. *Design in Offshore Structures.* Thos. Telford, London.

Faulkner, D. (1984) Reliability of offshore structures. *SERC Marine Tech. Seminar, Science and Eng. Rsch. Council.* London.

FIP (Commission on Concrete Sea Structures) (1985) *Design and Construction of Concrete Sea Structures, 4th edn.* Thos. Telford, London.

Fitzpatrick, J. (1984) Tarsiut concrete caissons. *Proc. FIP/CPCI Symp. Concrete Sea Structures in Arctic Regions, Canadian Prestressed Concrete Inst.* Ottawa.

Fjeld, S. (1983) Structures in ice covered waters. *Cold Regions Science & Technology,* **7**, 201–216 (Proc. Comité Arctique Int. Conf. on Arctic Energy Resources).

Forrest, J.C.M. (1986) The whole structure *Proc. Int. Conf. Design of Structures against Fire.* (ed. R.D. Anchor, H.L. Malhotra and J.A. Purkiss) Elsevier, App. Sc. Pubs., London.

Furnes, O. (1978a) *Fixed offshore platforms—design, safety and performance, Det norske Veritas.* Oslo.

Furnes, O. (1978b) Instability of plane and curved concrete walls in the design of gravity offshore platforms. *Proc. Euro Offshore Pet. Conf.* SPE (UK), London.

Furnes, O. (1979) *Certification of offshore structures, WEGEMT Second Graduate School, Det norske Veritas.* Oslo.

Gerwick, B.C. (1975) *Prestressed Concrete Ocean Structures & Ships, Prestressed Conc. Inst.* Chicago.

Gerwick, B.C. (1976) The future of offshore concrete structures. *Proc. Int. Conf. Behaviour of Offshore Structures.* Norges Tekniske Høgskole, Trondheim.

Gerwick, B.C. (1986) *Construction of Offshore Structures.* Wiley, New York.

Green, C.H. and Brown, R. (1978) *Life safety: what is it and how much is it worth? Current Paper CP52/78, Building Rsch. Estab.* (*Fire Rsch. Stn.*) Borehamwood.

de Graff, F.F.M., Koster, J.P. and Winden, M.v.d. (1981) *Prestressing steel and systems used at the piers of the storm surge barrier in the Eastern Scheldt in the Netherlands, FIP Notes* **91**. March/April.

Halliwell, L., Oakes, F. and Slater, D.H. (1979) Hazards, risks and social responsibilities. *Consulting Engineer, June:* 75–80.

Harlow, E.H. (1986) Closing address; Break waters. *Developments in Breakwaters. Proc. ICE Conf.* Thos. Telford, London.

Hawkins, N.M. and Shah, S.P. (1982) American Concrete Institute considerations for fatigue. *Proc. IABSE Coll. Fatigue of Steel and Concrete Structures.* ETH-Hönggerberg, Zürich.

Høland, I. (1976) Ultimate capacity of concrete shells. *Proc. Int. Conf. Behaviour of Offshore Concrete Structures.* Norges Tekniske Høgskole, Trondheim.

HRS (Hydraulics Research Station) (undated) *Breakwater Armour Blocks* (*Wallchart*), Wallingford.

ICE/MEG (1985a) Research Requirements in Coastal Engineering. *A report prepared by the Coastal Engineering Research Panel and Working Parties, Inst. of Civil Eng.* London.

ICE/MEG (1985b) *Coastal Engineering Research.* Thos. Telford, London.

Jakobsen, B., Eikenes, A. and Olsen, T.O. (1987) Recent development and potentials for high-strength offshore concrete platforms. *Proc. Symp. Utilisation of High Strength Concrete.* Tapir, Trondheim.

Jarlan, G.E. (1981) A prestressed concrete fixed drilling and production platform for the Hibernia oilfield development. *Proc. Symp. Production and Transportation Systems for the Hibernia*

Discovery. Petroleum Directorate, Govt. Newfoundland & Labrador, St. John's Newfoundland.

JCSS (Inter-Association Joint Committee for Structural Safety) (1978) Common Unified Rules for Different Types of Construction and Material. *CEB-FIP Int. Recommendations 3rd. edn.* CEB, Paris.

Jefferies, M.G., Stewart, H.R., Thomson, R.A.A. and Rogers, B.T. (1985) Molikpaq development at Tarsiut P-45. *Civil Engineering in the Arctic Offshore, Arctic '85.* ASCE, New York.

Jenssen, D.N. (1977) The Condeep gravity platform. *Proc. Symp. Safety of Offshore Structures. Eurotest, Int. Scientific Assocn.* Brussels.

Kakuta, Y., Okamura, H. and Kohno, M. (1982) New concepts for concrete fatigue design procedures in Japan. *Proc. IABSE Coll. Fatigue of Steel and Concrete Structures.* ETH-Hönggerberg, Zürich.

Kiyomiya, O. and Yokota, H. (1985) Non linear properties of arch-shaped concrete members. *Report of the Port & Harbour Rsch. Inst., Ministry of Transport.* Japan, **24(1)**, March.

Knecht, H.I., Stubbs, S.B. and Erb, P.R. (1979) Floating caisson vessel system for oil production in deep water. *Proc. 11th Annual Offshore Tech. Conf.* Houston.

Kjekstad, O. and Stub, F. (1978) Installation of the Elf TCP2 Condeep platform at the Frigg Field. *Proc. Euro. Offshore Pet. Conf., Vol 1.* SPE (UK), London.

Larsen, L.G. and Fosker, T. (1979) Brent platform B, design and installation. *Shell Brent 'B' Instrumentation Project, Soc. for Underwater Tech.* London.

Lee, G.C. (1982) Design and construction of deep water jacket platforms. *Proc. Int. Conf. Behaviour of Offshore Structures.* (ed. C. Chryssostomidis and J.J. Connor). Hemisphere, Washington.

Leivestad, S., Vik, B. and Ekeberg, P.K. (1987) The utilisation of high strength concrete: a survey of international codes and regulations. *Proc. Conf. Utilisation of High Strength Concrete.* Tapir, Trondheim.

Ligteringen, H., Mol, A. and Groeneveld, R.L. (1986) Criteria and procedures for the structural design of concrete armour units. *Developments in Breakwaters, Proc. ICE Conf.* Thos. Telford, London.

Lundrigan, H. and Lindgren, J. (1981) Potential concrete structures for Hibernia. *Proc. symp. Production & Transportation Systems for the Hibernia Discovery.* Petroleum Directorate, Govt. Newfoundland and Labrador, St. John's Newfoundland.

McIntyre, N.F. Environmental forces on a fixed platform and the ability of the platform to resist them, ibid.

Mast, R.F., Cichanski, W.J. and Magura, D.D. (1985) Cost-effective Arctic concrete structures. *Civil Eng. in the Arctic Offshore* (*Proc. Conf. Arctic '85*) ASCE.

Maxwell-Cook, P.V. (ed.) (1973) Concrete Sea Structures. *Proc. FIP Symp.* Tbilisi, FIP, London.

Mitchell, A.O. and Wainwright, D.A. (1983) Prestressed concrete sill units for Thames Barrier Project rising sector gates. *FIP Notes 3*: 18–23.

Moe, J. (1981) Technological development and cost uncertainties—as seen from the 'outside'—based on investigations on cost development in the North Sea. *Proc. Conf. Offshore Göteberg 81.*

Moses, F. (1976) Reliability of structural systems. *Proc. Int. Conf. Behaviour of Offshore Structures.* Norges Tekniske Høgskole, Trondheim.

Norwegian Contractors (1987) *NC News No. 1, June* Norwegian Contractors, Stabekk.

Ohtsuki, N. (1982) Curved slit caisson breakwaters. *Proc. 9th. Int. Congress FIP, Vol. 1,* FIP, Slough.

Olbjørn, E.H. and Foss, I. (1980) Certification of new concepts. *Proc. Oceanology International,* Brighton.

Olsen, O. (1978) Design of Condeep. *Concrete Structures, Proc. Inge Lyse Symp.* Tapir, Trondheim.

Owen, M.W. (1983) The hydraulic design of sea-wall profiles. *Proc. Conf. Shoreline Protection.* Thos. Telford, London.

Parat, J.L. (1980) Préfabrication lourde—une application originale à la construction de la platforme centrale du champ de Ninian. *Proc. Coll. Int. Tenue des Ouvrages en Béton en Mer.* CNEXO, Brest.

Parat, J.L. and Maher, D.R.H. (1978) Ninian field concrete gravity platform. *Proc. Euro. Offshore Pet. Conf.* SPE (UK), London.

Price, W.I.J., Tricklebank, A.H., and Hambly, E.C. (1982) Fatigue considerations in the design of conrete offshore structures. *Proc. IABSE Coll. Fatigue of Steel and Concrete Structures.* ETH-Hönggerberg, Zürich.

Ridley, T. (1978) Foundations of concrete gravity structures in the North Sea. *Proc. 8th. FIP Congress, Part 1, Cement & Conc. Assocn.* London.

Rowe, W.D. (1983) Acceptable levels of risk for technological undertakings. *Proc. Symp. Ship collision with Bridges & Offshore Structures*, IABSE.

Slagter, J.C. (1978) Concrete constructions in coastal areas and the problems involved. *Proc. 8th Cong. FIP, Cement & Conc. Assocn.* London.

Slagter, J.C. (1982) The Delta project, Keynote speech. *Proc. 9th Cong. FIP.* FIP, Slough.

Stahl, B. (1986) Reliability engineering and risk analysis. *Planning and Design of Fixed Offshore Platforms.* (ed. B. McClelland and M.D. Reifel) Van Nostrand, Reinhold, New York.

Steddum, R.E., Hershey, E.E. and Bruce, J.C. (1985) SHADS: Mobile Arctic drilling platform. *Civil Engineering in the Arctic Offshore, Arctic '85.* ASCE, New York.

Subra, F. (1980) Analyse comparative des différentes recommendations et règles en usage pour la realisation de structures en haute mer. *Proc. Coll. Int. sur la Tenue des Ouvrages en Béton en Mer, Actes de Coll.* no. **11**. CNEXO, Brest.

Tambs-Lyche, P. (1982) Submerged floating concrete tube across a Norwegian fjord. *Nordisk Betong*, **2–4**: 172–174.

Taylor, H.R. and Marsden, A.E. (1983) Some sea defence works in Eastern England. *Proc. Conf. Shoreline Protection.* Thos. Telford, London.

Thielen, G. (1982) Reflections on the presentation of fatigue in design codes. *Proc. IABSE Coll. Fatigue of Steel & Concrete Structures.* ETH-Hönggerberg, Zürich.

van Tongeren, H. (1986) The Saudi Arabia-Bahrain Causeway: large bridge construction between Saudi Arabia and Bahrain. *FIP Notes*, **4**: 7–11.

Tricklebank, A.H., Price, W.I.J. and Hambly, E.C. (1982) Implications of fatigue in the design of some concrete offshore structures. *Proc. Int. Conf. Behaviour of Offshore Structures, Vol. 2.* (ed. C. Chryssostomidis and J.J. Connor) Hemisphere, Washington.

UEG (1982) *Applications for Concrete Offshore. Report UR20, CIRIA-UEG.* London.

UEG (1984) *A management approach to weight engineering, Report UR24, CIRIA-UEG.* London.

Uherkovich, I. and Lindgren, J. (1983) Friction and elongation tests at Condeep site, Stavanger, Norway. *FIP Notes* **3**: 14–17.

Veritas, Det norske (1981) Design against accidental loads. *Tech. Note Fixed Offshore Installations, TNA 101.* Oslo.

Veritas, Det norske (1981) Impact loads from boats. *Tech. Note Fixed Offshore Installations, TNA 202.* Oslo.

Waagaard, K. (1982) Veritas' design recommendation for offshore concrete structures—background for special requirements for offshore structures. *Nordisk Betong*, **2–3**: 117–119.

Waagaard, K. (1982) Design recommendations for offshore concrete structures. *Proc. IABSE Coll. Fatigue of Steel and Concrete Structures.* ETH-Hönggerberg, Zürich.

Wang, J.L. and Peters, D.B. (1985) Overview of artificial island design and construction in the Arctic. *Civil Engineering in the Arctic Offshore, Proc. Conf. Arctic '85.* ASCE, New York.

Watt, B.J. (1984) Ice load considerations for concrete structures. *Proc. Arctic Offshore Tech. Conf.* Calgary.

Weir, F.V. (1981) The comparative environmental risks associated with fixed platforms and floating platforms and with tankers and pipelines. *Proc. Symp. Production and Transportation Systems for the Hibernia Discovery*, Petroleum Directorate, Govt. Newfoundland and Labrador, St. John's Newfoundland.

Wetmore, S.B. and Borchardt, D.R. (1984) The design, construction and deployment of a concrete island drilling system, the Glomar Beaufort Sea 1. *Proc. Arctic. Offshore Tech. Conf.* Calgary.

Yee, A. A. (1975) Design considerations and configurations: the Yee concept. *Proc. Conf. Concrete Ships and Floating Structures.* (ed. B.C. Gerwick) Univ. Calif. Berkeley.

Yee, A.A., Masuda, F.R., Kim, C.N. and Doi, D.A. (1984) Concrete module for the Global Marine concrete island drilling system. *Proc. FIP/CPCI Symp. Concrete Sea Structures in Arctic Regions, Canadian Prestressed Concrete Inst.* Ottawa.

7 Other matters

The problem with a book such as this lies as much in deciding what to omit as what to include. Consequently, concluding with a chapter devoted to 'other matters' represents an opportunity to touch on important aspects which may have been barely referred to previously. Two in particular ought to be discussed a little further although to some extent that is again a personal preference. They concern floating structures and inspection, maintenance and repair, the last three topics being commonly unified.

7.1 Floating structures

That there has been considerable interest in floating concrete is evidenced by the conferences devoted to the subject over the years as well as the papers and proposals elsewhere. Four conferences in particular set the tone: in Tbilisi (1972), California (1975), London (1977) and Rotterdam (1979). Additionally, the classification societies have formulated rules or proposed criteria and, most recently, the American Concrete Institute has produced a state-of-the-art report on barge-like concrete structures with many useful references. There is thus a substantial body of literature extant covering many aspects of the subject, ranging from direct experience to imaginative concepts for liquefied refrigerated gas storage and ocean thermal energy conversion. While there seems little doubt that successful economic application is feasible, there is evidence to suggest that the experience of users has not always been as happy as some protagonists would have us believe and it is also clear that the innovative steps proposed in a number of cases have been too great for the financial risk takers. However, a simple technical analysis does indicate, albeit in a fairly elementary way, that floating concrete is becoming more competitive with other materials. In view of the depressed state of the world ship-building industry, it is not likely to compete successfully in conventional markets where there is over-provision of capacity. On the other hand, for specialised applications where the material has potential *technical* advantages, opportunities may well develop.

7.1.1 *Floatability of concrete*

It is by no means unusual to encounter scepticism about the floatability of concrete, despite the many examples of concrete structures and components moved by water. This arises from the perceived massiveness of the structures and it might therefore be worthwhile outlining a fairly simple but possibly indicative analysis (Marshall, 1988).

Consider unit length of an immersed horizontal cylinder sufficiently long to allow the ends to be ignored. For an internal diameter d, wall thickness t and unit weight γ.

$$\text{displacement} = \frac{\pi}{4}(d + 2t)^2\gamma_w \tag{7.1}$$

and

$$\text{dead weight} = \pi(d + t)t\gamma$$

where γ_w is the unit weight for water, so that the specific gravity of the concrete $s = \gamma/\gamma_w$. Let $r = d/t$. Then, for the (empty) cylinder to float,

$$\text{dead weight} \leqslant \text{displacement}$$

or

$$\pi(d + t)t\gamma \leqslant \frac{\pi}{4}(d + 2t)^2\gamma_w$$

which leads to

$$r \geqslant 2(s - 1)\left[1 + \sqrt{\frac{s}{s - 1}}\right] \tag{7.2}$$

Assuming that, for concrete, the range of values of s is 1.5 (lightweight) to 2.5 (normal) then Table 7.1 can be compiled:

Table 7.1 Floatability; the variation in ratio internal diameter: wall thickness with changing specific gravity

Material	s	r
concrete	1.50	2.73
	1.75	3.79
	2.00	4.83
	2.25	5.85
	2.50	6.87
Steel	7.8	28.17

For practical purposes this approximates to a straight line relationship with minor error:

$$r = 4.02s - 3.22 \tag{7.3}$$

[As a rule of thumb, for a cylinder to float, the ratio of internal diameter d to wall thickness t must then be at least (4 × specific gravity minus 3.] Assuming that the load on the cylinder wall is due to hydrostatic pressure p, thin cylinder theory gives the wall stress σ from

$$p(d+2t)=2\sigma t$$

i.e.

$$r=2\left(\frac{\sigma}{p}-1\right)$$

or

$$r\approx\frac{2\sigma}{p}\quad\text{for }\sigma\gg p \tag{7.4}$$

and $p \approx 0.1\ \text{N/mm}^2$ at 10 m depth.

Hence at a nominal depth of 20 m, $r = 10\sigma$ and at 50 m, $r = 4\sigma$.

7.1.2 *Flotation efficiency*

This hypothetical floating cylinder has to support two loads; its own weight and all surcharge loads (or 'cargo' plus dead weights from additional parts of the structure, deck, equipment, and so on). Its flotation efficiency index F_e might then be defined by the ratio of 'additional' or 'cargo' weight to self weight (i.e. dead weight), where 'cargo' weight is the difference between displacement and dead weight.

$$F_e=\frac{\pi/4(d+2t)^2\gamma_w-\pi(d+t)t\gamma}{\pi(d+t)t\gamma}$$

(Strictly speaking the displacement should also allow for any displaced volume between the cylinder and the water surface but that is being neglected for present purposes). This then leads to

$$F_e=\frac{r'}{s}-1 \tag{7.5}$$

where

$$r'=\frac{(r+2)^2}{4(r+1)}$$

Substituting $r = 10\sigma$ and 4σ from above and approximating slightly, it will be found that r' is proportional to σ and so from eq. (7.5) it is clear that F_e is a simple linear function of σ/s which is the so-called specific strength. In other words, the flotation efficiency is proportional to the specific strength (or strength to weight ratio). Substituting appropriate values enables compilation of Table 7.2.

The criteria adopted in this analysis are very limited and many other factors have to be considered but the table indicates that high strength, lightweight concrete, on this basis at least, is competitive with mild steel.

Table 7.2 Variation in flotation index F_e with construction material

Material and Strengths		Index F_e at Nominal pressure head 20 m	50 m
Normal weight concrete ($s = 2.5$)	35 N/mm^2	34	13
	50 N/mm^2	49	19
Lightweight concrete ($s = 1.8$)	35 N/mm^2	47.6	18.4
	50 N/mm^2	68.4	26.8
	60 N/mm^2	82.3	32.3
Steel ($s = 7.8$) yield strength	200 N/mm^2	63.1	24.6
	250 N/mm^2	79.1	31.1

That is, it is just as 'floatable'. Moreover, while such concretes may not be commonplace, they are now certainly frequently manufactured on a production scale so that attainability is not in question.

Cost is an important consideration, of course, and just because concrete is, or may be, as floatable as steel does not mean that it is either cheaper or dearer; that will be decided by circumstance.

In some situations it may be fruitful also to consider the reciprocal of the flotation index which can then be regarded as a measure of the weight of structure required to support unit weight of cargo. As stated previously, the stability and motion response of floating structures are outside the scope of this book. However, it may be noted that they govern the configuration and general arrangement, including depth of immersion, of such structures. 'Light' cargoes require relatively heavier structures (with appropriate ballast capacity) to float at the hydrodynamically most suitable depth, LNG (liquefied natural gas) being a conspicuous example with a specific gravity of around 0.45. In this case, at least for a twin-hull semi-submersible, it has been observed (Marshall, 1988) that substantial ballast would be required for an appropriate steel vessel as compared with one made from concrete. In short, the concrete is a more efficient carrier of the light-weight cargo than steel, as one might perhaps expect from examining the inverted index.

7.1.3 *Structures afloat*

The scope for floating concrete is too great to attempt review in a few paragraphs but it may suffice to illustrate it with two 'extreme' applications, one intended for the open ocean at substantial depths and the other, very close to shore. Between, there is a plethora of uses such as tethered buoyant platforms for oil production, floating docks, barges for industrial plants, and so on.

OTEC (Ocean Thermal Energy Conversion) has been mentioned already, the objective being to operate a thermal power station based on the temperature difference between the warm tropical surface water layer (about

100 m thick) at around 30°C and the water at 800–1000 m depth, a few degrees above freezing. For this a pipe is required to bring the cold water up to the surface providing a major structural engineering challenge. The design problem is outlined (Maris and Paulling, 1980) as involving a pipe 1000 m long, from 1.5 to 30 m diameter according to plant capacity, suspended from a floating structure in the open sea, exposed to dynamic excitation from wave action on the platform as well as the upper portion of the pipe itself. It cannot be retracted easily so must be able to withstand the worst storm conditions and be durable, requiring no maintenance in effect. Ideally the structural material to be used must be flexible but strong enough to accommodate the inevitable flexural loading, and it must have a low density to reduce axial loads but be sufficiently heavy to restrict motion. Various possibilities have been and are being studied but prestressed concrete appears

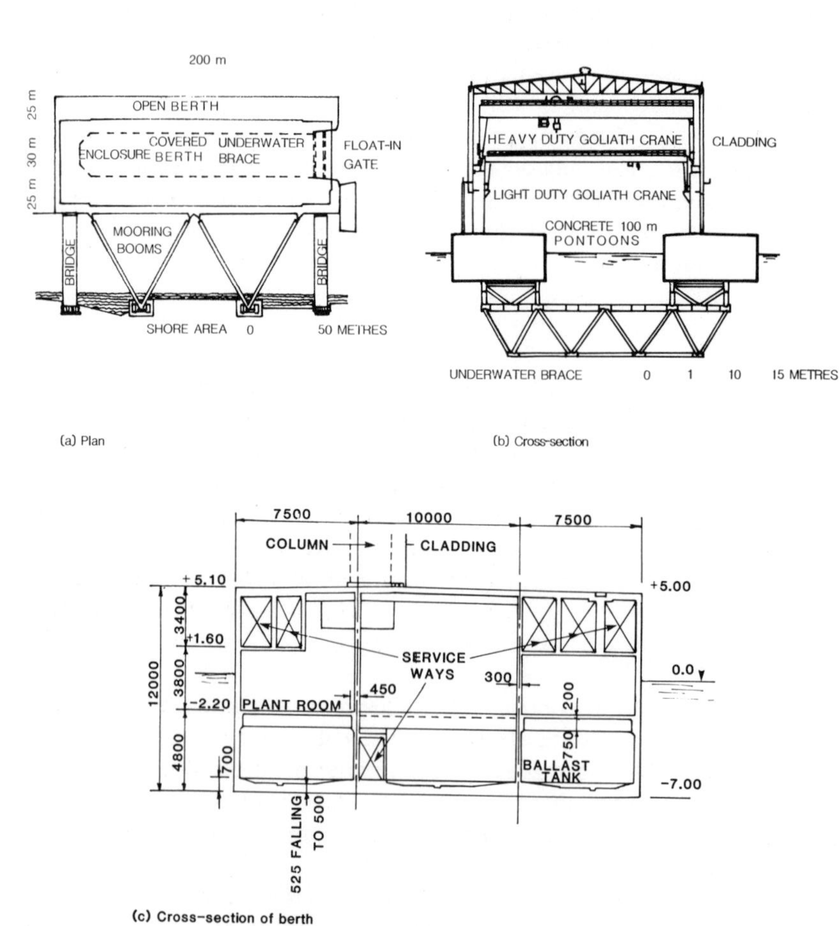

Figure 7.1 Twin pontoon construction

to possess most of the required attributes—its strength, stability and fatigue- and corrosion-resistant properties should enable production of a pipe which will last for the life of the plant. Of course its construction would be a major logistical exercise but if slip forming can be used to build large structures upwards from a fixed base it is surely not unreasonable to visualise a structure 'extruded' downwards in similar fashion, step-wise if necessary.

At the other 'extreme', a floating berth has been described (Warmington et al, 1988) which is only 60 m or so from the shore. The water depth and bed-rock slope preclude piling so the twin pontoon construction of Figure 7.1 was chosen. Of course the pontoons are quite substantial structures in themselves, being each 200 m × 25 m with a draft of 7 m, freeboard of 5 m, and displacement of 70 000T. The floating jetty is permanently moored by articulated tubular steel booms and provided with steel box girder bridges, the maximum tidal range being about 4 m. While the structure might, in most ways, now be described as fairly conventional, it is nevertheless an imaginative solution to a problem presenting many practical difficulties had the more usual fixed configuration been adopted.

In demonstrating the potential for floating concrete it is interesting also to note that these two concepts are based upon the imaginative use of well-proven technology: the major advance lies in ingenious adaptation; in evolution rather than revolution.

7.2 Inspection, maintenance and repair

Although it is convenient in discussion to separate these three topics, it is natural and logical to bracket them together. For example, inspection has little point unless there is an intention to take appropriate remedial action should it be found necessary. Similarly, the need for maintenance or repair cannot be properly assessed without prior inspection. Moreover, this alone is capable of a wide variety of interpretations ranging from the statutory requirements of hydrocarbon production structures to the infrequent glance over, say, a small jetty or quayside. It is indeed arguable that many more structures ought to be subject to more rigorous and more frequent inspections.

An Operator (of an oil production platform) must protect his personnel, protect his assets, and ensure continuity of production; part of the assurance of which is that 'the installation's structure is in a sound condition and fit for continued operation.' (Faulds, 1982). For this it is necessary to carry out periodic inspections. Generally there is also a statutory obligation to meet minimum requirements for the safety, health and welfare of personnel on board the installation. The *duty* or *responsibility* for safety, however, rests entirely upon the Operator, so that although guidance may be provided, it is for the Operator to determine and implement appropriate procedures and practice. Arising from this procedures have been developed for the inspection of offshore structures which might well provide a useful model for other structures, albeit in most cases rather less exacting. It is therefore worth while examining the practice in more detail.

7.2.1 *Structural inspection*

Structural inspections can be subdivided into four general categories (Henn and Spencer, 1984):

(a) inspection of new construction and repairs;
(b) periodic inspections for continued service;
(c) inspection factors related to types of materials;
(d) casualty inspections.

Although derived primarily from experience with ships they may be applied to any type of structure. Categories (a) and (d) are familiar enough, bearing in mind that (a) applies *during* and immediately after construction, but the other two may require some further explanation. For offshore structures it is often a statutory requirement that they have a (limited) annual inspection and a 'full' inspection every five years, so that their Certificate of Fitness (or Approval of Construction) can be renewed. Apart from that, however, a planned maintenance programme would normally demand regular inspection as a pre-requisite. Category (c) implies specialist knowledge on the part of personnel carrying out the inspection. For example a crack in steel has different connotations from a crack in concrete and the inspector should be aware of its potential significance in whichever case.

Certification of structures is by no means new, of course, it has been a feature of the operation of major dams for a good number of years, for example. However, offshore structures are typically subject to a more severe environment and unremitting dynamic loading. Experience is also rather more limited. Coupled with the problems (and expense) of access, these factors underline the necessity for a systematic approach, particularly in view of the size of some of the structures involved. Certain aspects indeed have to be incorporated or catered for at the design stage and it may even be found in some cases that it is cheaper to provide structural redundancy than to provide for inspection.

7.2.1.1 *Condition records* What has been called by one certification agency a Structure Condition Record (SCR) has to be established at the outset and updated on the basis of regular inspection during operation (Sletten et al, 1977). It is based on a so-called DFI-résumé (design, fabrication and installation résumé) which in essence is the as-built record of the structure. The design résumé would contain, for example, the design criteria and specifications and would identify critical or significant areas such as the location of stress concentrations, fatigue hot-spots, areas particularly subject to exposure and potential degradation, instrumentation, and so on. Included in the fabrication résumé would be quality control records, *actual* dimensions, areas of questionable workmanship, repairs, location of steel components fixed to the concrete structure, commissioning of instrumentation. The installation résumé would provide the *actual* location of the structure and records from towing and installation, including the foundations.

Although the terminology may vary from one operator or certification authority to another, the general practice is likely to be similar since the objective should be the same: safe operation. An important point to bear in mind is that the operational and inspection staff are unlikely to be the same people, or to be employed by the same organisations, as those responsible for design and installation: in some cases they may be scattered across several continents, in many they will be from more than one country. Continuity of contact is therefore doubtful and so it is essential that the structure's owner have adequate information if proper monitoring is to be achieved.

Since the structures to be inspected generally are large, frequent detailed inspection of their entirety is unrealistic and so representative areas may be selected in addition to the special ones indicated above. However, if such areas are chosen in advance of construction (so that their *actual* construction is incorporated in the record) it would seem advisable not to notify the construction personnel until after the event. Human nature being what it is, preselection can imply special attention which then results in non-representativeness. The primary aim here is, or should be, for objective observation in the interests of safety and its sustenance rather than the maintenance of standards which is achievable in other ways.

7.2.1.2 *Inspection levels* Significant or critical areas and what to look for are indicated in Table 7.3 (Sletten et al, 1977). Clearly this order of inspection needs to be very detailed and, where access and conditions are difficult, will be expensive. However, by definition, the same standard is not required everywhere and typically three classes or levels have been adopted as outlined in Table 7.4 (Browne et al, 1981). Noteworthy are the references to visual records and non-destructive testing and the sub-division into underwater and atmospheric inspection.

7.2.1.3 *Fault description* Obviously underwater inspection is a great deal more demanding in resources than that above the surface although the fact that part of a structure is in the atmosphere does not necessarily mean that it is easy to reach. Where divers are involved it means that their skills probably lie elsewhere than in concrete inspection and their observations have to be in a form which can be interpreted by engineers, for example. Photographic and video records are invaluable but an important additional aid has been produced by the UK Department of Energy (1985), subsequently

Table 7.3 Groups of significant areas and what to look for on concrete structures

What to look for \ Groups of significant areas	Coatings	Fixing plates and flange connections	Repaired areas	Connections deck/shafts	Areas subject to abrasion	Areas of inferior construction	Areas of stress concentration	Zone subject to wetting and drying generally	Construction joints	Scour protection	Instrumentation
Cracks											
Corrosion				×	×	×	×	×	×		
Distress				×			×				
Shrinkage			×			×		×			
Spalling											
Corrosion		×	×	×	×	×	×	×	×		
Distress		×		×	×		×	×			
Freezing and thawing								×			
Saltcrystallisation						×		×	×		
Inferior concrete	×		×		×	×		×	×		
Abrasion	×	×			×			×		×	
Deteriorated concrete											
Leaching of $CaOH_2$		×			×			×	×		
Efflorescence								×	×		
Freezing and thawing								×			
Chemical deterioration	×		×					×	×		
Corrosion of reinforcement											
Rust leaching		×	×	×		×		×	×		
Foundation										×	×

Table 7.4 Standard of inspection adopted for different areas of inspection

	Class 1	Class 2	Class 3
Standard	High standard of detailed inspection on a small number of specified critical areas	High standard of general inspection on wide surface within some specified areas	Good standard of general inspection covering most of the surface such that only massive structural damage will be detected
Objective	Detection of cracks > 0.5 mm of other defects > 20 mm (higher standards are sometimes required in critical areas)	Detectlon of cracks > 5 mm of other defects > 50 mm	
Underwater inspection	Cleaning and artificial lighting	Cleaning if necessary artificial lighting	No cleaning
	General video record of the area and surrounding areas	100% video survey at low speed,	General artificial lighting
	100% still photography	Still photography if needed	General visual observation with systematic record
	NDT if necessary	No NDT except possibly electropotential measurements	Short but systematic report
	Detailed report	Detailed report	
Atmospheric inspection	Cleaning if necessary	No cleaning	
	Artificial lighting	Artificial lighting if necessary	
	Complete visual inspection of the areas and surrounding areas (binoculars if necessary)	Complete visual inspection (binoculars if necessary)	
	100% still photography	Still photography if necessary	
	NDT if necessary	No NDT	
	Detailed report	Detailed report	

extended. In essence this is a photographic catalogue of 'blemishes' (which in fact include defects such as cracks and spalling). Transparent overlays identify various characteristics of the particular fault so that the photograph itself remains unobstructed. It is accompanied by a brief verbal description and an outline of probable causes and an index of key words is also provided to aid in identification and description. Perhaps a fair anology would be the rogues' gallery familiar from film and TV crime series.

7.2.1.4 *Zoning* Instead of or as well as divers, remotely operated vehicles (ROV's) may be employed, equipped with appropriate lighting and cameras. However, whatever survey system is adopted, it must be complemented by a suitable system for determining location. The structure may be zoned and subzoned, with four levels of division being adopted in one particular case (Browne et al, 1981):

1. primary zones representing different environmental exposures,
2. primary zone members of similar structural configuration,
3. individual structural members of the secondary zones,
4. members gridded or subdivided into areas $\not> 50\,m^2$.
   (Division 3 zones have been used for Class 2 inspections, details being plotted on 1:125 scale proformas (stylised location drawings) and Division 4 zones for Class 1 inspections using 1:50 scale proformas.)

A permanent reference system on the structure itself should be provided to enable the location of specific areas to be determined.

7.2.1.5 *Inspection frequency* The normal period for renewal of a Certificate of Approval is 5 years but that does not mean that the major inspection should take place every fifth year. It appears to be customary to spread the load by having a full inspection of 20% of the structure each year or, preferably, 25% in each of four years leaving the fifth for any necessary remedial work following an evaluation of the condition of the whole platform. Yet another pattern is to inspect 50% in year 2 and the remaining 50% in year 4: in essence, the specific time schedule depends upon the individual operator.

7.2.1.6 *Additional provision* It is fairly obvious that the requirements of inspection will have an impact upon design and a list of recommendations has been drawn up (CXJB, 1978). A number have been brought out in points made above but several are connected with the provision of a diving system as an integral part of the structure. The central platform of the Ekofisk field in the North Sea is a good example of this where a diving bell controlled by a 'van' on a monorail was installed (Bullington and Loper, 1977; Djortouian, 1980).

7.2.1.7 *Planning and evaluation* Emphasis has been given to the need for a coherent philosophy on which to base an inspection plan (Faulds, 1982). Where major extrapolations have been made from past experience, inspections should continue until the designs have been confirmed as sound. Only then can vigilance be relaxed, bearing in mind that it is stability of

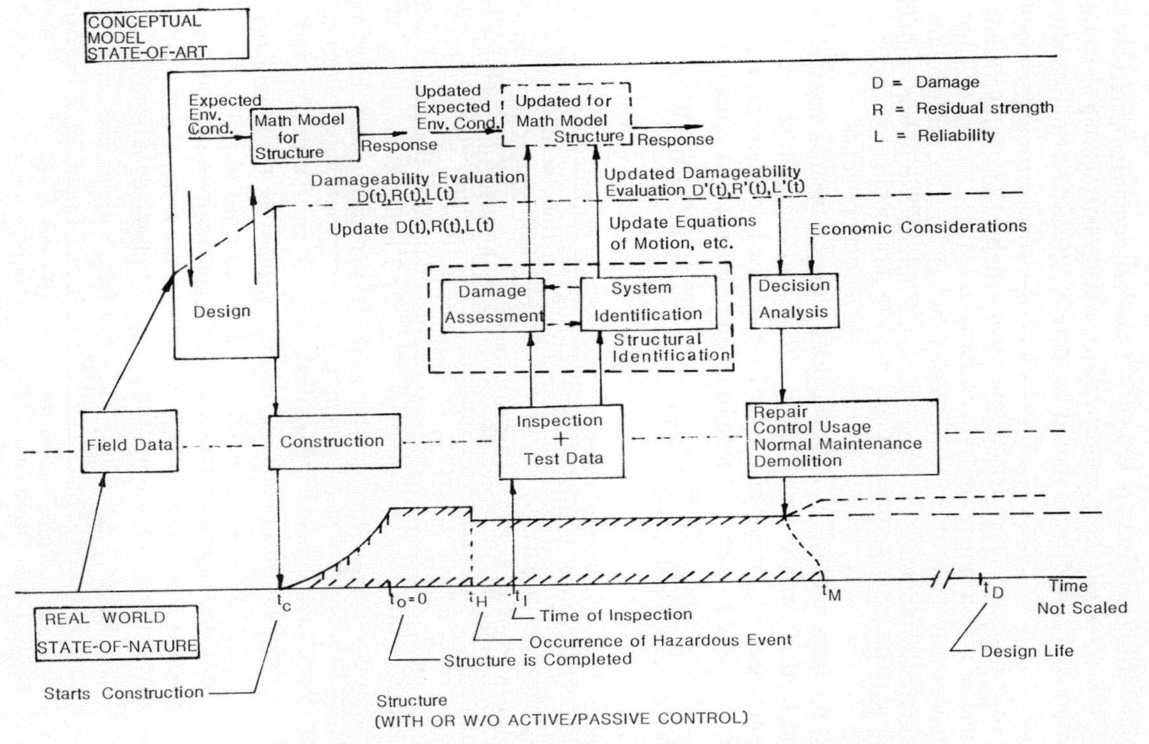

Figure 7.2 Schematic representation of accidental damage effects.

condition which is sought, whether it be scour, accumulation of debris, corrosion (or more properly its absence!), fatigue, marine growth, or whatever. Accidental damage is in a class of its own, of course, and the process of evaluation of its effects is shown schematically in Figure 7.2 (Moses and Yao, 1983). The results of consequent inspection and tests help in determining whether and how repairs should be carried out or if restrictions should be placed on use. In some cases of course, particularly for oil storage platforms, the consequences of *not* carrying out repairs may be too severe for any other course of action but restoration to be adopted. On the other hand many instances may arise where restriction is the simplest course to follow or, in the limit, demolition. Size, ease of access, alternative provision, will all play their part in resolving the problem.

7.2.1.8 *Inspection of 'minor' structures* So far the discussion has been concerned primarily with major structures where there are statutory requirements for inspection etc, and it has been suggested that a model has been provided on which procedures might be based for less critical structures. A 'cost-effective procedure' has been set out, in fact, for steel and concrete pier facilities (Bakun, 1988). This involves a *preliminary inspection* (at a frequency of one to three years) intended to assess visually the general condition of the structure and identify follow-up areas requiring further action. Then comes a *detailed inspection* to quantify deterioration where it has occurred, succeeded by a *structural analysis* to establish where the support loads can safely be carried. This ensures that *repair* is effected only on those parts of the structure which require it, thereby minimising costs. If further deterioration would render parts unsafe then preventive maintenance is carried out. The sort of criteria adopted in deciding what action is necessary are:

1. mild sporadic concrete spalling —no immediate action required
2. soft crumbling concrete, spalled concrete with observable metal loss in rebar (> 1.0 mm) —detailed inspection and possible analysis
3. shear and diagonal cracks in concrete beams, open cracks in concrete decks —immediate analysis
4. deteriorated expansion joints —immediate repair

Table 7.5 Suggested frequencies of inspection

Useful residual life of structure	Frequency of inspection
25 years	Every 5 years
16–25	4
9–15	3
4–8	2
4 years and all ferry terminals	Every year

(This list and Table 7.5 (Shah, 1988) can be compared with Table 4.5 in Chapter 4 for concrete oil production platforms.)

Both from this and the more elaborate procedures outlined earlier it is clear that an important element in minimising costs is the avoidance of time (and resource) wastage in examining sound concrete. To do this successfully requires an ability to 'read' the structure concerned with the implication that critical areas be readily identified—high stress zones, areas of severe exposure, and so on. The key factors to be considered are the relationship of deterioration to residual strength and its relationship in turn to the load to be carried. In fact it has been found in many cases where expensive upgrading repairs were being considered that adequate load carrying capacity remained due to initial excess capacity as a result of features such as higher safety margins or standardisation of member sizes or provision for weathering, etc.

Snah (1988) (which considers Canadian practice) provides a possible equation for calculating the useful residual life (URF):

$$URF = [TUF \times (100\% - WC) \times CF] - AA$$

where TUF is the theoretical useful life, WC is a weighting coefficient, CF a compensation factor and AA the actual age. The difficulty here, of course, lies in assigning values to WC and CF (TUF could be the original design life while AA is a matter of record). A possible maximum value of WC is put at 30 and a minimum for CF of 0.7 while a quantification model for them is given in Figure 7.3 (Shah, 1988). However, this still requires the estimation of values for insertion and a fairly considerable record base which one would imagine to be most easily compiled by large owners such as substantial public authorities, major industrial organisations (such as oil companies), and so on.

7.2.2 *Repair of structures*

Repair of concrete structures is generally regarded as a significant growth industry ranging from restoration and replacement of parts of structures or structural components down to patching of surfaces. Whatever the scale, however, one factor is common to them all other than in plain concrete: if the original concrete is damaged or has deteriorated to the point where it is no longer carrying load, then its replacement will not carry load either—unless live or surcharge loads are absent during the course of repair work and measures are taken to otherwise support as much of the dead load as is possible. An analogous distinction is that between applying a bandage to stop bleeding from a wound and a splint to support a fractured limb.

Repair is as simple or as complex a question as one cares to make it. However, unless proper consideration is given to the various factors involved, the practices and procedures adopted may simply result in further trouble at a later stage. For example, the repair itself may not be inherently durable

Types of structures	Theor. useful life	Influencing factors						Compensation factors	
Steel sheet piles Pipe piles		Variables	Description	Weighting coefficient: Steel	Concrete	Timber	Rock	Actual condition of structure	Compensation factors
			–Normal						
Concrete caissons		Use	–Heavy					Normal condition	
			–Abusive						
Timber caissons		Exposure to salinity and pollution	–None						
–treated			–Alternating						
–untreated			–Concentrated						
°above low water									
°below low water			–Mild (0′ to 3′)						
		Sea conditions	–Average (3′ to 8′)					Average deterioration	
			–Severe (≥ 8′)						
Concrete walls on timber caissons			–Good						
		Ice and waves	–Fair						
			–Inadequate						
Structures on untreated timber piles			–Good						
		Fender systems	–Fair					Considerable deterioration	
			–Inadequate						
			–Excellent						
On untreated timber piles		Foundations	–Fair						
			–Problems						
Structures on steel H piles		Construction and design	–Excellent						
			–Good						
			–Weak					Severe deterioration	
			–None						
		Biological attack	–Some						
Breakwater (rock)			–Advanced						

Structure: Useful residual life = {[(Theoretical useful life)} × (100% – Weighting coefficient)] × Compensation Factor} – Actual age.

Figure 7.3 Quantification model, marine structure: theoretical versus residual useful life

enough, it may lock in (and conceal) further corrosion, the materials used may not be appropriate to the circumstances, it may help to readjust load in a manner not envisaged by the designer, and so on.

7.2.2.1 *System categories* Six categories of repair systems have been suggested (Concrete Society, 1984):

1. re-casting,
2. sprayed concrete and mortar,
3. hand-applied cementitious mortars,
4. hand-applied resin-based mortars,
5. repair of cracks,
6. coatings.

Obviously these are linked to particular situations and the source provides the data from which Table 7.6 has been extracted.

Structural crack *repair* can be effected using low viscosity epoxy resins while cracks can be *sealed* using materials such as SBR (styrene butadiene rubber), acrylic and copolymer emulsions as well as low viscosity polyester and acrylic resins. Of course modifications to materials and development of new systems mean that there is no absolute division between categories. Generally, however, expense and the exothermic reactions involved confine resins to small volume repairs. 'Water friendliness' is also an important factor for marine structures—although some epoxy resins are moisture tolerant making them useful as bonding aids, for example. Clearly cementitious materials are suitable for use in moist or wet conditions with the problem at times becoming one of ensuring that the binder in particular is not washed out. In marine structures especially, measures may have to be taken to combat surface contamination of the area to be repaired by chlorides, algae and other marine growth (CIRIA/UEG, 1986). Other factors to be considered in materials selection include ease and location of mixing (e.g. whether in the wet or dry), strength and flexibility, permeability, and so on (Humphrys, 1980).

Table 7.6 Type and extent of concrete repairs

Possible materials for repair	Spalling large areas cover (mm) 25	12–25	6–12	Small areas cover (mm) 12–25	6–12
concrete, sprayed concrete, cement-sand mortars	*				
polymer-modified cementitious mortars		*		*	
epoxy resin mortars			*		*
polyester resin mortars					*

7.2.2.2 *Some cost implications* The methodology of repair is necessarily site-specific and may depend on considerations such as whether or not the remedial work can be carried out in the dry, with intermittent immersion or has to be done when totally submerged. There is an inevitable cost escalation which bears no relation to the volume of material. For example, remedying the damage to the dome of a storage cell of an oil production platform required a concrete patch 2.6 m × 2.45 m × 0.5 m deep with top and bottom reinforcement. Including inspection and testing it cost £1.4 m at 1981 prices (Faults et al, 1983), a concrete-in-place rate of £440 000 per cubic metre. That required divers. Another, involving use of a coffer dam, cost almost £1 m to repair an area 2.2 m × 0.8 m of a 400 mm thick tower wall. (In both these cases the damage was accidental, the first due to a falling pipe weighing 10 tonnes and the second due to collision from a supply vessel.)

7.2.2.3 *Fouling* Various aspects of marine structure repair are considered in a number of sources (see Gerwick, 1986, Steijaert and de Kreuk, 1985, Billington, 1979 as well as those already quoted, for example), and concrete repair in general is discussed more extensively elsewhere (Allen, 1986). However, it is worth repeating and giving further emphasis here to the effect of fouling which appears to take place quite rapidly, to the extent that ideally casting should take place within 1 to 2 days of the cleaning of the surface to be repaired. This is by no means always possible so that an anti-fouling paint may be required, otherwise the effectiveness of the work may be seriously impaired. This is an aspect which does not seem to have been fully appreciated previously (Steijaert and de Kreuk, 1985).

7.3 Concluding comments

One of the customary functions of a conclusion is to draw together the threads of the preceding text. Since a summary has been provided in the introduction, however, it might be more useful here instead to give a reminder of one or two aspects which so far have had to be left out. Omission in itself should not be regarded as a criterion of significance; rather, it is a reflection of the need to be selective if encyclopaedic proportions are to be avoided.

It has been possible to give foundations only scant attention in the present text. The bearing capacity of the sea bed has a clear influence on the design of the base of an offshore gravity platform, for example. Indeed, the conditions may determine the type of structure to be adopted. The incidence of pock marks is a contributory factor and may govern the precise location of the platform. These are important aspects warranting extended discussion but which are beyond our present scope. Antiscour measures may be required to prevent undermining from the action of waves or of currents. These might range from sheet piles at the toe of a sea wall, to the dumping of material from barges to form a berm for a platform. The possibility of subsidence may have to be investigated in view of experience at Ekofisk in the North Sea. In some geographic regions, primary and secondary seismic effects have to be taken into account. Such potential problems have to be reckoned with and can have a major influence on structural design.

Mention of the sea bed is a reminder that subsea structures may need to be protected against damage from fishing gear and dragging anchors, possibly even the grounding of vessels. Concrete may be of value in providing a defence for vulnerable pipelines for example. The risks have to be assessed and appropriate measures taken. Underwater concreting is perhaps better regarded as a construction technique and to be outside the scope of the present discussion.

Ferro-cement was suggested for inclusion, although regarded by some as a material in its own right, separate from the 'mainstream' materials dealt with here. Certainly it has been used successfully, particularly for the hulls of smaller vessels such as yachts and fishing craft, i.e. for thin shells. It might be bracketed with fibre-reinforced concrete (FRC) as a specialised application. While the latter could prove too expensive for widespread use, it could nevertheless have value in areas with a high risk of impact damage, such as around the water line of vessels for instance. Here one pictures the fibres supplementing rather than replacing conventional reinforcement. They may also have some merit for tetrapods or dolosse.

Turning to the constituents of concrete, it is evident that blast furnace slag cements have enjoyed considerable success in the production of durable structures. Some cement replacement materials clearly warrant serious consideration, although possible longer setting and hardening times might constrain their *in situ* use where inundation cannot be avoided or 'cushioned'. If additives are used obviously they must be compatible with sea water. As for aggregates, it is pre-supposed that sound materials are always employed. They must be durable and non-reactive and generally notable for their physical and chemical 'inertness'. Thermal compatibility with cement paste/mortar and reinforcement may be a very desirable characteristic which is not always considered. This may be particularly important in the tidal zone where there is cycling between solar radiation and immersion as well as wave action. Over the life of the structure, durability problems may be exacerbated as a result. For major structures such considerations are highly unlikely to be missed, but if maintenance of, say, coastal structures on a national scale is economically significant, much smaller structures could require more intense scrutiny. Choosing good quality, durable aggregates might therefore be the particular concern. For large floating and floated structures, on the other hand, weight reduction is a primary factor and so lightweight aggregates assume greater importance. They have been used successfully and in higher strength concretes but their potential durability has to be assessed very carefully. This again is a subject which would merit more than the passing mention which is possible here.

Certain other aspects of the behaviour of concrete might also be examined more extensively: the inter-relations between crack propagation, micro-cracking and fatigue for example. Although in many cases fatigue will be perceived as not being a problem, its relation to material degradation and durability may have long-term implications. At the other end of the structural scale, collapse mechanisms and non-linear behaviour govern response to collision and might be profitably discussed at some length.

Experience in the laboratory and in the field—which is where it matters ultimately—have taught us many lessons but we are still a long way short of being able to predict with absolute certainty. We can estimate what the environment is *likely* to be, we can assess how a structure is *likely* to respond, but we cannot know *for sure*. That is what makes the sea so intriguing and its structures so interesting for the engineer. There always has been and always will be a challenge and scope for initiative, enterprise and ingenuity. With these, go enquiry and curiosity and hopefully some of their directions have been indicated here.

References

Bakun, G.B. (1986) Cost-effective approach to pier inspection and maintenance. *Proc. 2nd. Int. Conf. Maintenance of Maritime & Offshore Structures*, Thos. Telford, London.

Browne, R.D., Doyle, V.J. and Papworth, F. (1981) Inspection of concrete offshore structures. *J. Pet. Tech.*, 2243–2251.

Bullington, A.R. and Loper, R.L. (1977) Underwater structural inspection, Greater Ekofisk, North Sea. *Proc. 9th. Ann. Offshore Tech. Conf.*, Houston.

CIRIA/UEG (1986) The influence of methods and materials on the durability of repairs to concrete coastal and offshore structures, UEG, London.

Concrete Society (1984) Repair of concrete damaged by reinforcement corrosion. *Conc. Soc. Tech. Report*, No. 26, London.

CXJB Underwater Engineers, Underwater inspection of offshore installations: guidance for designers. *Report UR10, CIRIA/UEG*, London.

Djourtouian, C. (1980) Système de plongée d'inspection pour zone annulaire de la plateforme centrale Ekofisk. *Proc. Coll. Int. Tenue des Ouvrages en Béton en Mer, Actes de colloques no. 11*, CNEXO, Brest.

Faulds, E.C. (1982) Structural inspection and maintenance in a North Sea environment. *Proc. 14th. Ann. Offshore Tech. Conf.*, Houston.

FIP (1982) *State of the Art Report*—the inspection, maintenance and repair of concrete structures, FIP, Slough.

Henn, A.E. and Spencer, J.S. (1984) A government perspective on the safety of marine structures. *Proc. Int. Symp. Role of Design, Inspection and Redundancy in Marine Structural Reliability*, National Academy Press, Washington.

McAlpine Sea Services Ltd. (for Dept. of Energy) (1985) Classification and identification of typical blemishes visible on the surface of concrete under water. *Conc. in the Oceans Tech. Report*, No. 9 (Offshore Tech. Report OTH 84 206), HMSO, London.

Maris, A.T. and Paulling, J.R. (1980) Analysis and design of the cold-water pipe (CWP) for the OTEC system with application to OTEC-1, *Marine Tech.* **17**(3), 281–289.

Marshall, A.L. (1988) Concrete marine storage of LNG—with particular reference to floating platforms. *Proc. 13th. Int. LNG/LPG Conf.*, Kuala Lumpur, Gastech Ltd., Rickmansworth.

Moses, F. and Yao, J.T.P. (1984) Safety evaluation of buildings and bridges. *Proc. Int. Symp. Role of Design, Inspection and Redundancy in Marine Structural Reliability*, National Academy Press, Washington.

Shah, V.K. (1986) Outline of guidelines for inspection and maintenance of marine facilities. *Proc. 2nd. Int. Conf. Maintenance of Maritime and Offshore Structures*, Thos. Telford, London.

Sletten, R. (1977) Underwater inspection of North Sea Structures. *Proc. Symp. Safety of Offshore Structures, Assoc. Int. à But Scientifique*, Brussels.

Sletten, R., Fjeld, S. and Røland, B. (1977) In-service inspection of North Sea structures. *Proc. 9th. Ann. Offshore Tech. Conf.*, Houston.

Warmington, J.R., Wainwright, D.A. and Clark, P.W.J. (1988) Design of a floating berth, *FIP Notes*, 1988/2, 5–7.

Appendix

Constants used in Fifth order theory

Note that $s = \sinh kh$ and $c = \cosh kh$

$$C_0^2 = g(\tanh kh)$$

$$A_{11} = 1/s$$

$$A_{13} = \frac{-c^2(5c^2+1)}{8s^5}$$

$$A_{15} = \frac{-(1184c^{10} - 1440c^8 - 1992c^6 + 2641c^4 - 249c^2 + 18)}{1536s^{11}}$$

$$A_{22} = \frac{3}{8s^4}$$

$$A_{24} = \frac{(192c^8 - 424c^6 - 312c^4 + 480c^2 - 17)}{768s^{10}}$$

$$A_{33} = \frac{(13 - 4c^2)}{64s^7}$$

$$A_{35} = \frac{(512c^{12} + 4224c^{10} - 6800c^8 - 12\,808c^6 + 16\,704c^4 - 3154c^2 + 107)}{4096s^{13}(6c^2 - 1)}$$

$$A_{44} = \frac{(80c^6 - 816c^4 + 1338c^2 - 197)}{1536s^{10}(6c^2 - 1)}$$

$$A_{55} = \frac{-(2880c^{10} - 72\,480c^8 + 324\,000c^6 - 432\,000c^4 + 163\,470c^2 - 16\,245)}{61\,440s^{11}(6c^2 - 1)(8c^4 - 11c^2 + 3)}$$

$$B_{22} = \frac{(2c^2 + 1)c}{4s^3}$$

$$B_{24} = \frac{c(272c^8 - 504c^6 - 192c^4 + 322c^2 + 21)}{384s^9}$$

$$B_{33} = \frac{3(8c^6 + 1)}{64s^6}$$

$$B_{35} = \frac{(88\,128c^{14} - 208\,224c^{12} + 70\,848c^{10} + 54\,000c^8 - 21\,816c^6 + 6264c^4 - 54c^2 - 81)}{12\,288s^{12}(6c^2 - 1)}$$

$$B_{44} = \frac{c(768c^{10} - 448c^8 - 48c^6 + 48c^4 + 106c^2 - 21)}{384s^9(6c^2 - 1)}$$

$$B_{55} = \frac{(192\,000c^{16} - 262\,720c^{14} + 83\,680c^{12} + 20\,160c^{10} - 7280c^8 + 7160c^6 - 1800c^4 - 1050c^2 + 225)}{12\,288s^{10}(6c^2 - 1)(8c^4 - 11c^2 + 3)}$$

$$C_1 = \frac{(8c^4 - 8c^2 + 9)}{8s^4}$$

$$C_2 = \frac{(3840c^{12} - 4096c^{10} + 2592c^8 - 1008c^6 + 5944c^4 - 1830c^2 + 147)}{512s^{10}(6c^2 - 1)}$$

$$C_3 = -\frac{1}{4sc}$$

$$C_4 = \frac{(12c^8 + 36c^6 - 162c^4 + 141c^2 - 27)}{192cs^9}$$

In deep water (i.e. $h/L \geqslant 0.5$) the coefficients simplify to the following values:

A_{13}	$\frac{5}{8s}$	B_{22}	$\frac{1}{2}$	C_1	1
A_{15}	$\frac{37}{48s}$	B_{24}	$\frac{17}{24}$	C_2	$\frac{5}{4}$
A_{22}	0	B_{33}	$\frac{3}{8}$	C_3	$-\frac{1}{4sc}$
A_{24}	0	B_{35}	$\frac{153}{128}$	C_4	$\frac{1}{16s^2}$
A_{33}	0	B_{44}	$\frac{1}{3}$		
A_{35}	0	B_{55}	$\frac{125}{384}$		
A_{44}	0				
A_{55}	0				

Index